A Textbook of
WELDING TECHNOLOGY

By
Er. G.D. Garg
B.Sc., B.E. (Mechanical) Hons.
Ex-General Manager,
M/s. Bharat Heavy Electricals Ltd., Haridwar
Ex-Chairman,
Indian Institute of Welding (Delhi Branch)

S.K. KATARIA & SONS®
Publisher of Engineering Books
4885/109, Prakash Mahal, Dr. Subhash Bhargav Lane,
Opposite Delhi Medical Association, Daryaganj, New Delhi-110002
Phone: +91-11-43551243, Mobile: +91-9871775858, 7678425209
E-mail: katariabook@gmail.com,katariabooks@yahoo.com
Website : www.skkatariaandsons.com
(e-books, subscriptions and bulk buying also available on our website)

Published by :
S.K. Kataria & Sons®
4885/109, Prakash Mahal, Dr. Subhash Bhargav Lane,
Opposite Delhi Medical Association, Daryaganj, New Delhi-110002
Phone: +91-11-43551243; Mobile: +91-9871775858, 7678425209
e-mail: katariabook@gmail.com, katariabooks@yahoo.com

First Edition: 2011, Reprint: 2026

Laser Typesetting by : Stride Computers

Printed at : New A.S. Offset

PREFACE

Welding is an ocean in which every time you dip you find new jewels. These jewels are invention of new materials, development of their welding processes and making use of them in important applications. Therefore the basic knowledge of welding technology and its applicability is very important. This book has been written keeping in view the complete syllabus of engineering students. This covers introduction to joining, welding processes, cutting processes, surfacing processes, various techniques, weldability of different materials and so on and so forth. This book describes the various topics which can be understood easily by engineering students and beginers.

Moreover, the welding industry in India has reached a cross roads where the higher productivity and quality processes have suddenly shown a tremendous spurt and are being very widely adopted by all sectors of industry. This necessitates automation in welding processes.

I feel the book very useful to students to give deep knowledge of the subject.

With all best wishes.

—AUTHOR

Dedication

The book is dedicated to my humble and beloved wife Smt. Suman Garg, whose inspirations and dedication towards family are unforgettable.

I wish that our daughters Nidhi, Sruti, Shikha and Neha will keep the candle enlightened in her sweet memory.

—G.D. GARG

CONTENTS

1

CHAPTER

INTRODUCTION TO WELDING

❑ INSIDE THIS CHAPTER

1.1 Background, 1.2 Welding, 1.3 Classification, 1.4 Features.

1.1 BACKGROUND

Welding is one of the latest and the most versatile process of manufacture and has very rapidly gained in stature and importance. The process is very simple in the hands of knowledgeable persons and for a technologist, easy of application. In some of its forms the process is as simple as gumming close the flap of an envelope. The ease of application can be gauged from the fact that massive and complicated structures can be fabricated and erected in very little time using variety of easily available raw materials.

The process of welding was more or less unknown till the turn of the current century. It's impact was felt in early years of the second decade when commercially accepted coated electrodes were developed. In the first world war when the importance of the time of manufacture was recognised as an independent factor the process received its necessity. The second world war completed its boost and the process with its inherent advantages acquired the necessary importance and today the process stands at the apex of the structure comprising various processes of manufacture.

The welding process, when all is set for the technologist, to proceed, it becomes so simple that a feeling has grown throughout the industry that nothing much is needed to be known beyond the operation and use of the plant. The process, no doubt is simple for the technologist. The process involves considerable understanding of metallurgy involving high heating and cooling rates, the impact of ambient conditions on the quality of the joint and the effect of thermal shock on the component. These all factors will be dealt in details in the coming chapters.

1.2 WELDING

Welding is a fabrication process of joining of materials wherein coalescence is produced by the application of heat with or without the application of pressure and with or without the use of filler metal.

Thus, from the definition, it is seen that in the process heat is always applied. Heat is an essential factor in welding and various processes of welding are usually named after the source of heat. The term material is used in the definition because the welding process is now-a-days is used for welding of plastics in addition to various metals.

1.3 CLASSIFICATION

The welding is broadly classified into two categories :

(a) Fusion welding.

(b) Pressure welding.

In fusion welding the heat is applied till the edges of parent metals to be joined melt and along with the molten filter metal flow together and form a pool which on solidification produces the required coalescence.

In pressure welding the edges of the metals to be joined are heated to a temperature at which coalescence is achieved by the application of suitable mechanical pressure.

From the above explaination it is clear that metal edges are raised to a higher temperature is fusion welding and to a comparatively lower temperatures in pressure welding.

1.4 FEATURES

Following are essential features in a welding process :

(a) Stability of the heat source.

(b) Uniformity of the heat input.

(c) Rapid rate of heat input.

(d) Shielding the molten filler metal (if used).

(e) Shielding the weldment against atmospheric contamination.

(f) Provide scavengers and deoxidisers in the weld-metal.

(g) Refine the weld metal.

(h) Slow and appropriate cooling rate of weldment.

❑❑❑

2
CHAPTER

WELDING PROCESSES

❑ **INSIDE THIS CHAPTER**

2.1 Oxy-acetylene welding, 2.2 Shielded metal arc welding (SMAN), 2.3 Gas arc welding processes, 2.4 Gas tungsten arc welding, 2.5 Gas metal arc and flux-cored arc welding, 2.6 Submerged arc welding, 2.7 Thermit welding, 2.8 Electroslag welding

2.1 OXY-ACETYLENE WELDING

Oxy-acetylene welding is the most widely used gas welding process. It is a manual welding process. In this process oxygen and acetylene gases stored in separate cylinders are led to the mixing chamber in the torch. The gases, usually in equal proportions, issue from the nozzle of the torch at a fixed velocity of approximately 110 meters per second. When ignited the flame impinges on the edges of the metals and heats to melt them. When the base metal reaches the melting point or is near to it, the filler metal is added. The torch is light in weight and is designed to give the welder complete control of the flame.

2.1.1 Plant, Accessories and Materials

The oxy-acetylene welding plant comprises of :

(i) Cylinder

There are two cylinders, one for oxygen which is coloured black and the other is for acetylene coloured maroon. Normally these cylinders are available in various capacities but for welding, standard cylinders of 6.2 cu. m (220 cu. ft) capacity are universally used. The pressure of the gas in O_2 cylinder is approximately 140 kg/cm^2 when full whereas in acetylene cylinders is approximately 25 kg/cm^2. The oxygen cylinder is solid drawn.

(ii) Regulator and Indicator Gauge

There is one gauge at each cylinder to indicate the pressure of the gas in the cylinder as well as in the pipeline leading to the mixing chamber of the torch. In addition the gauge regulates the discharge pressure of the gas from the cylinder. The threads in the oxygen cylinder gauge are normal right hand and the gauge is painted black. The threads in the acetylene gauge are left hand and the gauge is painted maroon. This is to avoid interchange of gauges. The pressure reduction in the gauge is sizeable and the process continues for a sufficient length of time, at a stretch. Hence, invariably two stage regulator gauges are used.

(iii) Welding Torch

Welding torch is a device to mix the two gases and to feed the mixture at the suitable point on the surface of the work price. The torch comprises two threaded ends to feed oxygen and acetylene gases, two sets of valves to control the gases, a mixing chamber and the tip. From the point of view of the welder, the tip is important.

A tip is that portion of the torch through which gases pass just prior to their ignition and burning. The tips which are numbered have proper arifice size. The numbering system varies from manufacturer to manufacturer. In the system adopted by Indian Oxygen Limited, it is based on the volume in cu. ft. of each of the gases discharged in one hour at the recommended pressure, for example, for a discharge of cu. ft. of acetylene and 1 cu. ft. of oxygen per hour. The nozzle is numbered 1 and for 5 cu. ft. of acetylene and 5 cu. ft. of oxygen the number of the nozzle is 5. Tips are some cases made integral with mixing chamber.

Tips are generally made of copper and copper alloys which are relatively soft materials and those should be guarded against damage. When cleaning the tips, tip clearance designed for the purpose should be used.

(iv) Hose Pipe

Hose pipes are of rubber strengthened with canvas. The hose pipes are coloured black for oxygen and red for acetylene. The pipe for oxygen is thick walled.

2.1.2 Welding Materials

Oxy-acetylene welding is a fusion welding process and to cover the gaps between the edges to be joined filler metal is invariably used. Though oxy-acetylene flame has an excellent protection against atmospheric oxidation, yet for some metals use of flux is utmost necessary.

(i) Welding Rods

Welding rods supply the filler metal to the joint. The properties of the metal deposited during welding should closely match those of the base metal (or, the required properties of the deposit). Because of these requirements, welding rods of various chemical compositions have been developed for welding of ferrous and non-ferrous metals. Obviously, it is important that correct welding rod is selected.

The quantity and the rate of supply of filler metal required increases with increasing thickness. To achieve this welding rods of varying diameters are manufactured and used.

The composition analysis of the weld deposit must be within specified limits. Owing to the intense heat of the flame some of the alloying elements in the rod may be burnt and hence the analysis of the rods must also be within specified limits to compensate for the expected loss.

(ii) Fluxes

Prevention of oxidation of weld pool and the removal of the oxide layer from the work surfaces is essential in welding. Unless the oxides are removed and its formation in prevented, fusion may be difficult and inclusions may be present with the result that the joint may lack strength. The oxides when present on the work surface will not flow out from the weld zone but will remain adhered to it and get entrapped inthe solidifying metal, interfering with the bonding of the filler metal. In fusion welding these conditions usually occur when the oxides have a melting point higher than the base metal. Hence means must be found out to remove the oxides and also to prevent their formation. Fluxes are used for this purpose. A flux acts by forming fusible slag with the oxides. The slag has a lower melting point and solidifies later than the melt and as such the process of solidification of the melts the slag floats out and thus, the oxides existing on the surface of the metal (or of the filler metal rod) is removed. The flux may be used as a dry powder, as a paste or a thick solution. The powder fluxes are applied by heating the end of the welding rod and dipping it into the can containing the powder flux. The fluxes in paste form are painted on the work surfaces.

The common metals requiring the use of flux are aluminium, bronze, brass, cast iron and stainless steel. Oxides of low carbon steel have lower melting points and hence for low-carbon steel, flux is not necessary.

2.1.3 Accessories

A large variety of accessories are used in oxy-acetylene welding.

- Goggles are a must in welding with oxy-acetylene flame and for that matter in any gas welding. Proper shade of goggles must be chosen.
- Next article invariably required is the friction lighter. In certain cases the operator uses smouldering rags for the purpose of lighting the torch. This practice is hazardous and should be discouraged. Friction lighter is very convenient and a quick means of lighting the flame.
- Other common accessories are tip cleaner, cylinder trucks, clamps, jigs and fixtures, gas economisers (these include hooks for holding the torch when not in use), aprons and gloves. Insome cases devices are entraining liquid flux into gas stream is also required.

In large shops engaged in oxy-acetylene welding or oxygen cutting gas manifolds and delivery pipe lines are used for the supply of gases – oxygen and acetylene.

2.1.4 Operation of the Plant

Prior to the commencement of the operation of welding the plant is to be set. The setting includes the adjustment of regulator gauge for the required discharge pressure of the gases, suitable size of the nozzle, the adjustment of the flame and the selection of proper composition and the size of the welding rod. All these have a bearing on the type and thickness of the metal to be welded.

Flame

The adjustment of the flame is the most important factor in the success of the welding operation. The adjustment of the flame means controlling the quantity of discharge of gases – oxygen and acetylene and the adjusting ratio of these gases.

The gases must discharge from the nozzle at a velocity slightly above the velocity corresponding to the flame propagation rate of oxy-acetylene flame which is approximately 110 m/second. The discharge velocity is controlled by the length of the hose pipe, the size of the nozzle and the pressure.

There are three types of flames, neutral (also called normal), carburising and oxidising depending on whether the gases are in equal proportions, excess of acetylene or excess of oxygen respectively. Excess of oxygen or acetylene in the flame is small. These three types of flames can be distinguished by the visual observation of the flame.

The final combustion reaction between oxygen and acetylene is

$$2C_2H_2 + 5O_2 = 4CO_2 + 2H_2O$$

Considering this reactions a neutral or normal flame will be one in which for every volume of acetylene there are $2\frac{1}{2}$ volumes of oxygen. But this is not the case for simple reason that the reaction between oxygen and acetylene takes place in two stages as under :

Stage I. $\quad C_2H_2 + O_2 = 2CO + H_2$

Stage II. $\quad 2CO + O_2 = 2CO_2$

$$1H_2 + \frac{1}{2}O_2 = H_2O$$

The oxygen for the second stage of re-action is obtained from the atmospheric air. It should be noted that in the first stage CO and H_2 are produced. In case there is excess of either of the two gases issuing from the nozzle the products of combustion will be CO, H_2 and the excess quantity of either of the two gases.

In case of excess of acetylene, C_2H_2 is split into its constituents C_2 and H_2, H_2 will combine with H_2 from the reaction and C_2, in nascent form will be available. Since carbon can be readily absorbed in metals, particularly, steels there is a tendency to carburise the melt and hence the flame is called carburising. On the other hand when there is excess of oxygen, from oxygen O_2 will be available, momentarily though, along with C_2 and H_2. This O_2 will oxidise the metals and hence the flame is referred to as oxidising flame.

This two stage reaction between oxygen and acetylene is at the same time, the strength and the weakness of the oxy-acetylene process. The strength lies in the fact that the products of combustion are of reducing nature and provide not only excellent protection of the weld pool but also reduce, atleast in part, the oxides present on the surface of the rod and the base metal. The weakness lies in the fact that the total heat is generated in two stages. In the first stage, which is the effective stage for the purpose of welding, only 1/3 heat is produced (every cu. ft. of acetylene produces 1500 B.Th. US of heat). The remaining 2/3 of heat is produced when CO and H_2 combine with atmospheric oxygen. Only a small part of this heat is utilised for the purpose of welding.

2.1.5 Base Metal Prepration

The base metal prepration comprises cleaning of the surfaces to be joined and suitably shaping the edges. The basic object of the two being to ensure effective coalescence along the entire edge. The cleanliness of the surfaces (edges) means removing dirt, oil and oxides. These are often referred to as contaminants. The contaminants cause blow-holes, porosity, slag-inclusions and incomplete fusion and on the whole either prevent welding entirely or give reduced strength. The edge prepration is necessary to subject the edges, simultaneously and uniformly, to the influence of the flame.

It will be noted that the flame impinges on the surface of the base metal and when the two base metals are placed side by side the heating of the edges can be only indirect by virtue of conduction. This is likely to be not only time consuming but will give lower temperature in the core of the metal. Hence the metal edges are cut to suitable angles to ensure direct access to the flame. The square edges which do not require any shaping of the edges is the easiest (used upto 8 mm in downhand welding and upto 12.5 mm is vertical welding with two operators) and require smaller quantity of filler metal. In other cases suitable edges are prepared.

2.1.6 Welding Positions

Welding can be performed in any position but its use is avoided in overhead position. As far as possible welding should be carried out in downhand positions, though it is often carried out in vertical position with equally good results. Vertical welding has the advantage that square edges can be used even upto 12.5 mm thickness and no additional shaping of edges is involved. In the downhand welding there are two techniques, forward welding also called left-ward welding and backward welding also called right-ward welding (The use of the term left ward and right ward be avoided since they are dependent on the operator whether he is left-handed or right-handed).

2.1.7 Effects of Oxy-acetylene Welding

The characteristic of all the welding processes is the steep temperature gradients and one of the greatest problems is to prevent the deterioration in the properties of the metals which are affected by the intense heat of welding. Besides there is a problem of distortion. Thus heat which is essential factor in welding is the prime cause of various problems of welding. Therefore, precautions are necessary to avoid adverse effects of heat. Correct design of the joint, the shape of the edges, suitable heating and controlled cooling reduces the above mentioned adverse effects. In oxy-acetylene welding the temperature gradient through steep is comparatively less but the heat affected zone and the heat input, because of the lower temperature and the spread of the flame on the edges and surfaces, is rather large leading to greater distortion. Because of the greater heat input and larger heat affected zone there is greater chance of changes in the properties of the metal. The oxy-acetylene flame allows better control of the operator to regulate the heat input and because of the different type of flame the carbon content can be regulated to certain extent. When an oxidising flame is used rapid reaction results between the oxygen and the carbon of the metal. Some of the carbon is eliminated in the form of carbon monooxide (CO) and the steel itself and the other constituents are oxidised. When a carburising flame is used, carbon is added to the weld puddle.

2.2 SHIELDED METAL ARC WELDING PROCESSES (SMAW)

There are large number of welding processes which use an electric arc as the source of heat. Amongst these, Shielded Metal Arc Welding (SMAW) is the msot widely used process. This process was originally referred to, simply as Metal Arc Welding but currently the term shielded metal arc welding has been substituted.

Shielded Metal Arc Welding is like oxy-acetylene welding, a fusion welding process. The heat source is an electric arc estabilished between a metal rod, referred to as electrode and the base metal. The electrode serves an additional purpose of providing the filler metal. The electrode is always coated with a thick layer of flux and is referred to as coated Electrode. The shielding, which protects the weld and the weld metal against oxidation is obtained from decomposition of coating. The composition of the coating meterial and the electrode material control the mechanical, chemical and metallurgical properties of the weld.

Both A.C. and D.C. currents are used in shielding metal arc welding. The current required is high and voltage low. The current generally varies from 50 to 500 amperes and the voltage varies from 25 to 30 volts. In manual shielded metal arc welding the arc is formed by striking between the base metal and electrode. The temperature of the arc is of the order of 6000°C as against the temperature of the oxy-acetylene flame which is generally help of this. The intense heat of the arc instantaneously melts the surface of the base metal and also the tip of the electrode and the welding commences instantaneously.

2.2.1 Plant, Accessories and Materials

The shileded metal arc welding plant compries :

(a) Welding Generator/Transformer

The welding generator provides D.C. power whereas the welding transformer provides A.C. power. A.C. as well as D.C. power is used in welding but a study of the nature of the component and of the material of component is necessary to determine whether A.C. or D.C. is to be used for the job.

The power source can be classified into three hypes—A.C., D.C. or A.C./D.C. D.C. power derived from A.C./D.C. power source to the extent that the instantaneous voltage of A.C./D.C. power source fluctuates between zero and maximum whereas the voltage of D.C. power source is almost constant. The characteristics of the power source may be either constant voltage or constant current. Generally constant current characteristic are perferred, particularly in shielded metal arc welding, since it is difficult for the welder to hold a constant arc length. Welding is performed in steep portion of the current-voltage curve to minimise the change in current during manual metal arc welding. The open circuit voltage of the power source is of the order of 65–70 volts and the voltage on load reduces to 25–30 volts.

(b) Welding Cables

Welding cabsle should be of copper and of suitable size to minimise the drop in voltage and reduce the heating effect. The normal length is about 15 m.

(c) Electrode Holder

Electrode holder is simply a clamping device for holding and transferring the welding current into the electrode. The handle of electrode holder is insulated. The jaws which grip the electrode should be in good condition and kept unconteminated to provide a good conducting path and to reduce heating of the jaws. The holding should be designed to hold the electrode vertical as well as at an angle to the workpiece. It should permit quick and easy change of electrode and it should be light in weight.

(d) Ground Clamp

The ground clamp is a device for connecting one of the leads to the workpiece and it should permit an easy, good and firm electrical contact.

2.2.2 Welding Materials

The coated electrode is the only material used in the shielded metal arc welding. The electrode used in the shielded metal arc welding is invariably coated with a thick layer of flux. The thickness however varies. Usually thickly coated electrodes in addition to establishing the arc and supplying the filler metal for the weld deposit, introduce other materials in the arc and around the arc. Depending upon the type of arc used, the coating performs the following functions :

(a) Stablises the arc.

(b) Shields the molten metal against atmospheric contaniminations.

(c) Provides scavengers and deoxidisers to refine the grain structure of the weld metal.

(d) Provides a slag blanket over the deposited (and solidifying) weld metal to retarted the cooling rate.

A variety of electrodes are available in the market and are designated in various ways from the point of view of the material of the electrodes, the type and thickness of coating and the power source including the polarity to which it is to be connected. Some electrodes can be connected in any manner and with any power source. These are called universal electrodes.

The exact composition of the electrode rod, the nature and composition of flux are the trade secrets of the manufactures. The technologist has only his own experience and the data furnished by the manufactures to make a selection of the electrode. Very often the electrodes are not what they are shown to be. The experience of the user goes a long way in selecting the electrode., The electrodes are sold under a code number stardardised under IS : 815 : 66.

The electrodes are available in various gauge sizes. Three S.W.G. sizes 10, 6, 8 approximately 1/8", 5/32" and 3/16" (corresponding sizes in metric system 3.15, 4 and 5 mm.) are most common. The average current used for these electrodes is 100, 150 and 200 amperes. When worked out, current density is approximately 7000 amps/in^2 (1000 amps/cm^2) and is constant. Larger currents adversely affect the coatings.

The arc welding was attempted easly in the century but its application was not considered possible till the coated electrodes were developed in 1912. The arc welding process became a success immediately thereafter. Though various processes using the arc energy have been developed and introduced, the shielded metal arc welding, still maintain its supermancy larger on account of its simplicity and versatility. Of the seven essential functions to be performed in any welding process, the coated electrodes helps to perform the following functions :

(i) Rapid rate of heat intput
(ii) Provides scavengers and deoxidisers
(iii) Stability of heat source
(iv) Shielding
(v) Reparding cooling rate.

(i) Rapid rate of heat input

The rapid rate of heat input arises from the nature of the arc. The arc has a very high temperature of the order of 6000°C and heat source is concentrated. Barring plasma arc, electron beam and some gas arcs, the shielded metal arc has the highest temperature and most concentrated heat source. Besides the nature of the voltage drop and consequently the generation of the heat maxes this source ideal for welding.

(ii) Scavengers and deoxidisers

These are provided from the coating. The coating also generates the inert atmosphere to protect the weld metal against oxidation but it is inadequate. Hence scavengers and deoxidisers are used to prevenet the harmful effects of oxidation.

Incidentally, the oxidation being an exothermic process, the heat of oxidation improves fluidity of the weld pool.

(iii) Stability of heat source

The arc is described as the sustained electrical discharge through an ionised gas. In shielded metal arc welding, the gas is the atmospheric air which comprises, mainly, oxygen and nitrogen. The ionisation in the atmospheric air (in any gas for that matter) unlike in an electrolyte, vanishes instantaneously when the applied voltage reduces. In A.C. as well as rectified D.C. the voltage fluctuates and every hundredth of a second it becomes zero, hence there is a tendency of ionisation to vanish and this tends to extinguish the arc. The coating creates ionisation thermionically and sustains the electrical discharge and hence provides stability to an arc.

(iv) Shielding

The flux in the coating melts and intermingles with the molten metal protecting it against oxidations. The flux forming slag with the oxide deposites on the surface of the bead and protects the bead, while hot from oxidation.

(v) Cooling rate

Cooling rate in shielded metal arc welding is always high. The blanket of slag tends to reduce it but only to a limited extent. To a limited extent it refines the grain structure as well.

2.2.3 Accessories

The accessories required are :

(i) Protective clothing

The globules of molten metal and some spanus are always thrown out from the arc and heat is intense. To ensure protection against burns and to prevent exposure to ultraviolet rays the operator should wear leather gloves (not rubber) and a protective apron and a leather jacket. Bare legs should be particularly covered when welding in sitting position.

(ii) Face shield

The face shild incorporating a dark glass is a must in welding. In oxy-acetylene welding, goggles were enough but in shilded metal arc welding where the heat is intense, there is emission of ultra violet rays and a large number of arc sparks are flying, the face and head shield is a must. The dark glass should be protected on the outside by a clear glass. The face and heat shields are usually constructed from processed fibres to make them light and fire-proof. The shield is either held in hand or head mounted.

2.2.4 Arc as a Source of Heat

In shielded metal arc welding vis-a-vis the oxy-acetylene welding, the arc replaces oxygen and acetylene as the heat source. The arc has been described as the sustained electrical discharge between the electrodes separated by a column of ionised gas. In shielded metal arc welding atmospheric air is present in this column and the length of the column is approximately equal to the diameter of the electrode and ranges from 2.5 mm to 6 mm. The electrical dischange is from the cathode (–ve electrode) through the column and into the anode (+ve eletrode) and through the cable back to the source. The heat is liberated at the tip of the two electrodes and in the column. The column generates only a small amount of heat of the order of 20%. The rest of the heat is generated at the electrodes. The heat at the +ve electrode is theoretically estimated to be 70% of the total heat at the electrodes. Practically, however it is determined to be of the order of 60%. In short the heat at the +ve electrode is more than at the (–)ve electrode.

The discharge commences from the cathode, spreads out in the form of a cone such that the arc column represents the frustum of a cone. When using A.C. the electrode and the work-piece are alternatively –ve and +ve and it makes no difference how the work-piece and the electrode are connected. The position is different when either rectified D.C. or D.C. (pure) is used. In rectified D.C. the cathode is at the zero potential whereas the anode reaches the maximum voltage and falls to zero every 1/100th of a second. In pure D.C. also the cathode is at zero potential but the anode is always at the maximum voltage.

When the electrode is connected to the negative pole (when using D.C. or rectified D.C.) and the work-piece is connected to the positive pole it is called **straight polarity** welding. When the connections are reversed it is called **reverse polarity** welding. These is no such difference when welding with A.C.

As explained earlier there is more heat at the positive pole and less at the negative pole. Hence the amount of heat input into the electrode and the work-piece can be controlled, there being more heat input into the work-piece is straight polarity welding. As stated earlier the arc forms a frustum of a cone. In stragith polarity welding (work-piece positive) the larger end of the cone impinges on the work-piece. Hence there is relativiely, less concentration of heat in the work-piece. Hence, on the whole, the variation in intensity of heat on the area when the arc impinges is reduced to an extent, with two different types of connections. The arc has some cleaning action as well. The cleaning action is on the cathode. Often, the cleaning aspect of the arc determines the polarity to be used.

2.2.5 Operation

In operating the plant the machine is set on the required current and the electrode holder and the work-piece are suitably connected to the required pole. The electrode held in the holder is made to tough (scaratch against) the work-piece and the arc struck. The welding commences immediately. The work can be carried out in any position. The current is suitabsly adjusted. The deposit is in the form of small beads. Usually the electrodes are 45 cms long (18") and are deposited in one and a half minute. The length of the beads deposited is usually 1/3rd the length of the electrode. While depositing the arc-gap maintained should be as small as possible.

Generally the arc-gap (arc-length) is equal to the diameter of the core wire of the electrode.

Before commencing the welding the work-piece is to be prepared as in case of oxy-acetylene welding. In this case angle 'V' is smaller. As compared to oxy-acetylene welding the temperature of the source in shielded metal arc welding is high-twice as much. Hence the rate of heat transfer is high-twice as much. Hence the rate of heat transfer is high. Besides, the source of heat is concentrated which further increases the rate of heating. As a consequence the heat input and heat affected zone are smaller and there is deeper penetration and less distortion in SMAW. Deeper penetration has its effect on the edge-design. Generally the edges have less included angle and larger shoulder width which reduce the quantity of filler metal required.

In SMAW the coating provides protection against oxidation but it is never adequate. Oxidation of the weld pool occurs and oxides are formed. These oxides are removed in the form of slag. The slag serves a useful purpose of protecting the welding and retards the cooling rate of the weld.

2.3 GAS ARC WELDING PROCESSES

Oxy-acetylene welding and shielded metal arc processes have their advantages and disadvantages. Hence a process which will combine the advantages of both the processes and reduce/eliminate the disadvantage is evidently called for. Gas arc welding processes arc such processes which have revolutioned the techniques of welding.

Gas arc welding also called Gas shielded metal arc welding is an arc welding process in which the arc is protected against atmospheric contaimation by a shield of a protective gas. In the process arc is formed between the end of a metal electrode, either consumable or non-consumable and the component to be welded. The shielding gas is usually inert though it may be active. Usually filler metal is used.

The gas shielded arc welding process are divided into two catagories :

(i) Metal Inert Gas Welding (MIG)

(ii) Tungsten Inert Gas Welding (TIG)

In shielded metal arc welding the arc is struck by scratching the electrode against the work piece (momentarily causing a short-circuit). On the other hand in inert gas shielded arc welding the arc is usually struck by high frequency spark. In mannual welding under which TIG process is classified, the electrode hold in the electrode holder, is held at an angle of approximately 75° to the surface of the weld puddle. The filler metal is provided from a separate filler metal rod. TIG welding is usually used for sheets.

MIG is the mechanised form of shielded metal arc welding. The electrode which serves as the filler metal as well, is mechanically fed into the weld pool at the desired feed. The electrode is in the form of a spool. MIG welding is used for welding thicker sheets.

A.C. as well as D.C. is used for shielded metal arc welding. Generally for TIG welding either A.C. or D.C. which electrode connected to (–)ve pole (straight polarity) is used. When using reverse polarity a larger size of electrode is taken. The power source has drooping characteristics.

Non-consumable electrodes used for gas shielded metal arc welding usually are tung sten (ii) tungsten alloy with 1 or 2% thorium and (iii) tungsten alloyed with 0.3 to 0.5 percent of zirconium. Tungsten electrodes (99.5% purity) are less expensive and are generally used in less critical operations. Tungsten electrode has a relatively low current carrying capacty and a low resistance to contamination. Tungsten electrodes with 1–2% thorium (called thoriated electrodes) are superior to tungsten electrodes is several respects. They have higher electron emmissivity, better current carrying capacity, longer life and greater resistance to contamination. With these electrodes the arc starting is easier and the arc is more stable. Tungsten electrodes with 0.3–0.5% zirconium have properties that generally fall between those of tungsten and thoriated electrodes.

The importance and the utility of gas shielded arc welding arises from the fact that the process combines in itself the good features of oxy-acetylene and shielded metal arc welding. Oxy-acetylene offers excellent protection to the weld but its rate of heat input is low and not concentrated enough and because of this heat affected zone (HAZ) is large. In shielded metal arc welding because of high temperature and concentrated nature of source, the rate of heat input is high and therefore heat affected zone (HAZ) is small.

Amongst the two shielded inert gas arc welding processes namely TIG and MIG, the TIG process is less economical owing to the fact that feeding the filler rod is slow like in oxy-acetlylene welding and the anode/cathode drops of voltage and consequently the heat developed at these poles is not effectively utilised. In addition TIG process being manual it is preforce less economical.

2.3.1 Plasma Arc Welding

The term plasma refers to the ionised state of a gas. This state of gas is sometimes called the fourth state of matter, the other three being solid, liquid and gas. The term "Plasma Arc" is used to describe an electric arc which is constricted. The constriction of the electric arc, usually accomplished by passing the arc through a water cooled copper orifice, is intended to increase the energy density of the arc stream.

The plasma arc was introduced by Schonherr in 1909 who used it to study arc phenomenon. Schonherr used gas vortex to concentrate the arc and achieve higher current densities and higher temperature. Gage attempting to use Plasma Arc to melt refractory metals observed similarity between the plasma arc and the gas flame and noted the intense temperature of the, so called flame. Efforts to control the heat intensity and velocity of the "Plasma gas" let to the development of cutting of metals in 1955. Similarlity of this "new tool" with "tungsten arc" was soon realised and in 1963, let to the use of Plasma Arc for welding of metals.

It may be pointed out that plasma exists in any arc occurrence-welding arc carbon arc and in arc furnaces. The expression "plasma arc" is used only where the plasma is constricted and thereby its temperature and stability increases. The temperature of the plasma arc can go very high even beyond 18000°. Another important feature of the plasma arc is that it can have very long length.

In plasma arc the heat and velocity of gas can be varied over a wide range. With very low currents of the order of 0.1 amp, it can be used to light a cigaretle and with higher currents to melt highest temperature metals. The flame can be very soft having a very low velocity or can be very harsh with very high velocity. The velocity of the palsma gas is obtained from the conservation of momentum provided by arc constriction. The gas issuing from the plasma arc can acquire supersonic velocity which is several thousand feet per second. The plasma arc system is similar to the conventional tungsten inert gas welding process with the exception that the electrode is kept inside the copper orifice and the arc is constricted. The arc is struck with a pilot arc and like in TIG, tungsten electrode is used. The filler metal is also provided in the same way. The currents used arc in the range of 100 to 400 amps. The power source has drooping characterics and the open circuit voltage is of the order of 70 to 80 volts. Protection against oxidation is good and longer lengthsd of arc can be maintained.

2.4 GAS TUNGSTEN ARC WELDING

Gas tungsten arc welding (GTAW), also called tungsten inert gas welding (TIG), is a process that uses an arc between a non-consumable tungsten electrode and the base metal. During welding an inert gas shielding is used for protecting the weld metal, the electrode and hot areas of base metal. Filler metal may or may not be added in the weld metal. (Fig. 2.1). In GTAW, the arc is struck between the tungsten electrode and the base metal and the heat developed by the arc is used for melting the base metal and filler metal. A shielding gas either argon or helium is used for eliminating the oxidations of the weld metal, base metal and filler metal. The shielding gas is supplied to the weld zone from the gas cylinder. This shielding gas also acts as medium for ionisation of establishing the arc. Once the arc is established and a steady weld pool is created (if required filler metal is added to the weld pool), the TIG torch is moved along the groove steadily to effect the welding and complete the joint.

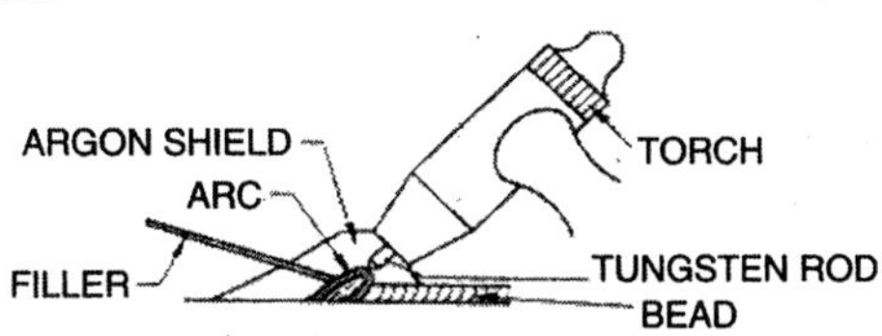

Fig. 2.1 GTA welding.

Advantages : The major advantages of GTAW process are as described below :

- **It produces a high quality weld :** The welds arc clean and free from spatters. Filler metal may or may not be added according to the thickness of the joint. Root pass welding can be controlled to produce a smooth contour in the under-bead. Autogenous (without filler wire) can be done for low thickness joints.
- **Welding can be done in all positions :** The process parameters can be controlled accurately to produce a consistent joint. Almost all metals and alloys can be welded. The process can be automated.

Limitations : Deposition rates are lower as compared to other processes, and hence completion time for weld joint is longer than in other processes. The welding cost is more because it uses argon gas for shielding. Compared to SMAW welder's skill requirement is more. Shielding becomes different in windy site welding condition.

2.4.1 GTAW System

The GTA welding system consists of :

(i) Power source.

(ii) Torch.

(iii) Shielding gas supply system.

(i) Power Source : (Fig. 2.2)

GTA welding uses a constant current type of power source which gives a more or less constant current even when there is a slight variation in the arc length. As the process is used mostly manually, the normal variations in the arc length due to the hand unsteadyness will not produce a drastic variation in the current. As the penetration in the joint is mainly controlled by the current, constant current will produce nearly inform penetration. Both D.C. and A.C. can be used for welding. In D.C. the electrode connection to the negative (DCEN) of the power source is the

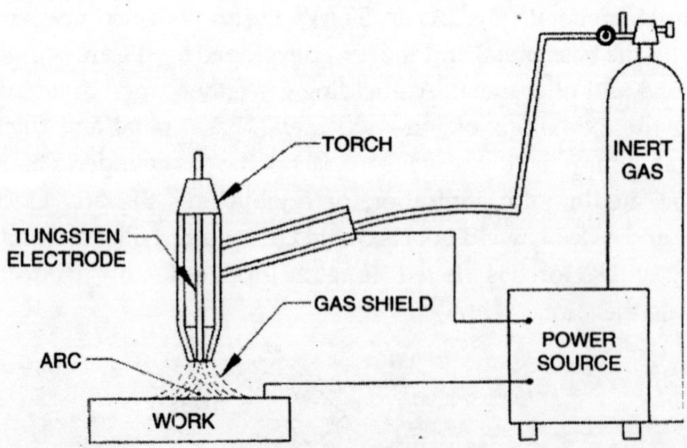

Fig. 2.2 GTAW system.

preferred type of connection, it is observed that the penetration is more for this type of connection, and hence widely used. The other type of connection DC electrode positive (DCEP) results in poor penetration and overheating of the torch (Fig. 2.3). Hence, normally this is not done except for metals line aluminium and magnesium. In the case of these metals a thin touch oxide layer is always present on the surface and as soon as it is removed by any cleaning methods, it forms again very quickly. During welding this layer can cause defects like oxide inclusions or lack of fusion. It is observed that this oxide layer can be cleaned by the arc itself in the DCEP. As already explained about this connection gives a poor penetrations.

But, in AC polarity is alternating and the electrode polarity goes through a half cycle during which the oxide is cleaned. Hence, for aluminium and magnesium and their alloys AC is preferred. Figure 2.3 and Table 2.1 shows that the different materials are welded with preferred polarity. Different types of power sources are available to provide a constant current either in DC or AC.

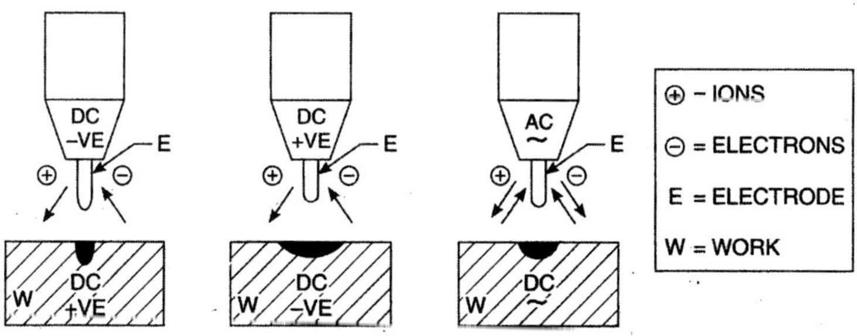

Fig. 2.3 Influence of current and polarity on bead geometry.

Table 2.1 : GTAW Parameters for Stainless Steel Sheets

Characteristics	DCEN	DCEP	AC
Penetration	Deep and narrow	Shallow and wide	Medium
Heat distribution	$E = 33\%, W = 67\%$	$E = 67\%, W = 33\%$	$E = 1/2, W = 1/2$
Cleaning of oxide	Not possible	Fully possible	Only in one half cycle
Electrode	—	—	—
Capacity	∅ 3.2 mm - 350 amps	∅ 6.3 mm - 120 amps	∅ 3.2 mm - 220 amps

Motor-generators, transformers-rectifiers, thyristorised and transistorised power sources are all suitable. Electronically-controlled power sources can be used with feedback control and can given very accurate control of current. They are however expensive. For the accurate control of the bead, a pulsed current attachment is used with generators and rectifiers.

(ii) GTAW Torches

The GTAW torch holds the tungsten electrode and directs the arc to the desired locations. It also provides a shielding gas supply to shield the arc zone. [Fig. 2.4(a)]. The tungsten electrode is held in a brass collet and the collet is screened to the rear cap of the torch. The torch body is cooled by water in the case of water cooled torches and by the shielding gas in the case of air-cooled or gas-cooled

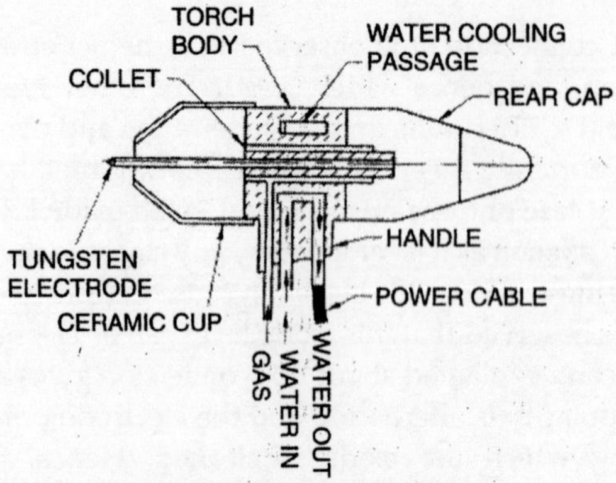

Fig. 2.4(a) TIG torch configuration.

torches. A gas nozzle located in the front part of the torch directs the shielding gas to the arc zone. The shielding gas issues out of the torch through the annular gap between the electrode and the gas nozzle.

(iii) GTAW Electrodes

The electrode is made of tungsten or an alloy of it and this provides easy arc starting and steady arcing. Tungsten is chosen as the electrode material for its high melting point of about 3410°C. It also has a good electrical and thermal conductivity and electron emmissivity. Tungsten rods are available in different diameters and lengths. The diameters are 0.5, 0.8, 1.0, 1.6, 2.24, 3.2, 4.0, 5.0 and 6.3 mm. The lengths are varying from 50 mm to 500 mm : 150 mm being the most commonly used. Electrodes are normally ground to a vertex angle of 60° [Fig. 2.4(b)] for DCEN polarity welding. Figure 2.4(c) shows the recommended tip shape for AC Tig welding. The common varieties of the tungsten electrode are :

(i) Pure tungsten.
(ii) 1% or 2% thorium oxide and tungsten.
(iii) 0.5% zirconium oxide and tungsten.

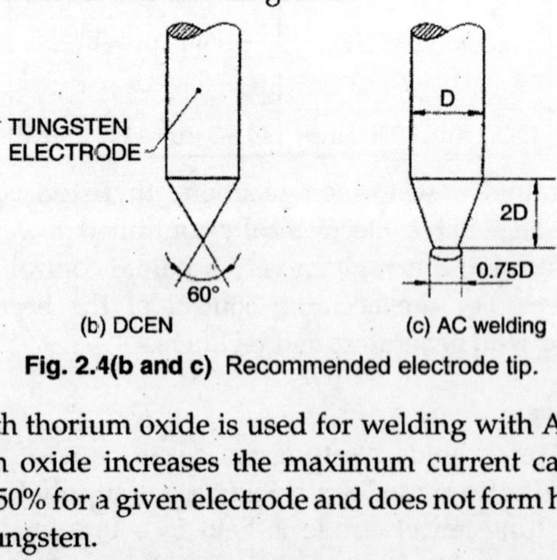

(b) DCEN (c) AC welding

Fig. 2.4(b and c) Recommended electrode tip.

Tungsten with thorium oxide is used for welding with AC. An additions of 1% or 2% thorium oxide increases the maximum current carrying capacity by approximately 45-50% for a given electrode and does not form hemispherical blobs as does the pure tungsten.

Thoriated tungsten electrodes are, however, preferred for DC as the arc wanders when used on AC. Red and yellow colour bands are used widely to indicate 2% and 1% alloyed thoriated tungsten electrodes. Zirconium alloyed tungsten is used in AC applications because of its high resistance to contaminations as well as good arcing characteristics. The current carrying capacity depends on the type of shielding gas (whether argon or helium), the length of electrode sticking out of the collet, the cooling (water or gas) of the holder, the position of welding and polarity used. With larger electrodes for a specified current, the arc becomes erratic and welding becomes difficult. The current, the arc becomes erratic and welding becomes difficult. The selection of smaller diameter rods for a particular current, however, would increase chances of the electrode melting. Argon and helium are the common inert gases selected for gas tungsten arc welding. Argon gas is used to shield the arc because of the following reasons :

(i) It smoothness and quietens arc action.

(ii) Permits lower arc voltage.

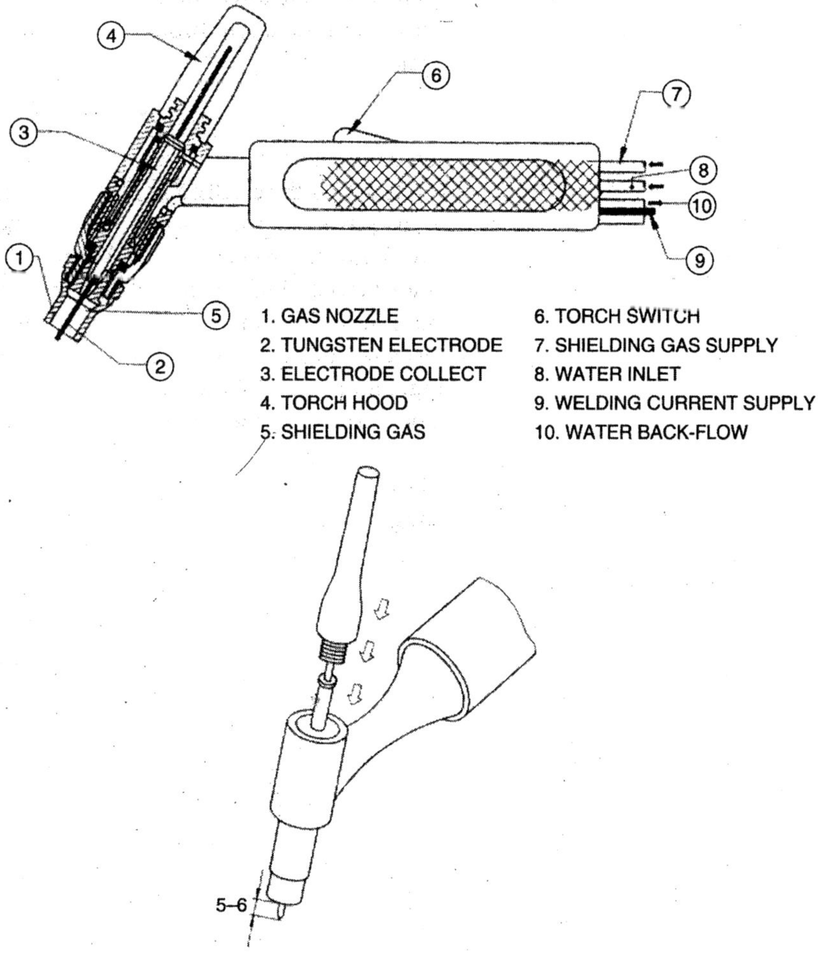

Fig. 2.5 TIG torch and its accessories.

(iii) Its cheapness and good availability.

(iv) Lower flow rates are enough for good shielding.

The welding parameters for argon shielded gas tungsten arc welding varies with respect to the type and thickness of the material and the welding positions involved. Table 2.2 gives selection of GTAW shielding gases for different materials. Table 2.3 and 2.4 give the GTA welding parameters for stainless steel and aluminium. Figure 2.5 shows the TIG torch and its accessories.

Table 2.2 : Selection of GTAW Shielding Gases

Gas or Gas Mixture	Application
Argon 99.5% pure	Almost all materials.
Argon + 5% hydrogen (occasionally upto 15% H_2	Stainless steel, nickel and nickel alloys.
Argon + helium (75% Ar + 25% He)	Aluminium alloys.
Helium 99.99% pure	Almost all materials.
Nitrogen 99.9% pure	Copper and some of the copper alloys.
Argon + Nitrogen (85% Ar + 15% N_2)	Copper and some of the copper alloys.

Table 2.3 : GTAW Parameters for Stainless Steel Sheets

Sheet thickn-ess nm	Welding current (amps)			Elect-rode ∅ mm	Filler wire mm	Welding speed nm/min.	Argon lit./min	Num-ber of passes
	Hori-zontal	Verti-cal	Over-hand					
0.8	15-30	14-28	13-27	1	–	300-350	3	1
1.0 B	25-60	23-55	22-54	1	1	250-300	4	1
1.0 F	55	51	50	1	1.5	250-300	4	1
1.5 B	25-60	23-55	22-54	1	1.5	250-300	4	1
1.5 F	90	85	80	1	2	250-300	4	1
2 B	80-110	75-100	70-100	1.5-2	1.5-2	175-225	4	1
2 F	105	98	95	1.5-2	2	175-225	4	1
3 B	120-200	110-185	110-180	2-3	2	125-175	5	1
3 B	125	115	110	2-3	3	125-175	5	1
4 B	120-200	110-185	110-180	2	3	100-150	5	1
4 F	180	165	160	2	2	100-150	5	1
5 B	150-250	140-230	135-225	2-3	3-4		5	1
5 F	300	280	270	3	4		4	1

Table 2.4 : GTAW Parameters for Aluminium and Aluminium Allow Sheets

Sheet thickness nm	Welding current (amps)			Electrode ∅ mm	Filler wire mm	Welding speed nm/min.	Argon lit./min	Number of passes
	Horizontal	Vertical	Overhand					
1.0 B	35-60	35-60	35-60	1-1.6	1.6-2	270-330	4	1
1.0 F	45-70	30-50	40-70	1-1.6	1.6-2	245-295	5	1
1.5 B	60-80	60-80	60-80	1.6	1.6-2	255-305	6	1
1.5 F	70-90	70-90	70-90	1.6	1.6-2	220-255	5	1
2.5 B	75-100	75-100	75-100	1.6-2.4	2-2.4	255-270	5	1
2.5 F	65-90	65-90	65-90	1.6-2.4	2-4	220-245	6	1
4.0 B	150-180	140-170	125-160	2.4	2.4-3.2	255-280	7	1
4.0 F	160-200	150-190	145-180	2.4	2.4-3.2	205-255	8	1
5.0 B	190-220	190-220	180-210	3.2	3.2-4	220-255	8-9	1
5.0 F	210-240	190-220	180-210	3.2	3.2-4	170-200	8-9	1

2.4.2 Plasma Welding

(a) Process Fundamentals

Plasma welding uses a non-consumable tungsten electrode as in TIG welding. The torch has a nozzle that creates a gas chamber surrounding the electrode. An inert gas supply called the plasma gas gets ionised in this chamber and issues, out through an orifice in the nozzle. In the case of the transferred arcs the arc is maintained between the electrode and the job. In the case of non-transferred arcs the arc is struck between the copper nozzle and the tungsten electrode. The arc in plasma welding is more concentrated and is of higher temperature compared to TIG arc. (Fig. 2.6). This produces a deeper penetration in the welds and provides for faster welding speeds. In some cases welding may be effected without any filler metal. Figure 2.7 indicates the schematics of the process.

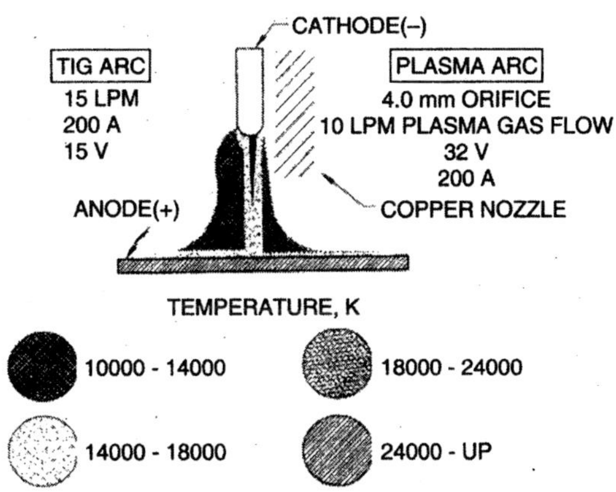

Fig. 2.6 Typical effect of arc constriction on temperature.

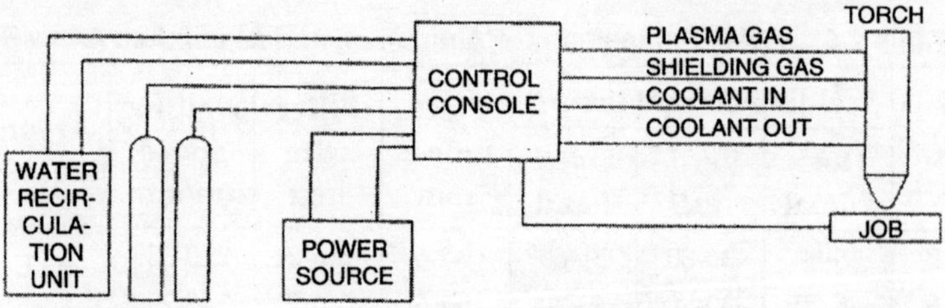

Fig. 2.7 Plasma welding system.

Advantages :

(i) Plasma arc welding can be used to weld almost all metals and alloys.

(ii) It provides a deep penetration resulting in reduced welding time.

(iii) The arc is very stable and is less affected by external disturbances such as magnetic field and wind. This enables the welder to have better directional are control.

(iv) As the electrode is placed inside the chamber, the electrode can not come in control with the molten metal and contaminate the weld and the filler metal cannot touch the electrode to cause a disturbed arc.

Disadvantages :

(i) The equipment is costly.

(ii) The process is complex and hence it requires more operator training to understand the process parameters interaction and use the process-effectively.

(b) Process Techniques

The process can be operated under two modes :

(i) Keyhole mode (ii) Melt-in mode.

(i) Keyhole Mode

In plasma welding of certain ranges of metal thicknesses, appropriate combinations of plasma gas flow, arc current and weld travel speed will produce a relatively small weld pool with a hole, penetrating completely through the base metal. This is called a keyhole. Plasma welding is generally performed in the down- hand welding positions on material thicknesses ranging from 3.15 mm to 9.5 mm. In the keyhole operation, the molten metal is displaced to the top bead surface by the arc to form the keyhole. As the torch is moved along the weld joint, the metal melted at the front side of the keyhole is pushed by the arc around to its back where the molten metal gradually solidifes to form the weld. The main advantage of the keyhole welding is the making of the weld in a single pass. As the parameters specifically, the speed arc to be controlled accurately, this mode is used in automatic welding only.

(ii) Melt-in Mode

In the melt-in mode the welding is done as in the TIG welding. The arc does not have as much force as in the keyhole mode and the molten metal formed by the arc heat conducts further heat to the bottom and the root is melted to effect the joint. The penetration is also not as deep as in the keyhole mode and the bead is also wider.

(c) Equipments

The process uses a constant current type of power source as in TIG welding and in manual metal arc welding. Both DC and AC can be used. As in TIG welding, DCEN is always used in all the alloys except aluminium where AC is used. The power source can be transformer rectifier, thyristor or transistor type. In some cases the pulsed current welding is also performed for producing a consistent joint. The arc initiation in plasma arc welding is effected by the pilot arc. A pilot arc is

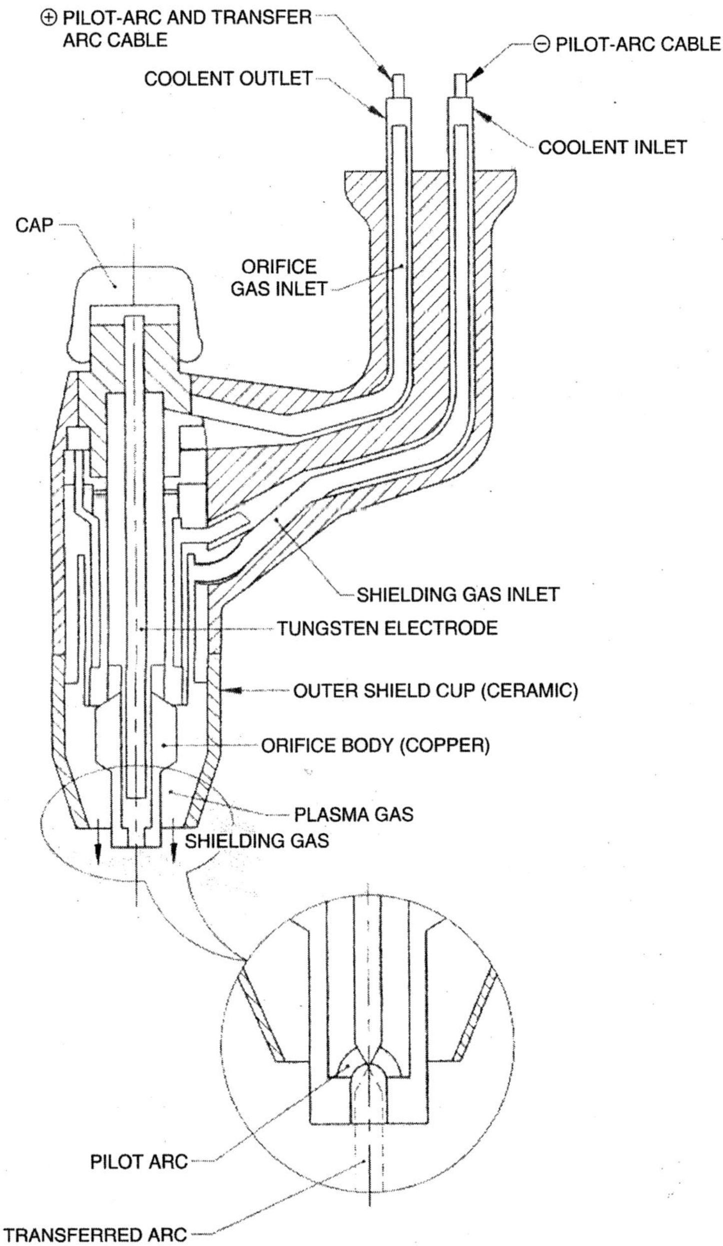

Fig. 2.8 Typical manual plasma torch head design.

initiated by the H.F. spark between the tungsten electrode and the copper nozzle using a low current pilot power source. The pilot arc, issuing out of the nozzle through the orifice, acts as a conducting medium for the arc. The pilot supply is switched off once the arc is established. In some cases such as in the surfacing applications, the pilot arc is continuously switched on to maintain the arc.

(d) Plasma Welding Torch

The plasma welding process utilizes mostly the transferred mode of arc, and hence the arc is between the electrode and the job. The torch is a complex one in construction and consists of following components. (Fig. 2.8) Electrode holder and electrode, nozzle and nozzle holder, torch body, shielding gas cup and the gas lens. The electrode made of thoriated tungsten is held in a brass collet which is held to the torch body. The nozzle is screwed to the nozzle holder which is water cooled. The gas cup is screwed to the front of the nozzle. The annular space between the electrode and the nozzle acts as a chamber for ionization. The annular space between the copper nozzle and the shielding gas cup with a gas lens acts as a passage for the shielding gas. The copper nozzle is continuously water cooled since it is in touch with the plasma. It is an essential requirement and controls are provided to switch off the current if the water supply is not present. Touches are water cooled even for the lowest current for the above reason. The copper nozzle has orifices of different diameters depending on the current used. The torch connections are :

(i) Water inlet with pilot power supply cable.
(ii) Water outlet with the power supply cable connected to the negative of the power supply.
(iii) Plasma gas supply.
(iv) Shielding gas supply.

(e) Control Console

A control console is a unit which houses the various essential controls for the process. Arc current control, plasma gas supply, shielding gas supply, pilot power supply controls are all provided in the console. Normally an option is provided for the pilot power supply to be on continuously or switched it off automatically once the main arc is established. Additionally, interlocks are provided for ensuring the safety of the torch. If the gas or water supply is absent then the system is designed for power shut-off Fig. 2.9 shows the plasma welding torch.

Fig. 2.9 Micro plasma welding torch.

(f) Gases used in Plasma Welding

The gases used are normally argon for plasma gas or argon and hydrogen mixture for the shielding gas. These are supplied from the gas cylinders and connected through the pressure regulators and flow meters. The final flow is from 1 to 10 liters per minute for the plasma gas and 6 to 50 liters per minute for the shielding gas. The pressure is to be set at the same level for both the gases. The hydrogen-argon (5% hydrogen) mixtures can be mixed in the mixing unit before use. The argon used is of 99.95% purity. The gas flow rates are chosen as per the guidelines of the plasma torch manufacturer.

(g) Filler Metals

The filler metal may be either in wire form or powder form for the plasma welding. In the case of wire forms, the fillers used for TIG welding can be used. The feeding can be either manual using rods or wires from spods in the case of mechanised welding. The diameters used range from 0.5 to 6 mm. In the case of powders the consumables are in the power form, only for surfacing applications as these alloy cannot be coiled on to the spods or drawn to wires. Separate powder feeders are available for the feeding of these powders and fed into the arc zone through a special nozzle. For these applications of powder feeding automatic systems are used.

2.4.3 Micro-plasma Welding

Micro-plasma welding is a low current version of the plasma welding. With this process thin components of various metal combinations are welded. The current used can be from 0.2 Amp to 50 Amp. The thicknesses that can be welded using this process are 0.1 mm to 0.5 mm. The torch is smaller than the conventional torch. This torch also requires water cooling even through the current used could be as low as 1 Amp. Normally this process is used for welding of thin components such as bellows, sensors etc. current pulsing, ranging from 5 pulses to 5000 pulses per second is commonly applied for the welding of precision components.

2.4.4 Safety in Plasma Welding

As the plasma arc welding has a bright arc at a high temperature proper protection must be provided for the eyes and body of welder. Shield glasses of DIN 6 to DIN 15 types are used when plasma welding is done. The parts of the body exposed to the arc will be affected due to the severe ultraviolet and infrared radiation emanating from the arc. The parts must be covered by protective clothing. For persons working in the far vicinity of the welding area it is enough if light clothing preferably of light colours are worn. For the welder or the operator near the weld all parts of the body must be covered fully to avoid skin burns. For welding of metals from which can emanate injurious fumes proper exhaust form must be provided.

2.5 GAS METAL ARC AND FLUX-CORED ARC WELDING

Metal inert gas (MIG) welding is a process in which an electric arc is formed and maintained between a continuously fed consumable electrode wire and the work piece. The tip of the consumable electrode wire is melted and the molten metal is transferred across the arc into the weld pool. The arc and the weld pool are protected from atmospheric contamination by the shielding gas supplied through the nozzle. The selection of the shielding gas depends on the base metal that is being welded. Generally for welding of non-ferrons metals, inert gas like argon or helium is used. When such inert gases are used for welding, the process is called 'Metal Inert Gas Welding' (MIG). However, for welding of steel, CO_2 gas or argon

+ CO_2 mixture is used as the shielding gas. When CO_2 or argon + CO_2 mixture gas is used for shielding, then the process is called either as 'metal generally gas' (MAG) welding or CO_2 welding.

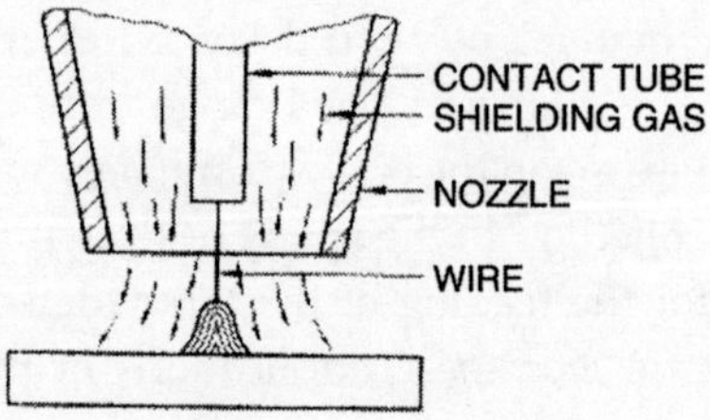

Fig. 2.10 Schematic view MIG welding gun.

The MIG welding process generally employs a DC power source. The consumable electrode wire is always connected to the positive of the power source and the work-piece is connected to the negative of the power, source. This type of connection is known as "Direct Current Reverse Polarity" (DCEP) or more explicity "Direct Current Electrode Positive" (DCEP). The consumable electrode wire is fed through a copper contact tube from which it picks up the current. The gas flows out through a nozzle surrounding the contact tube as shown in fig. 2.10. The gap between the nozzle and the work-piece is known as "stand off distance". The stand off distance should be maintained between 10 to 20 mm to ensure positive gas protection of the weld pool from atmospheric contamination. The distance between the tip of the contact tube to the arc is known as electrode extension or "stick out".

2.5.1 Equipment and Accessories

The MIG welding equipment comprises of a power source, a wire feeding unit, welding torch and accessories (Fig. 2.11). MIG welding equipment can be used either manually or automatically. In manual welding, the arc travel is performed by the welder whereas in automatic welding the arc travel is automatic and is controlled by the positioner.

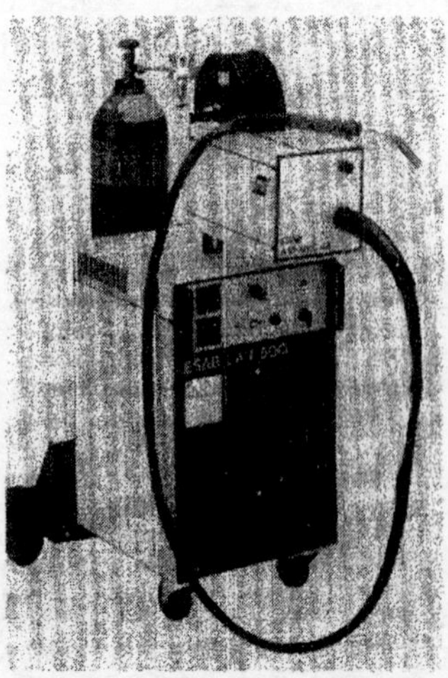

Fig. 2.11 MIG welding equipments.

2.5.2 Wire Feeders

The basic requirement of the wire feeder is to feed electrode wire continuously into the arc through the torch. The control provided in the wire feeder maintains the wire feed rate at a predetermined rate. The required wire feed rate is obtained by setting a dial on the control. The control also regulates the starting and stopping of the shielding gas flow and wire feed in response to a signal from the torch switch. Also the shielding gas, cooling water and welding power are delivered to the torch through the wire feeder control box. The pre-flow, post flow and energising the power supply contactor are also determined by the control. Conventional wire feedes have two or four roll drive for feeding the wire. The feed rolls have different sizes of V-grooves on its periphery to feed wires of different diameters. Based on the location of the wire feeder in relation to the spool and torch, they arc classified as push type, pull type and push-pull type.

2.5.3 MIG Torch

The welding torch guides the wire and shielding gas into the weld zone and also transmits current to the wire through the contact tube. Further, the torch must be flexible, light in weight and of ergonomic design. Based on the type of cooling used, the MIG torches can be classified as gas-cooled or water-cooled torches. Gas-cooled torches used the shield gas to assist in cooling the torch during operation. Figure 2.12 shows the construction of a gas-cooled torch. It is important to note that the shield gas is channeled along an extended surface of the inner conductor, which removes the heat conducted away from the contact tip area and also pre-heats the shield gas. Also the outer case of the neck is electrically isolated from the central conductor, allowing direct contact between the gas nozzle and its finned seating area. Based on the type of cooling used water cooled torches are also available. The temperature of the gas nozzle is prevented from rising to excessive

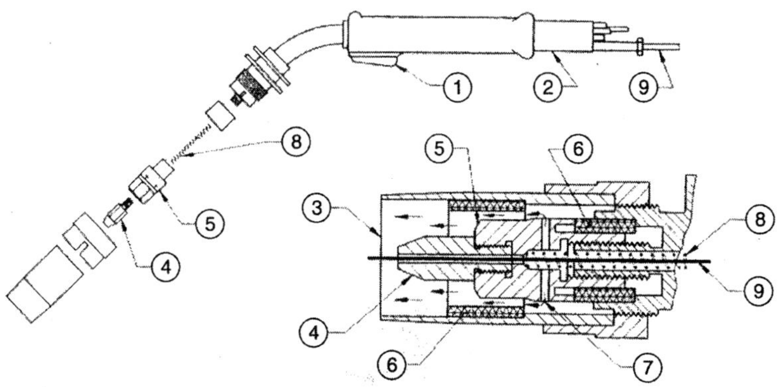

1. TORCH SWITCH
2. HOSE ASSEMBLY
3. GAS NOZZLE
4. CURRENT CONTACT NOZZLE
5. CURRENT CONTACT HOLDER
6. INSULATORS
7. SHIELDING GAS
8. SPIRALLY WOUND WIRE ELECTRODE GUIDE
9. WIRE ELECTRODE

Fig. 2.12 MIG gas cooled torch.

levels during welding. Gas-cooled torches can be used for current upto 400 amps. in the semi-automatic mode.

Water-cooled torches require a pump for circulating the water through the torch. The water cools the inner conductor and gas nozzle seating and on its return it cools the power cable, resulting in a considerable weight reaction of the cable. The major disadvantage of the water cooled torch is that a lack of coolant will result in a rapid failure of the torch cable. Therefore, for the safe operation of the water cooled torch water pressure switches must be provided so that welding cannot take place without water flow. Water-cooled torches are used for currents greater than 400 amps and for mechanized welding applications. The contact tube is made of copper in order to transmit the current effectively. The bore diameter of the contact tube is important because the wire must be fed easily through the contact tube and also should make good electrical contact. Uniform feeding of the wire is necessary for good arc stability. To ensure uniform feeding the wire is fed through a flexible spiral of the conduit with a liner provided within the torch. The inner diameter of the spiral conduit is again very important. Different conduits are available to suit different diameters of wires. When using steel wires a steel spiral conduit is used and for aluminium wires, nylon liners are uses.

2.5.4 Wire Burn-off Characteristics

A linear relationship exists between the rate at which the wire is fed into the arc and the current required to melt it off to maintain a constant arc length. This is known as the burn-off characteristics and it differs for different filler wire composition and diameters. Figure 2.13 and 2.14 illustrate the typical burn-off characteristics of

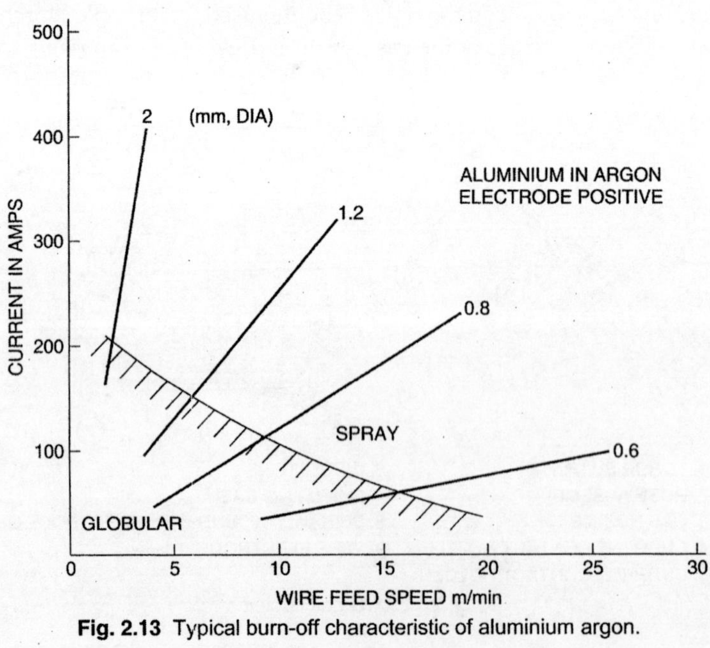

Fig. 2.13 Typical burn-off characteristic of aluminium argon.

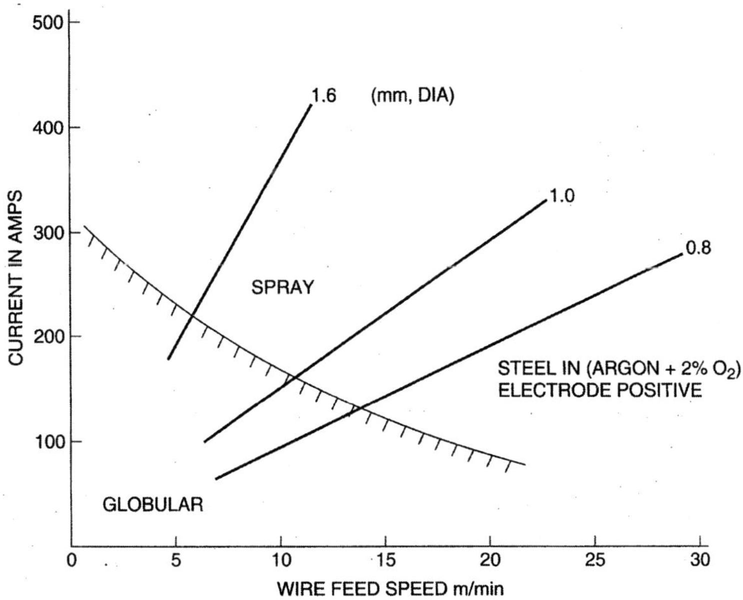

Fig. 2.14 Typical burn-off characteristic of steel in CO_2.

aluminium in argon and steel in CO_2 respectively. It can be observed from the figures that to alter the welding current a proportionate increase in wire feed rate is required. The amount of current drawn by the arc depends on the wire diameter, wire material and the feed rate. For a given wire material, at the same wire feed rate, the smaller diameter wires drawless current compared to the bigger diameter wires.

2.5.5 Power Source-Characteristics and Features

A power source with constant voltage type characteristic is used for MIG welding. The constant voltage type power source is used because it provides "Self regulated arc" and is also able to meet special demands of the short-circuit welding. Self-regulation of the arc means maintaining the arc length at a constant level throughout. This requires a balance between the wire feed rate and wire melting rate. Since in MIG welding wire feed rate is preset and kept constant, any fluctuation in the arc length can be corrected only by wire melting rate. The wire melting rate is a function of the welding current. That means to vary wire melting rate, the welding current must be varied. This variation of the wire melting rate with current is best achieved in the constant voltage power source compared to a constant current power source. The constant voltage power source provides best self-regulation of the arc length, this can be explained as shown is Fig. 2.15. The operating point is determined by the output characteristics of the power source and the arc characteristics. That means the point at which the power source characteristics and the arc characteristics meets become the operating point. Any fluctuation in the arc length will shift the operating point to right or left. For instance, if the arc length is increased, for whatever reason, the operating point will shift to the left, resulting in slower burn-off rate and consequently the arc length

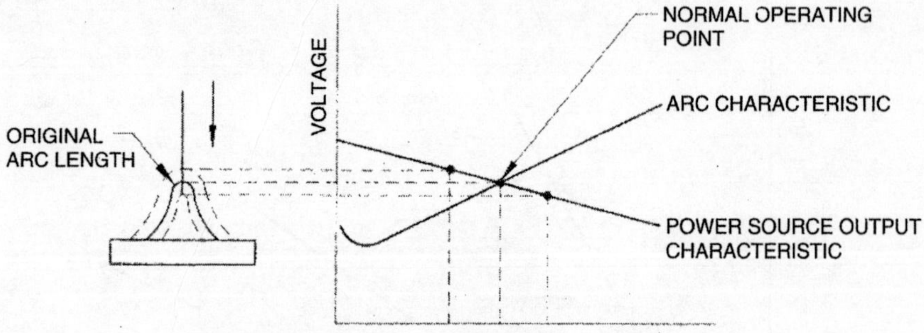

Fig. 2.15 Principle of self regulating of the arc length.

will be adjusted to the original level. On the other hand if the arc length is shortened, *i.e.*, less than the arc length corresponding to the original operating point, the operating point will shift to the right, resulting in length will be adjusted to the original level. This self-adjustment will variations, is sufficiently large to produce in current, produced by voltage variation, is sufficiently large to produce a large change in the burn-off rate. From this point of view the smaller diameter wires (*i.e.*, 1.0, 1.2, 1.6 mm diameter) lend themselves easily to better self-adjustment of the arc, compared to the larger diameter wires.

2.5.6 Shielding Gases

The principal functions of the shielding gases are :

(i) Protection of the weld zone from atmospheric contamination.

(ii) Providing stable arc.

(iii) Providing good bead geometry.

(iv) Providing improved weld metal properties.

The most common gases used in GMAW are argon, helium and CO_2.

(a) Argon

It is an inert gas which produces a stable arc. It has high density relative to air and therefore, gives good shielding efficiency. The low ionisation potential (15.8 eV) facilitates easy arc striking. However, when pure argon is used for welding of steel, it results in arc wandering and poor bead wetting characteristics. As a result pure argon is seldom used for welding of steel. Ar + 20% Co_2 or Ar + 1% oxygen mixtures are generally used for welding of steel. For welding of non-ferrous metals such as aluminuium, copper etc. pure argon can be used for shielding.

(b) Helium

Helium gas is chemically inert but it has the disadvantage of low density relative to air and also it is very costly. These features have restricted its use as shielding gas.

(c) Carbon dioxide

Carbon dioxide is used only for gas metal arc welding of steel. It is an active gas. In the welding arc, the CO_2 gas dissociates into carbon mono-oxide and free oxygen resulting in highly oxidising arc atmosphere. The oxidising tendency of the CO_2 gas is counteracted by adding sufficient deoxidisers such as Si, Mn, etc. in the the wire. Since these de-oxidising elements have higher affinity to oxygen than to iron or carbon, the above oxidising reactions arc suppressed and result in a sound weld metal.

(d) Shielding Gas Mixtures

In GMA welding, argon + CO_2 or argon + CO_2 + O_2 mixtures are used mainly to reduce the spatter and maintain arc stability. The gas mixtures containing less than 20% CO_2 given stable spray are transfer at high current with no spattering. It produces a finger type penetration profile. The gas mixture can also be used for thin sheet welding using short circuit transfer.

2.5.7 Metal Transfer

In MIG welding molten metal droplets are transferred from the continuously filler wire to the weld pool. The way in which the metal is transferred across the arc gap influences the spatter levels, stability, positional capabilities, bead shape and the weld quality. The metal transfer occurring in MIG welding can be broadly classified into three types : namely, dip transfer (short circuit transfer), globular tansfer and spray transfer.

(a) Dip Transfer

The dip or short-circuit transfer is characteristics by the repeated short-circuiting of the wire with the weld pool. The dip transfer occurs at low current and low voltage. Under such conditions the wire melting rate is insufficient to maintain a constant arc length. The arc gap gradually closes and the wire eventually touches the weld pool. In response to this electrical, short-circuiting, the current from the power source increases rapidly causing rupture of the bridges between the wire tip and the weld pool. During the process of rapture, a portion of the heated electrode tip is transferred to the weld pool and in the gap the arc is reestablished. The process is then repeated. A typical dip transfer welding produces very low heat input levels and small weld pool size. Fig. 2.16. Therefore, it is well suited for this sheet welding and all-position welding.

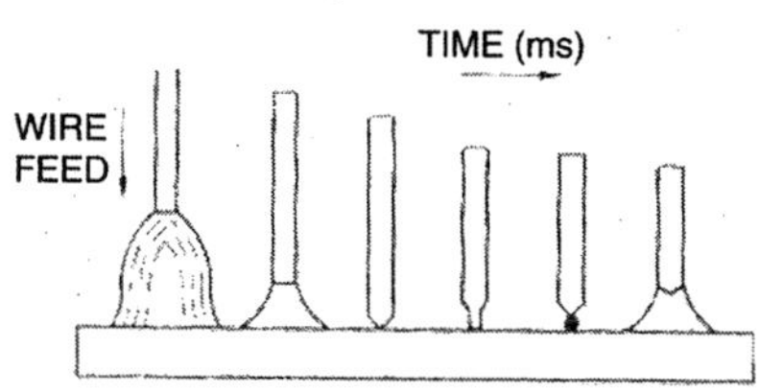

Fig. 2.16 Dip transfer.

(b) Globular Transfer

The globular mode of transfer occurs at medium range of current and voltage. In this mode droplet sizes are large, the rate of transfer is low and irregular. [Refer Fig. 2.17(a)]. The globular transfer being the transition from dip transfer to spray transfer, it occurs over a very narrow range of current. It can be used in down hand position only. Therefore, it has very limited application.

(c) Spray Transfer

The spray transfer occurs at higher currents and voltages. In this mode the size of the droplets at the end of the wire tip is decreased and is transferred across the arc as a fine spray. [Fig. 2.17(b)]. Since this mode of transfer can only be produced at high currents, the heat input is relatively high and the weld pool will be large. Therefore, this mode of transfer is suitable for welding of thick plates but only in downhand position. It cannot be applied for out of position welding. Spray transfer provides a stable arc, deep penetration and smooth weld bead profile without any spatter.

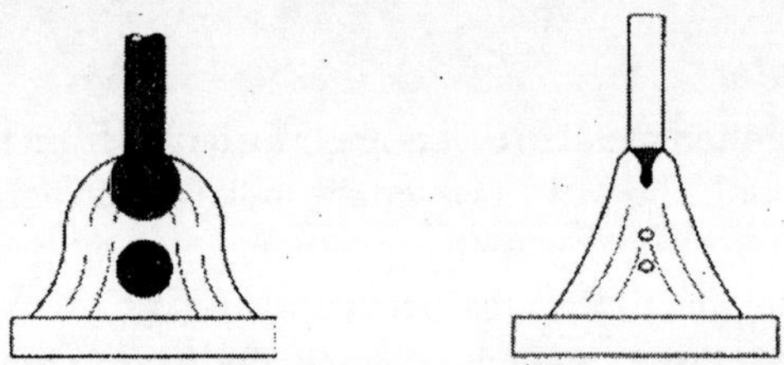

(a) Globular transfer (b) Spray transfer

Fig. 2.17 Metal transfer.

(d) Repelled Transfer

Repelled transfer also known as 'non-axial globular transfer' occurs generally in solid wires CO_2 welding where current and voltage are significantly above the range of short-circuiting transfer. The repulsion of the droplet is thought to be due to the electromagnetic force which acts between the plate and a point on the droplet surface to cause a reaction force towards the electrode wire tip, capable of lifting the molten droplet. (Fig. 2.18). The repelled transfer or non-axial transfer generally causes spattering problem. In CO_2 welding, spattering problem can be reduced by using the "burried arc technique". The burried arc technique involves using lower arc voltage for any given current. In so doing the arc is burried in a cavity created by the arc force and the spatter generated is trapped by the surrounding cavity. However, the welding speed has to be properly adjusted to achieve proper wetting of the weld bead.

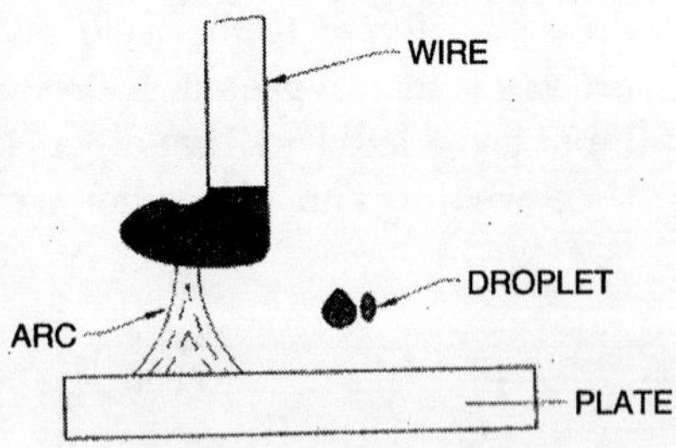

Fig. 2.18 Repelled transfer.

2.5.8 Classification of Consumables

The electrode wires used in MIG and MAG welding are quite small in diameter compared to submerged arc welding. The wire diameters commonly available are 0.8, 1.0, 1.2, 1.6 and 2.0 mm. However, 1.2 mm and 1.6 mm diameters wires are mostly used. These wires are available as continuous coils wound in standard spools. All the solid wires used in GMAW welding are coated with copper. The copper coating prevents the resting of the wire and improves the shelf life. It also improves the current pickup when the wire passes through the copper contact tube. Almost all the electrode wires used for GMAW of steel have deoxygenerating elements added to them to minimize the amount of porosity and improves the mechanical properties. The use of electrode wire switch appropriate amount of deoxygenerators is very important when using CO_2 or oxygen bearing shielding gas. Because of the smaller electrode wire size and relatively wide current range employed with each wire size, the feed ratio of wire normally ranges from about 2 nots/minute to 16 mts/minute. The smaller wire size results in high surface to volume ratio and therefore the cleanliness of wire becomes very important. Drawing compounds, rust, oil or other foreign matters at the surface of the electrode wire can cause weld metals defects like porosity etc.

2.5.9 Flux-cored Arc Welding

Flux-cored arc welding (FCAW) is an arc welding process in which the heat for welding is produces by an arc established between the flux-cored tubular consumable electrode wire and the work-piece. There are two major versions of the process, namely "self-shielded type" (in which the flux performs all the functions of shielding) and the "gas-shielded" type which requires additional gas shielding. The gas shielded type FCAW is widely employed for welding of carbon steel, low alloy steel and stainless steel in flat, horizontal and overhead positions. (Fig. 2.19). However, the self-shielded type FCAW is mainly used for carbon steel welding and the quality of weld produced by this type is generally inferior to that of welds made with gas shielded type.

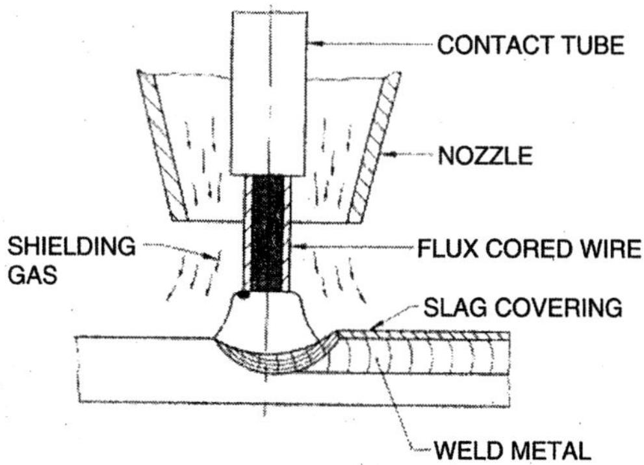

Fig. 2.19 Flux cored arc welding.

(a) Equipment

The noticeable differences between the equipment used for GMAW and FCAW are in the construction of welding torch and feed rollers. The welding torch used for self-shielding wire is very simple in construction as there is no need for the gas nozzle. Similarly, the feed rolllers used for flux-cored wires have to ensure positive feeding of the wire without applying to much pressure on the soft tubular wire.

(b) Metal Transfer in FCAW

The metal transfer in FCAW differs significantly GMAW process. FCAW process exhibits two distinctly different modes of metal transfer namely, large droplet transfer and small droplet transfer. However, both are classified as free flight transfer. The FCAW process does not produce a stable dip transfer as that of solid wire GMAW. The large droplet transfer occurs at the lower current-voltage ranges. At higher current-voltage ranges the transfer mode changes to smaller-droplet transfer. An important aspect to be observed during FCAW metal transfer is the presence of the 'flux-pole' at the core of the arc column protruding into the arc. The flux-appears only during welding with basic type flux-cored wire. However with rutile wire flux pole does not occur and the metal transfer is of spray type (Fig. 2.20).

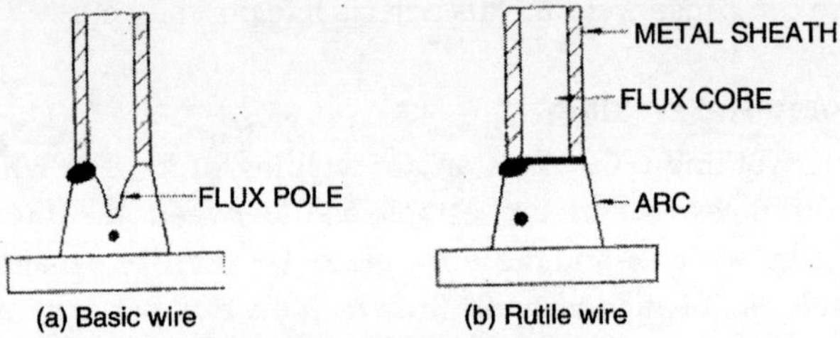

Fig. 2.20 Appearance of flux pole.

(c) Classification of Flux-cored Wire

The basic function of the flux contained within the tubular wire include providing protective slag on the weld bead, introducing the required alloying elements and deoxygenerators into the weld pool and providing stability to the arc besides producing the required shielding medium to protect the arc and the weld medium. Flux-cored wires are now available for welding of plain carbon steel, low-alloy steel and stainless steel and also for hard-facing applications. These wires based on the nature of the flux may be classified as rutile-gas shielded, basic gas shielded, metal cored and self-shielded. Retile gas shielded wires have extremely good arc running characteristics, excellent positional welding capability and good slag removal and mechanical properties. Basic gas shielded wires give reasonable arc characteristics, excellent tolerance to operating parameters and good mechanical properties. Metal-cored wires contain very little mineral flux, the major constituent being iron power and ferro alloys. These wires give smooth spray transfer in argon/CO_2 gas mixture. They generate minimum slag and are suitable for mechanised welding applications. Self-shielded wires are available for general purpose downhand welding.

The flux cored wires are available in both seamless and folded types. The seamless type is generally coated with copper, whereas the folded type of wires (*i.e.*, close butt and overlapped type) are treated with special compounds.

(d) Deposition Rate and Efficiency

Deposition rate is defined as the weight of metal deposited per unit time. The deposition efficiency is defined as the ratio of weight of weld metal effectively deposited to the weight of wire consumed. In GMAW welding the deposition efficiency is generally between 93 to 97% and in FCAW the corresponding figure is between 80-86%. These values are determined by the spatter losses and slag formation. The low depositions efficiency in the case of FCAW is due to the slag formation. Generally, the spatter loss can be minimised by using Argon/CO_2 mixed gas instead of CO_2 gas.

2.5.10 GMA Narrow Gap Welding

Narrow gap welding is a technique wherein welding of over thick plates is accomplished with parallel-sided butt joint preparation. As a result in narrow gap welding the weld volume is greatly reduced. Since narrow gap welding employs single parallel-sided butt joint, achieving adequate side wall penetration is the prime concern. In GMAW-NG welding a number of techniques are used to achieve the side wall penetration. These techniques are classified as :

(i) Curved fixed electrode narrow gas welding technique (Fig. 2.21)

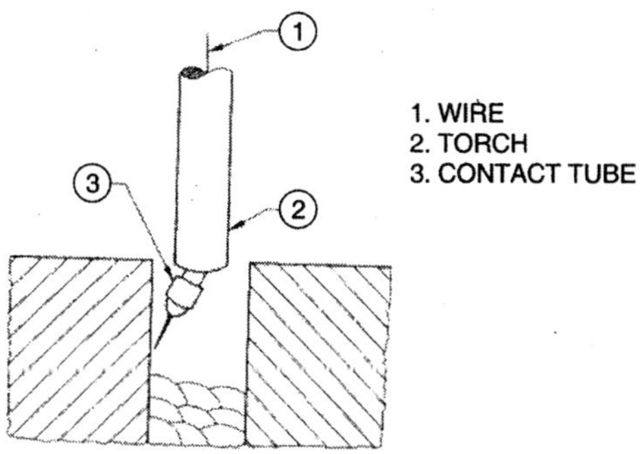

Fig. 2.21 Curved fixed electrode NGW.

(ii) Narrow gap welding technique with arc oscillation (Fig. 2.22)

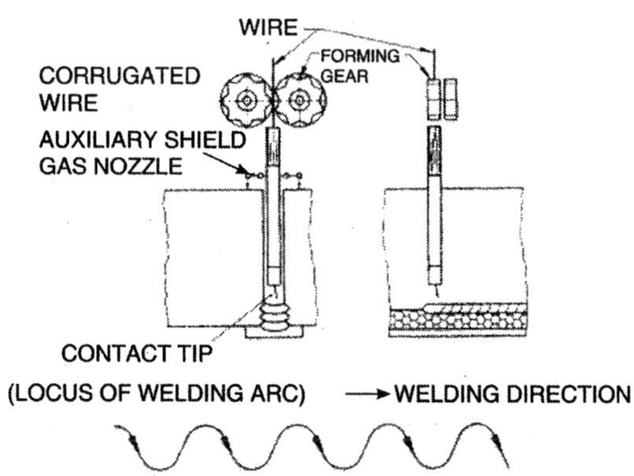

Fig. 2.22 NGW with arc oscillation.

(iii) Narrow gap welding technique with arc rotation. (Fig. 2.23)

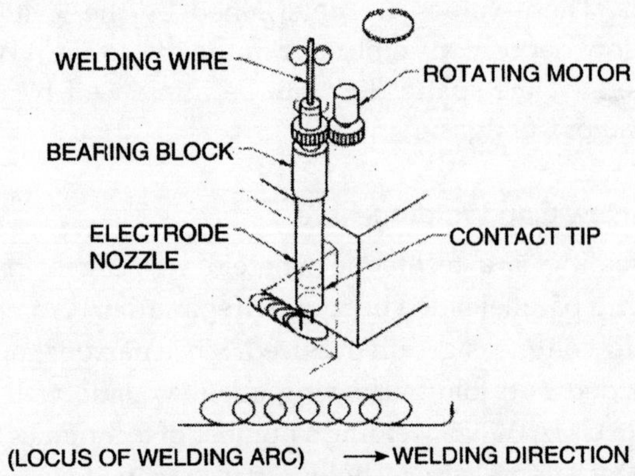

Fig. 2.23 NGW with arc rotation.

In all these techniques special mechanism are used for feeding the wire and supplying the shielding gas to the arc. This make the GMAW narrow gap torch appear quite different from the conventional GMAW torch. The main advantage of the narrow gap welding lies in simple joint preparation and reduced volume of the weld metal. The mechanical properties of GMAW narrow gap are able to meet the requirements of various codes and the GMAW narrow gap techniques are used in the industries for the fabrication of pressure vessels, penstocks, large beams and structures.

2.5.11 Electro Gas Welding

Electro gas welding (EGW) is an automatic method of GMA vertical butt welding using a solid or flux-cored consumable electrode wire. The molten weld metal created is confined by the water-colled copper shoes provided on both sides of the weld joint. The shielding gas required for protecting the arc and the molten weld metal is supplied concentrically through the torch as well as through the special hole provided in the copper shoes. The welding head comprises a special torch, a wire feeler along with a wire spool, an oscillation system, shielding gas and cooling water supply lines and the control unit (Fig. 2.24). During welding, the welding

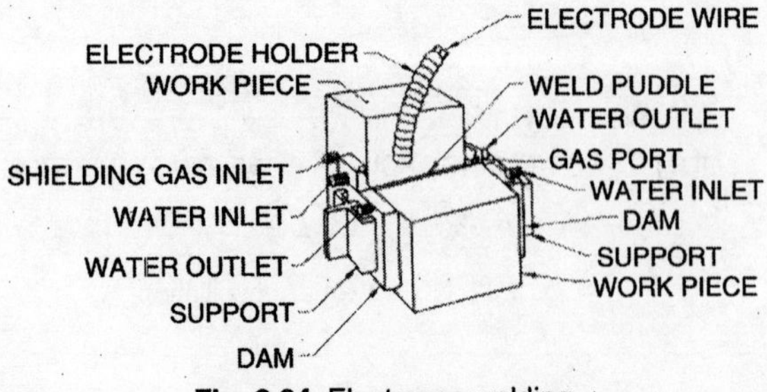

Fig. 2.24 Electrogas welding.

head comprising of all the above units moves as a single unit and the welding head should move at the same speed at which the melton pool is rising. This can be accomplished either manually by the operator or automatically using feed back controls. Although the axis of the weld is vertical, the process is actually flat position welding with vertical travel. The main advantages of electro gas welding are improved welded impact properties, relaxation of postweld heat treatment and easy restarting of the weld. Moreover, for welding of carbon steels no preheating is required. Electro gas welding is widely used in fabrication industries for welding of storage tanks, ships and other structures.

2.5.12 Synergic Pulsed — MIG Welding

The principle of adjusting all the parameters simultaneously from a single knob control is commonly known as "synergy" and synergic pulsed MIG welding has some form of algorithm to achieve this. Synergic pulsed MIG employs rectangular wave-current pulses. (Fig. 2.25). The metal transfer takes place only during the

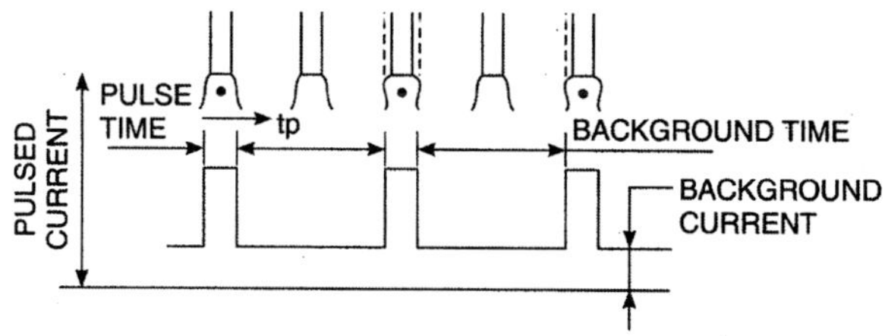

Fig. 2.25 Synergic pulsed MIG welding.

high pulse current period and the low background current helps to just maintain the arc. The main advantage of synergic pulsed MIG welding is its ability to provide "spray transfer from mean current as low as 70 amps to 300 amps." Thus this process can be successfully used for thin sheets as well as thick plates welding and also for out of position welding. Since synergic pulsed MIG operates in spray transfer mode at all currents, the operation in quite smooth and absolutely from spatter. The fume generation is very low. The deposition efficiency is quite high compared to conventional MIG welding.

2.5.13 Defects – Causes and Prevention

(a) Porosity

Porosity is defined as cavity type discontinuties formed by gas entrapment during solidification. Porosity can also be divided into two types : surface porosity which can be seen by the naked eye or visual inspection techniques and sub-surface porosity which can be established only X-ray or radiographic techniques. Causes and remedies are as follows :

(i) **Cause :** Poor Gas Coverage.
Remedy : Check gas flow rate. Use proper nozzle to work distance. Remove spatter from the nozzle. Eliminate leakage in the gas line. Check the CO_2 gas heater. Eliminate side wind or fan air blowing into the arc.

(ii) **Cause :** Moisture in the shielding gas.
Remedy : Use welding grade shielding gas. Check for water leaks in water-cooled torches.

(iii) **Cause :** Electrode and work piece contamination.
Remedy : Ensure that the electrode wire and work piece surfaces are free from oil, greese etc.

(b) Spatter

(i) **Cause :** Spattering in short-circuit transfer welding.
Remedy : Use proper inductance. Adjust the arc voltage. Use proper stand off distance.

(ii) **Cause :** Spattering in CO_2 open are welding, *i.e.,* in globular transfer.
Remedy : Decrease the are voltage or increase the wire feed rate to get burried arc condition and contain the spatter.

(c) Lack of Fusion

(i) **Cause :** Joint surface is contaminated.
Remedy : Clean all groove faces of any mill scale or oxides etc.

(ii) **Cause :** Too large a weld puddle width.
Remedy : Increase welding speed and reduce weaving.

(iii) **Cause :** Improper joint design.
Remedy : Check the included angle of the groove joint.

(iv) **Cause :** Improper welding technique.
Remedy : When using a welding technique increase the dwell time on the side walls of the groove.

(d) Lack of Penetration

(i) **Cause :** Improper joint preparation.
Remedy : Joint preparation must provide access to the bottom of the groove while maintaining proper stand off distance. Reduce the root face and increase the root gap.

(ii) **Cause :** Improper welding parameters.
Remedy : Increase the welding current and reduce welding speed. Maintain proper stand off distance.

(iii) **Cause :** Improper welding technique.
Remedy : Keep arc on leading edge of the weld puddle.

2.6 SUBMERGED ARC WELDING

Submerged Arc Welding (SAW) also called sub-arc welding is an arc welding process where the welding arc is submerged under a granulated flex. The arc is initiated between the electrode wire and the weld joint under the medium of flux. A wire feeding mechanism then feeds the electrode wire towards the joint at a controlled rate. A flux feeding mechanism feeds the flux around the electrode wire continuously to keep the arc under submerged condition. The electrode wire and the joint edges melt due to the heat of welding and form a weld pool to bridge the gap between the joints. Depending on the application, the welding arc can be stationary or moving to carry out the welding operation. Figure 2.26 shows a schematic view of submerge arc welding process.

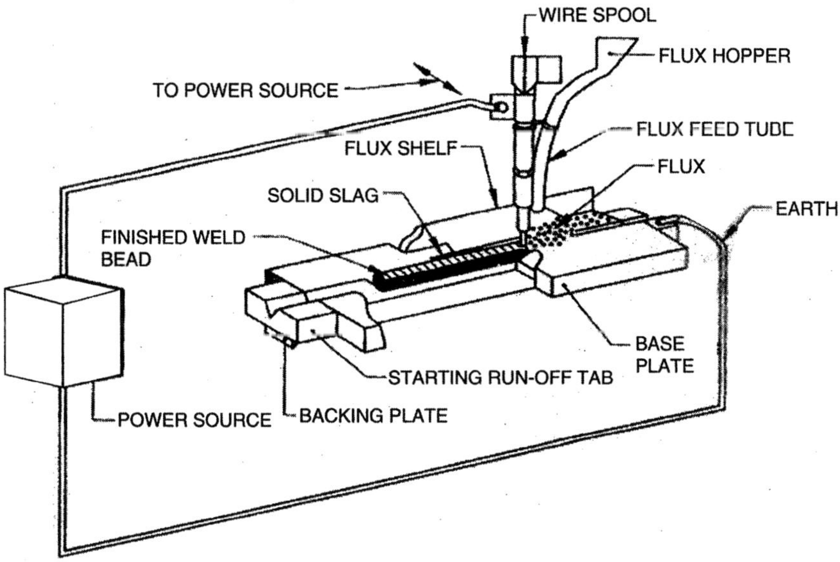

Fig. 2.26 Schematic view of submerged arc welding process.

2.6.1 Welding Methods

Submerged arc welding is carried out either in semi-automatic or fully automatic mode. Many types of fixtures and positioning equipments are available to mount and move the jobs so that the molten weld pool and the flux remain in position until they get solidified. Semi-automatic welding is done using a hand-held gun (welding) which delivers both the flux and the electrode. The electrode is driven by a wire feeder and the flux may be supplied by gravity hopper mounted on a gun or pressure fed through a hose. It is difficult to go in for a higher welding current and speed in this method.

Mechanised welding is done using an equipment that performs the operation without requiring a welding operator to continually monitor and control adjustments. In this method it is possible to use higher welding currents and speeds to achieve high production rates. Automatic welding is highly suitable for pipe welding using move than one electrode wire. Generally two or three wires are used in tandem. Fully automatic welding is done using equipment that is monitered by a welding operator to position the work, stop and start the welding and to adjust the controls. Machine welding is used in jobs having varying weld profiles.

Arc Starting Methods

The welding arc can be initiated using one of the six methods depending on the type of job, joint set-up and the importance of arc starting at a particular place on the joint. The methods are :

(i) Steel wool ball start.
(ii) Sharp wire start.
(iii) Scratch start.
(iv) Molten flux start.
(v) Wire retract start.
(vi) High frequency start.

Three most popular arc starting methods are described below :

- Steel wool ball start is done by positioning a ball (10 mm) of steel wool directly between the wire electrode and job. The flux is applied around the wire and welding is started. The steel wool ball melts rapidly and creates an arc.
- Sharp wire start is done by making the end of the wire sharp like a chiesel by using a wire cutter. The electrode is lowered to the job, the flux applied and welding started. The chiesel point melts rapidly and the arc is started.
- Wire retract start is a cost effective method when frequent starts have to be made and when starting locations is important. It is done by touching the electrode on to the job, covering with flux, turning on the current, and withdrawing the tip of the electrode from the surface of the work piece.

2.6.2 Equipment

For submerged arc welding following equipment is needed :

(i) Power source.
(ii) Electrode wire feeding mechanism.
(iii) Flux feeding system.
(iv) Travel arrangement.
(v) Process control system.

The accessories include positioning or manipulating the system. Figure 2.27 indicates the welding equipment.

Fig. 2.27 Wire feeding and flux feeding system.

(i) Power Sources

Both D.C. and A.C. power sources are used in SAW process. A D.C. power supply can be a transformer rectifier or a motor driven generator A.C. power supplies are generally transformer type. Most SAW power source are constant voltage types. For submerged arc welding 100 percent cycle power sources are always recommended.

(ii) Wire Feeding System

This consists of a wire feeder motor, gear assembly and feed rollers which feed the wire down through the nozzle assembly.

(iii) Flux Feeding System

This consists of a hopper, a control valve, a feeding tube and a cup at the end of the table to distribute the flux around the wire. A flux hopper can be filled manually or mechanically by a flux delivery system.

(iv) Travel Arrangement

A welding head with fire feed assembly and flux feed assembly usually may be mounted on a travelling trolley or on a boom fitted to a column.

(v) Process Control System

The control system in submerged arc welding is used to control the wire feed speed, welding voltage, welding current and the travel speed of the welding head.

2.6.3 Flux-recovery System

The flux recovery system is used to remove the unused flux from the joint for reuse. It consists of a vaccume generating unit, a recovery tube and flux collecting drum.

2.6.4 Job-positioners

A variety of positioners are available commercially to moment and rotate jobs is welding.

2.6.5 Process Application

Submerged arc welding is a versatile process and find its uses in various fabrication industries. It is employed in fabricating columns, beams, bridges and ship decks. Pressure vessel fabrication needs SAW for high rate of deposition. SAW also finds its use in hard facing and metal cladding.

(a) Operating Variables

For a given wire and flux the main operating variable are :

(i) Welding voltage.
(ii) Welding current.
(iii) Electrode wire extension.
(iv) Electrode wire extension.
(v) Depth of granular flux.

Welding Voltage

Variation of welding voltage alters the arc length. Increase in voltage increases the arc length. Increase in arc voltage increases the bead width, results in more flux consumption and reduces penetration into the base metal.

Welding Current

Welding current decides the depth of penetration and the melting rate of the wire. Increasing the current results in more depth of penetration and causes excessive reinforcement. Too high a current may cause undercuts or even puncture of the base plate. The different current ranges selected for different wire diameters are :

Wire diameter (in mm)	Current (in amps.)
2.4	230-700
3.2	300-900
5.0	480-1100
6.3	700-1600

Welding Speed

For a combination of current and voltage, the welding speed decides the heat input and metal input to the joint. If the travel speed is increased one would obtain a small weld bead with less reinforcement. Excessive speed will result in incomplete filling and undercut.

Table 2.5 : Comparison between SAW vs SMAW/FCAW-Welding Speed

Welding process	Arc speed (mm/min)
(i) Manual shielded metal arc	100
(ii) Semiautomatic flux cored	250
(iii) Automatic submerged arc-one wire	450
(iv) Twin arc submerged arc	520
(v) Tandem submerged arc-two wires	750

Electrode Wire Extension

This is the length of the wire extending out of the nozzle tip. Due to resistance heating (I^2R) the extended wire gets heated and becomes unstable. An electrode extension of approximately eight times the electrode wire diameter is recommended. Excessive extension will result in a wavy weld bead due to instability of the wire.

Depth of Granular Flux

The depth of the flux decides the appearance and soundness of the weld. If the granular layer is too deep, the gases will not escape and will result in irregular an distorted beads. If the granular layer is too shallow the arc will flash and spattering and porosity will occur. About 25 mm depth of flux is recommended for smooth operation.

2.6.6 Flux, Wires and Flux-wire Combinations

(a) Fluxes

Fluxes is SAW help in sustaining the arc, protecting the weld pool, cleaning the weld pool and modifying the chemistry of the weld. They also influence the shape of the weld beam. From the manufacturing point of view there are basically two types of fluxes are used :

(i) Fused flux (ii) Agglomerated or bonded flux.

Fused flux is manufactured by fusing the raw materials in an arc melting furnace and then quenching or granulating in water. The resultant glossy mass is furnace-dried, crushed, screened for size and packet. The fused fluxes have following advantages :

(a) Good chemical homogeneity.
(b) Non-hygroscopic in nature.
(c) Can be recycled without disintegration.

The disadvantage is that alloy additions cannot be done due to oxidation at high temperature melting.

Agglomerated flux is manufactured by first dry mixing the raw materials and then wet mixing with the addition of binders like potassium silicate or sodium silicate or the combination of both. After wet mixing mass in pelletised and based in a furnace. The pellets are then broken up, screened to size and packed. Agglomerated fluxes have the advantage of alloying additions for deoxidation and compensation. The disadvantages are that they absorb moisture easily and disintegrate while recyling.

(b) Wires

Submerged are welding wires are available to produce weld deposits matching carbon steels, low alloy steels, stainless steel and nickel alloys. Special alloy wires are available for hard facing applications. Carbon steel and low ally steel wires are copper coated to prevent resting. SAW wires are available in various sizes from 1.6 mm to 6.3 mm in diameter.

(c) Flux-wire Combination

The flux and wire should be matched to produce a weld of desired chemistry and mechanical properties, matching the base material. Test weld is made with a flux and wire as per a particular code and then recommended for use on a job. American welding society specification AWS 5.17 prescribes the requirements for electrodes and fluxes for SAW of carbon steels.

2.6.7 Weld Joints

The submerged arc welding process is employed in groove welding, fillet welding, plug welding and surfacing. Groove welds are usually made in flat position and fillet welds are made in flat and horizontal positions. Plug welding and surfacing are both carried out in flat position only. Groove joints are commonly butt joints with different types of grooves depending on the thickness of the plates welded. (V-joint double V, U or double U-joints). To minimise the weld deposit, plates of higher thickness require 'U' preparation. Edge preparation may be done on the wedge plate in fillet welds ensume good penetration.

2.6.8 Welding Procedures

(a) Joint Setup

To make a good welding joint, a proper procedure has to be followed. The plates be welded are edge-prepared using gas cutting and grinding or machining. The edge prepared plates are properly aligned and fitted up using recommended root gap. Sometimes to contain the weld metal at root, a backing strip is tack welded to the joint and it becomes an integral part of the joint.

(b) Electrode Alignment

The electrode wire a properly aligned in relation to the joint for proper penetration and joint completion.

(c) Electrical Connections

Generally, direct current electrode positive (DCEP) are used for welding joints by SAW. For surfacing applications, direct current electrode negative (DCEN) are used. Alternating current can also be employed in SAW, where weld quality can be compromised.

(d) Running the Arc

The operator has to see to it that the welding arc runs along the groove as per the wire alignments made. Minor adjustments may be required in process of laying the beads.

(e) Terminating the Arc

When the welding arc completes travelling the full length of the joint, it has to be terminated. The crater has to be filled before terminating the arc. This is done by stopping the travel and allowing the arc and wire feeding to continue for a controlled length of time.

(f) Slag-cleaning

After each bead is laid, the slag must be cleaned before laying the next bead. Slag particles sticking at the corners should be removed to prevent the slag inclusion. Table 2.6 and 2.7 indicates the typical procedural variable for submerged arc welding of 6 to 20 mm thick plates, butt and filled joints respectively.

Table 2.6 : Submerged Arc Welding Parameters for Single Electrode (Fully Mechanised)

Weld Size (mm)	5	6	8	10	12	16	20
Plate thickness (mm)	6	8	10	12	16	20	25
Electrode size (mm)	3.2	4	5	5	5	5	5
Current (amp) DC	520	620	720	800	870	920	970
Volts	30	32	34	36	38	39	40
Welding speed (m/min)	1.4	1	0.7	0.56	0.36	0.25	0.20
Electrode required (kg/m)	0.10	0.18	0.28	0.40	0.70	1.1	1.6
Flux required (kg/m)	0.05-0.09	0.75-0.12	0.14-0.18	0.18-0.27	0.33-0.45	0.53-0.75	0.83-1.2
Total time (hr/m for weld)	0.012	0.016	0.0024	0.03	0.47	0.67	0.09

Table 2.7 : Submerged Arc Welding Parameters for Single Electrode (Fully Mechanised)

Plate thickness (mm)	6		10		12		16		20	
Pass	1	2	1	2	1	2	1	2	1	2
Electrode size (mm)	5		5		5		5		5	
Current (amp) DC+	600	750	650	800	750	850	750	850	800	900
Volts	31	33	33	35	35	36	35	36	36	37
Welding speed (m/min)	1.8	1.8	1.2	1.2	0.9	0.9	0.6	0.6	0.5	0.5
Electrode consumed (kg/m)	0.13		0.23		0.35		0.56		0.63	
Flux consumed (kg/m)	0.14-0.16		0.19-0.25		0.3-0.4		0.5-0.65		0.55-0.72	
Total time (hr/m of weld)	0.019		0.028		0.38		0.59		0.6	

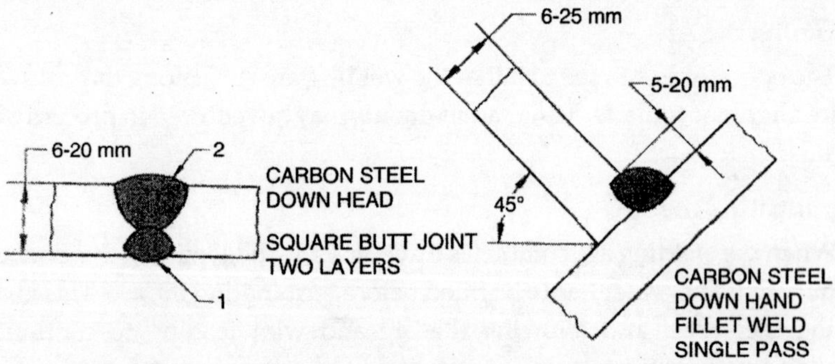

Fig. 2.28 Submerged arc welding parameters for single electrode.

(g) Weld Quality

The submerged arc weld should be free of defects like porosity, slag entrapments, improper penetration and cracks.

2.6.9 Developments in Submerged Arc Welding

A lot of development has taken place in the submerged arc welding process since it was used by the fabrication industries. Some of them are discussed below :

(a) Hot Wire Submerged Arc Welding

In addition to the main welding wire another wire is fed to the weld pool. The second wire is heated with the electrical resistance (I^2R) by using a separate AC transformer. The hot wire fed to the weld pool does not arc on the base material but is just fed in a red hot condition. It melts by using the head available from the main arc. The deposition rate can be increased by 30 to 40%. This process is well suited for pipe welding.

(b) Strip Cladding

Stainless steel deposits and high alloy steel deposits can be deposited over carbon steels and low alloy steels for corrosion resistance in chemical processing vessels and nuclear vessels. Instead of a electrode wire a strip is used in strip cladding process. Commonly a strip of 0.5 mm. Thickness and 60 mm width is used. A wide shallow penetration bead can be obtained in this process.

(c) Narrow-gap Submerged Arc Welding

When thick plates are welded, a considerable amount of weld deposit is made even with U-groove joints. To minimixe the weld metal deposited, a narrow gap joint with parallel faying edges is setup for welding. For a plate thickness of 100 mm, the weld metal deposited by narrow gap SAW is only 40% of the deposits required by a U-groove joint. The gap between the faying edges is kept at 18-20 mm. A special wire aligning system is used to weld in deep grooves. A special flux is required for easy slag removal and smooth bead formation in deep grooves. Normally two passes per layer are laid in narrow gap submerged arc welding.

2.7 THERMIT WELDING

Thermit welding is one of the oldest joining processes having similarities with welding in foundry. The process involves quickly creating hot molten metal in the prefabricated moulds or reaction chambers. The essential reaction is an exothermic reaction between metallic oxides and aluminium. The reaction ends in the reduction of metal with desired alloying. Preparation of hot molten metal without any electric or mechanical systems at the site itself, is the greatest advantage of this process.

Thermit reaction occurs at the constant pressure and is an instantaneous oxidation and reduction process which is driven by temperature. The chemical reaction takes place actually on the boundaries of individual solid particles called the interfaces. Metallic oxides of Cu, Cr, Ni and Mn are also often used instead or iron oxide (Fe_3O_4). The important oxide, however, is that of iron and is used with aluminum dust which acts as a reducing agent for most of the application. The heat

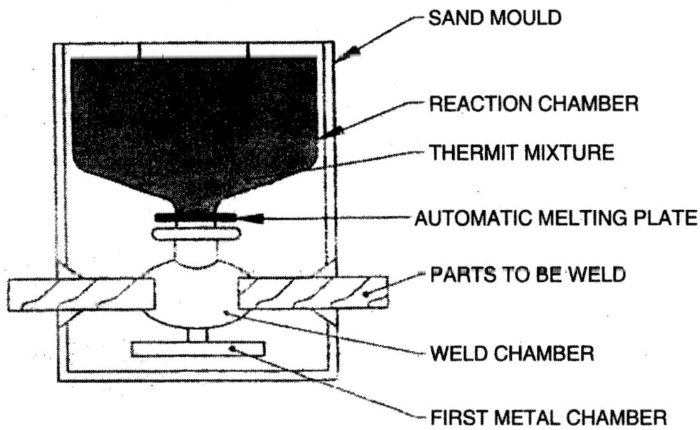

Fig. 2.29 Schematic arrangement of thermit welding.

formation for 1 kg of thermit mixture can reach a temperature of about 3000°C. Addition of alloying elements brings down the temperature further. The thermit mixture is not inflammable at ordinary temperature and can only be ignited with an initial temperature of 1200°C. Thermit reaction time in just around 7 to 10 seconds. A schematic diagram for thermit welding is shown in Fig. 2.29.

2.7.1 Types of Thermit Welding Processes

The thermit welding processes can be classified as follows :

(a) Ordinary Thermit Welding (OTW)

Preheating is carried out by inserting a burner into the heating gate provide at the bottom of the mould. The heating gate is closed by a metal plug lined with asbestos after the parts have achieved around 950°C temperature. Refer Fig. 2.30.

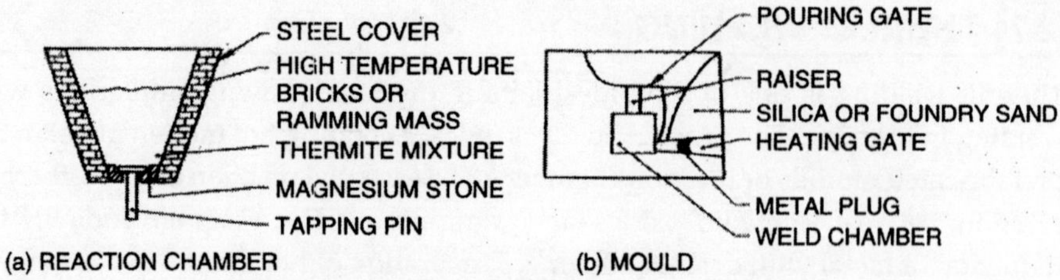

Fig. 2.30 Mould with separate reaction chamber.

(b) Quick Thermit Welding (QTW)

Preheating is carried out from the top of the mould and the flame is passed through the raiser channel located just above the parts to the joined. Unlike in OTW, the preheating time is restricted.

(c) SOV

SOV is the German abbreviation for weld without preheating. The gap between the parts to be joined is around 30 mm and the thermit portion requirement is more compared to ordinary thermit welding where the gaps are about 10 to 15 mm.

2.7.2 Types of Moulds

The moulds are usually prefabrication with or without reaction chambers. Higher volume weld metal moulds are made at the site itself and the joining is accomplished by using separate reaction chambers. Reaction chambers are fabricated from mild steel and are lined with high temperature bricks or magnesite ramming mass.

(a) Moulds for Non-repetitive Welds

Repair welds arc normally non-repititive and the moulds are prepared by wax pattern at the site itself. A heating gate is centered directly on the wax pattern. Vent holes are to be provided to facilitate the escape of moisture during preheating. Silica sand or boundary sand is used with water glass binder for the fabrication of the mould. Reaction chambers are used to raise the hot molter metal and the same is allowed to flow into the moulds, fitted with the parts to be joined.

(b) Moulds for Repititive Welds

The mound for this purpose is mostly prefabricated and is strengthened with CO_2 for rapid curing. Separate reaction chambers are not used and the same is incorporated in the mould itself as shown in Fig. 2.29. Suitable heating gates and risers are to be provided. Hollow chambers are provided to accomdate the preheat metal or first metal which passes through the weld area and does not have enough super heat to completely melt the parts to be welded. Suitable melting plates with linings are used in between the reaction chamber and flow gate which will automatically melt at the end of the reaction and allow the hot metal to flow into the weld chamber.

2.7.3 Preheating

The most common method of preheating is by using natural gas, oxy-acetylene gas, LPG or kerosene. Electric induction heating of the parts was not successful. Preheating temperatures are in the range of 750°C to 950°C. Preheating of the weld area may be accomplished by allowing a part of the hot metal to get into direct contact with the joint. It is then removed from the weld area by withdrawing the parts and then allowing the flow of final hot metal.

2.7.4 Sequence of Operation

(i) Align the parts to be joined with proper gap.
(ii) Fix the mould and pack the gap in between the moulds perfectly with asbestos pasts.
(iii) Preheat the mould and then the parts to be welded through the heating gate.
(iv) Load the thermit mixture in the reaction chamber.
(v) Ignite the thermit mixture by using special matches and magnesium tapes.
(vi) Tap the hot metal at the end of the reaction, *i.e.*, 20-25 seconds after ignition.
(vii) At the end of solidification, the moulds are to be dismantled and the excess weld metal is trimmed off by chiesel.

2.7.5 Applications

Thermit welding is mainly used in rail welding, concrete reinforcement rod welding, building up of steel mill wobbler ends and for electrical connections.

(a) Rail Welding

Rails are welded to create long rail ways for electrified and other fast tracks. This increases the passenger comfort and keeps the maintenance cost down. Spillage of goods like coal is minimised in mining industry.

(b) Reinforcement Steel Rod Welding

In big building projects a massive number of joints are to be carried out for joining of reinforcement steel in a short time. Thermit welding is applied by using prefabricated moulds with reaction chambers.

2.8 ELECTROSLAG WELDING

Electroslag welding is a fusion welding process where in the electrical resistivity of a molten flux pool produces continuous heat energy to melt the filler wire and the base metal. The molten flux is contained by a pair of water cooled copper shoes covering both the sides of the joint gap. This is a vertical welding process and is mainly used for heavy thickness joints. The electroslag process can be applied either on the longitudinal or circumferential joints.

2.8.1 Operation

The electroslag welding is a continuous process. Heat energy is supplied by the resistance heat developed by the slag puddle and not by the arc, except for a brief period in the beginning. A schematic diagram of electroslag welding is shown in Fig. 2.31. The starting sequence for electroslag welding is as follows :

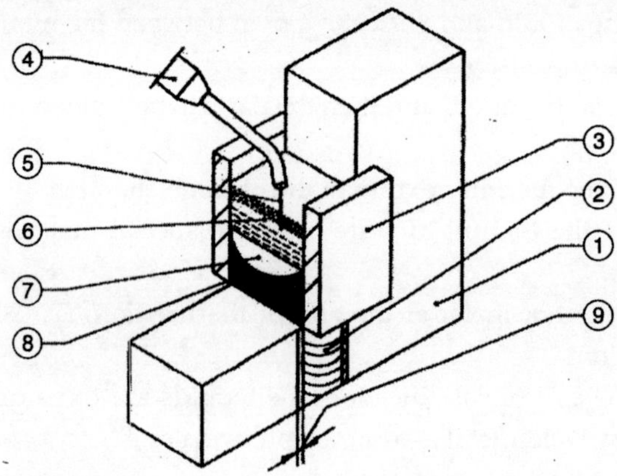

1. BASE METAL
2. ELECTRO SLAG WELDMENT
3. WATER COOLED COPPER SHOE
4. WIRE FEED NOZZLE
5. FILLER WIRE OR STRIP
6. MOLTEN SLAG POOL
7. MOLTEN METAL POOL
8. SOLIDIFIED WELD METAL
9. SIDE WALL FUSION

Fig. 2.31 Schematic diagram of electroslag welding.

(i) The plate/base material is kept in positions with a gap of 30-50 mm.

(ii) A start on plate/dam is welded to a length of 150-200 mm on to the base material at the bottom. This extension is closed with a 30 to 40 mm plate to contain the molten pool in the beginning for a start on the dam arrangement. (Fig. 2.32).

(iii) Water cooled copper shoes are mounted on either side and the welding head is positioned in the gap.

(iv) The filler wire is made to touch the start-on plate welded at the bottom wire and electroslag flux is added.

(v) The power source is switched on and the welding starts.

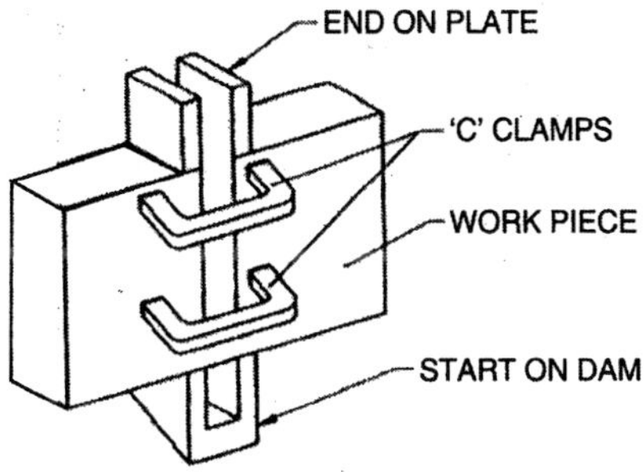

Fig. 2.32 Preparation for welding - Start on dam.

In the beginning an arc is created in between the start on plate and the filler wire. The arc energy melts the welding flux and a molten slag pool is created. Within a very short time, the arc gets quenched and no arcing will take place further. The filler wire carries electric power into the molten slag pool. Because of the resistivity of the molten slag pool, heat energy is created and same is utilised to melt the filler wire, flux and the base material. The filler wire tip should not touch the molten metal pool surface below the slag pool so as to avoid stray arcing. Figure 2.33. Shows molten pool parameters.

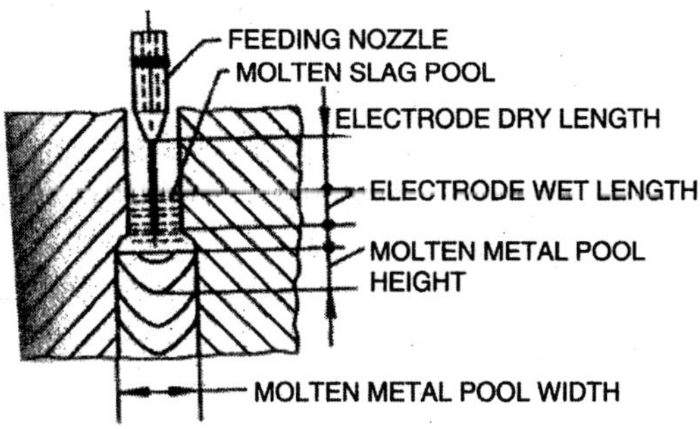

Fig. 2.33 Molten pool parameters.

2.8.2 Equipment

A universal column and boom is used to carry out electroslag welding. Basically vertical and transverse motion should be possible with the equipment. The welding head is mounted on the boom, on a cross-slide, to enable transverse movement of the welding head which is controlled by a separate motor. the copper shoes are connected to the boom and they move along with the welding head. An alternating current or the direct current with constant voltage power source is used for welding. The power sources should be having a rated capacity of 750-1000 amperes with 100% duty cycle at 60 volts.

Since multi-wires are employed in electroslag welding the power rating should be accordingly multiplied by the number of wires. Usually separated power sources are used for each wire. Some fabricators use single power source for a two-wire system with high power rating. Constant voltage transformer-rectifiers are extensively used for ease of control. Constant current transformers-rectifiers are difficult to control. For circumferential welding applications suitable fixtures and positions are used. The capacity of the positioner is selected based on the weight of the jobs to be welded and probable welding speed to be employed. Stand-by generator sets are a must in the event of anticipated power failure to avoid costly restarting/repairs.

2.8.3 Consumables

(a) Wires

Wires with low and medium carbon content as used in submerged arc welding could be used for welds in structures and other such application. However for high temperature applications such so joining of pressure vessels and chemical reactors, alloy steel wires with chromium and molybdenum additions are used. This is to ensure better creep properties at higher operating temperatures. Usually wires with 2.25 Cr, 0.5 Mo and 2.25 Cr 1 Mo composition and 3.15 mm diameter are used in most application. Higher diameter wires are difficult to straighten and as a result feeding problems are experienced. Flux-cored or tubular wires are used to produce specific weld metal composition. The amount of flux in these types of wires should be precisely controlled to avoid low side fusion. Strip electrodes are used in surfacing applications.

(b) Fluxes

The flux is the most critical element in the electroslag welding process. The fluxes are predominantly based on mineral oxides of manganese, silicon, calcium, titanium, aluminium and calcium flouride. The calcium fluoride based mineral is the key element is controlling the slag viscosity and resistivity at high temperatures. The welding fluxes should have optimum viscosity so as to avoid the scape of the molten slag pool through the gap in between the copper shoe and the base material. It the molten flux is of a higher viscosity, the stirring effect of the molten slag pool will be restricted leading to improper side wall fusion. The melting point of the flux should be lower than that of the material to be welded. The flux should conduct electricity without stray arcing and the temperature should be maintained homogenously throughout the molten pool. Mostly fused fluxes are used in electroslag welding.

2.8.4 Welding Operation

The materials to be welded are cut to the required sizes. The edges to be welded are neatly prepared by simple gas cutting and it is not necessary to machine the surface. After cleaning the surface for oxides the plates are kept in position. Strong reinforcement bars are provided and welding head is positioned. In the meantime, water cooled copper shoes are mounted and adjusted. Asbestos paste is applied along the moving surface edge to arrest leakage of the molten slag pool. Welding is then started and parameters are adjusted to stabilise the process. In the initial stage of starting, most of the heat energy is diverted to create a molten slag pool and less energy is available to melt the plate edges.

Since, the slag pool level is to be raised above the joint top to complete the welding, an end-on plate or dam is added to top as shown in Fig. 2.33. The elctroslag process is a continuous one and any interruption is not desirable.

2.8.5 Welding Parameters

The electroslag welding parameters can be catagorised into two groups, *i.e.*, main welding parameters which can be controlled directly and auxiliary welding parameters which are controlled by main welding parameters.

Main Welding Parameters :

Electrode feed rate	V_e m/hr
Welding voltage	V – Volts
Depth of molten slag pool	d – mm
Electrode dry length	e_d – mm
Electrode wet length	e_w – mm
Diameter of electrode	d_{ia} – mm
Cross transverse speed	V_t – m/hr
Cooling water flow rate	q – lt/min
Number of electrodes	n
Welding gap	b – mm

Auxiliary Welding Parameters :

Welding current	I – amp
Welding speed	V – m/hr

The electrode dry length is a distance between the tip of the nozzle to the surface of molten slag pool. The depth to which the electrode is to be submersed in the molten slag pool is known as electrode wet length. The rate of the electrode transverse movement in molten slag pool, which provides homogeneous temperature distribution throughout the molten slag pool is known as cross-transverse speed. The cooling water rate is set, based on the outgoing water temperature which is maintained around 70-80°C. Below are given optimum values of some parameters for welding 50-100 mm thick plates with single and double wires. (Table 2.8).

Table 2.8 : Electroslag Welding Parameters

Parameters	Values
(i) Welding voltage	40-45 volts
(ii) Depth of molten slag pool	35-45 mm
(iii) Electrode wet length	25-35 mm
(iv) Electrode dry length	30-40 mm
(v) Cross transverse speed	30-40 m/hr
(vi) Electrode dwell time	2-3 seconds
(vii) Weld joint gap	35-40 m/hr
(viii) Welding speed	2-4 m/hr

The other two important factors are as follows :

(a) Coefficient of form factor : This is the ratio between the molten pool width (b) and its its height (h) and is expressed as $= h/b$. The coefficient varies from 0.8 to 10 and the optimum value is around 1.5 to 4. It is necessary to restrict the value so as to avoid the central line cracking in the weld metal, during solidification.

(b) Mass balance equation : This equation calculates the wire-feed rate to be set on the machine, by equating the volume of the weld metal to be filled in the joint to the volume of electrode melted and is expressed as :

$$S_w \times V = S_e \times V_e$$

where, S_w is cross-section of weld.

V is welding speed.

S_e is cross-section area of electrode.

V_e is electrode feed rate.

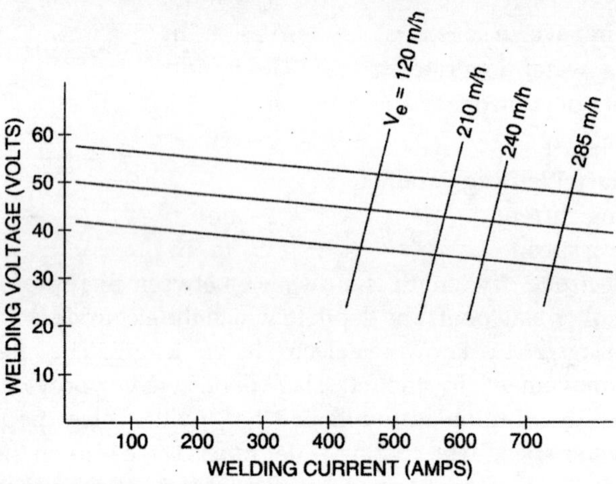

Fig. 2.34 Nomogram for wire feed rate.

From the calculated wire feed rate, the number of wires is decided depending on the machine capacity for the wire feed. Based on nomogram, shown in Fig. 2.34, the welding current value is chosen.

2.8.6 Effect of Welding Parameters

In electro-slag welding, the heat energy is obtained by the passage of the current through the molten slag pool. Only a small portion of the slag pool, around the immersed electrode, acts as a source of heat. The major portion of the current flows through the electrode tip and hence the maximum temperature is obtained around the electrode tip. The position of the electrode tip in the molten slag pool depends on the welding parameters practiced. Figure 2.35 shown the effect of current on the molten metal pool height and electrode tip position. Increase in the current level leads to higher feed rate and deposition. The side wall fusion decreases with increase in the current level. At higher current, the electrode wet length increases and may even touch the surface of the molten metal pool leading to stray arcing. Figure 2.36 illustrates the effect of variation in welding in voltage. A higher voltage increases the heat input in the molten slag pool leading to higher side wall fusion and prompting wider bead width. Figure 2.37 shows the effect of electrode-slag flux on the molten pool geometry. The bead width is higher while using a welding flux with higher electrical resistivity. Welding flux with a higher conductivity gives poor side wall fusion due to the low heat generated in the molten slag pool.

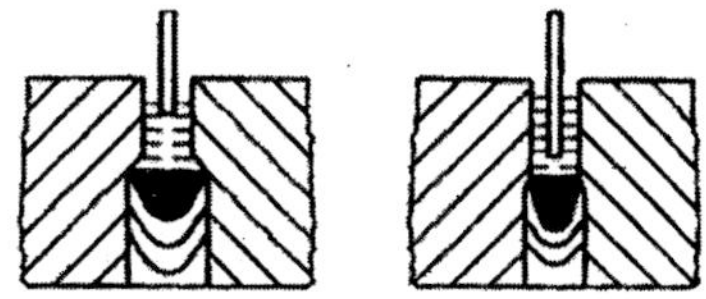

Fig. 2.35 Effect of welding current.

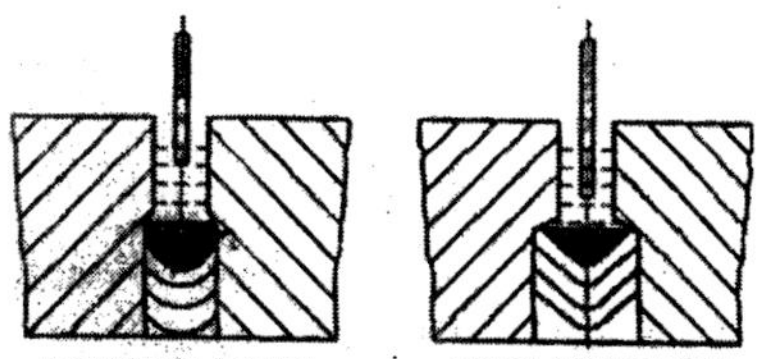

Fig. 2.36 Effect of welding voltage.

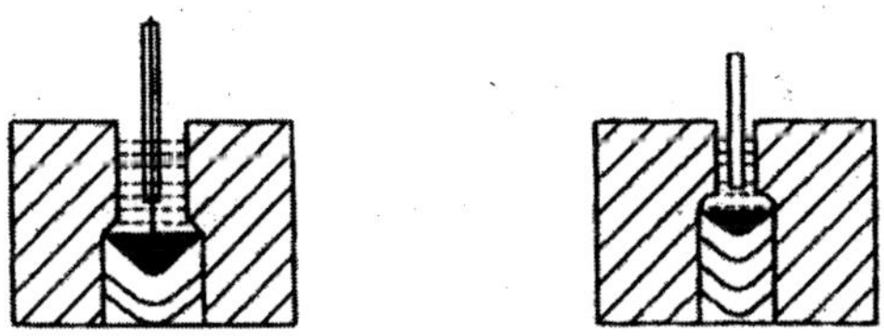

Fig. 2.37 Effect of flux characteristics.

2.8.7 Defects, Causes and Remedy

Usually, slag inclusion are rare in electroslag weldments except in the very beginning. Higher current and low voltage levels lead to low side wall fusion. Side wall fusion can be controlled by current and voltage. Restarting is necessitated due to interruption during welding. Prior to restarting, the area should be cleaned for residual slag. After the completion of welding, the restarted point is to be drilled out and filled again by manual welding or the consumable guide electroslag process. Centre line cracking can be avoided by selecting optimum welding parameters and controlling the form factor.

2.8.8 Process Limitation

The electroslag welding process is most suited for high thickness welding application since the heat input involved is high and the cooling rates are very low, leading to coarse grain structure. A coarse grain structure means inferior impact property so this process is more advantageous for statically loaded structures than for cyclically loaded ones. Careful selection of flux, avoidance of coarse grain structures and obtaining clean weld metal can overcome the above constraint. Heat treatment, stress-relieving and normalising are to be carried out without fail which increasing the production cost however.

2.8.9 Advantages

The electroslag process can be applied for any thickness with suitable support facility. Usually preheating is not needed for any thickness since it is being taken care of the process itself. Stirring action of the molten pool cleans the side walls, avoiding machining or cleaning after gas cutting. The flux consumption is only 50 gms per kg of weld metal deposited. This is highly productive process with deposition rates of 12 to 15 kg/hr per wire. Any standard column and boom and other auxiliary facilities can be utilised to keep the capital expenditure low for establishing electroslag welding setup. The operation is relatively easy and does not call for intensive training of the opertor.

2.8.10 Applications

Electroslag welding can be economically applied right from 30 mm to any higher thickness. Mostly, statically loaded structures like storage, ship hulls, flames are electroslag welded. Many bridges were welded by this process and the performance was satisfactory. In pressure vessel fabrication this process is predominantly used. For bigger machine parts like thick plates and shafts the electroslag welding process is applied successfully. Also, for rebuilding of worn out rollers in steel mills, crane wheels and crusher hammers, the elecroslag welding process is used advantageously.

❑❑❑

CHAPTER 3

FRICTION WELDING

INSIDE THIS CHAPTER

3.1 Friction welding, 3.2 Ultrasonic welding, 3.3 Diffusion bonding.

3.1 FRICTION WELDING

Friction welding is a solid phase welding process in which bonding is produced at a temperature lower than the melting temperature of the two parts being joined. This is a method of welding in which the welding heat is generated by sliding or rubbing together of two mating surfaces under pressure. Friction welding machines, available commercially, look like lathes. The schematic arrangement of a friction welding machine is shown in Fig. 3.1. There are two basic requirements which are to be satisfied before welding can take place between metals. There are :

(i) to produce two absolutely clean and uncontanimated surfaces.

(ii) to bring these clean surfaces in intimate contact with each other so as a metallic bond can be produced across the interface.

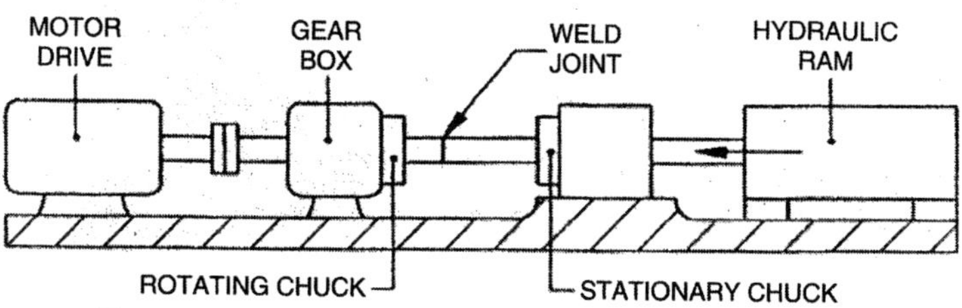

Fig. 3.1 Schematic arrangement of friction welding machine.

Principle of Operation

In the friction welding process, the two surfaces to be joined are made to rub against each other under the action of an axial force. Figure 3.2 illustrates a

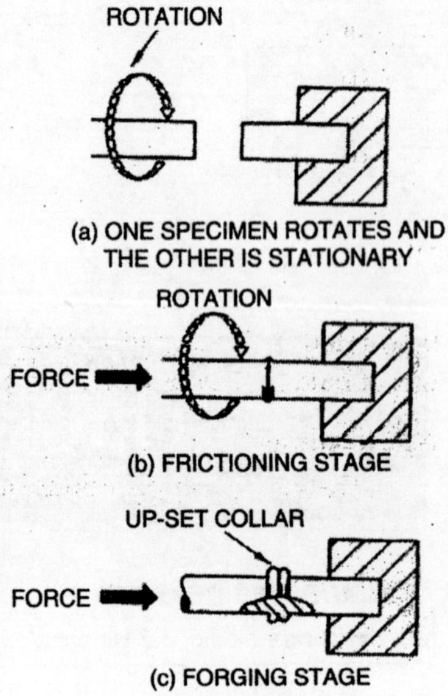

Fig. 3.2 The friction welding process.

schematic diagram of the friction welding operation. The sequence of operation necessary to produce a weld is as follows :

(i) The components to be joined are separately loaded and clamped into the rotating chuck and work table respectively.
(ii) The work table is moved forward to close the gap between the two components until they are almost touching.
(iii) The chuck is then rotated at a very high speed and the gap is closed so that the mating surfaces are rubbing together. This is the start of the heating cycle.
(iv) Axial force is simultaneously applied so that the rate of heating due to friction is increased.
(v) The components become heated near their interfaces and plastic deformation causes metal to be squeezed out. This is called "burn off".
(vi) After a predetermined amount of burn-off has occured the rotation of the chuck is abruptly arrested by a brake and an upset force is simultaneously applied.
(vii) The weld cools rapidly by conduction and upset force is then released.
(viii) The welded component is then removed from the machine by releasing the chuck and retracing the work-piece. The friction welding machine is shown in Fig. 3.3.

Fig. 3.3 Friction welding machine.

Process Parameters

(i) Spindle speed
(ii) Fractioning force/friction time.
(iii) Forging force/forge time.

Advantages

(i) High quality welds are produced consistently. Many combinations of dissimilar metals can be welded by this process.
(ii) High production rates can be achieved.
(iii) No fluxes or consumables are required.

Disadvantages

It is difficult to weld non-circular section. As one or both components have to be rotated, it is difficult to weld large and unsymmetrical sections because of the out of balance forces which are produced.

Applications

The various applications of friction welding are :

	Industry	Components
(i)	Automobile	— Axle cases
		— Engine exhaust values
		— Transmission Jacks.
		— Automobile Jacks.
(ii)	Cutting tools	— Twist drills
		— Tools bits
(iii)	Oil and gases	— Pump shafts
		— Control valve shafts
(iv)	Power	— Steel flanges of valves.

3.2 ULTRASONIC WELDING

Process Fundamentals

This is a method of welding which employs mechanical lateral vibrations of the order of 10 to 100 kilocycles per second to produce frictional effects which break up surface films. This joining process is mostly suitable for joining thin gauge materials and wires to a high standard of positional accuracy. The process is illustrated in Fig. 3.4. Basically the ultrasonic welding equipment consists of a high frequency power source connected to a transducer coupling system which converts the electrical energy into high frequency vibrations, typically 1-80 microns amplitude at 20-60 kilo hertz frequency.

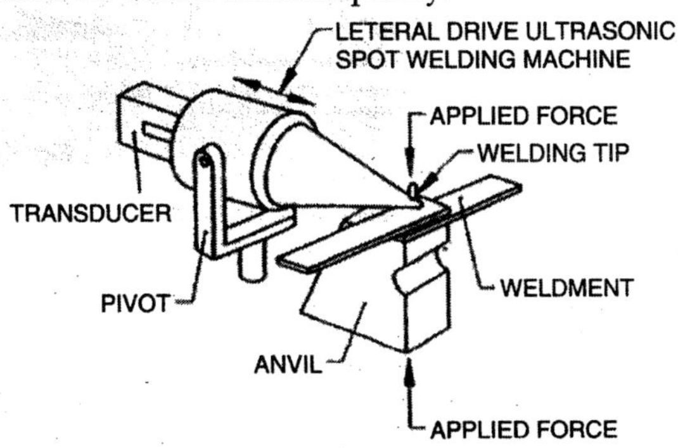

Fig. 3.4 Ultrasonic welding.

The stages of the welding operation are :

(i) Clamping the two components together under a moderate static pressure.

(ii) Subjecting the joint area to intense vibratory motion by way of an upper clamping tool termed as NOTRODE. After a period of approximately 2 seconds sufficient deformation is produced at the interface of the oxide film to fissure and disperse.

Process Parameters

The welding parameters include ultrasonic power, clamping force and weld time or rate. There must be specified and adjusted work-wise. Power required is more when hardness and/or thickness of the job is higher. Correct clamping force is to be established. Time interval ranges from 0.01 second for fine wires to about 1.0 second for heavy sections.

Advantages

(i) The main advantages of ultrasonic welding is the absence of heat and the fact that dissimilar metals can be joined to each other including metals to non-metal.

(ii) The weld are generally as strong as the parent material itself.

Disadvantages

(i) The range of application is limited to spot and similar welds in thin shuts and foils.

(ii) There is a limitation of thickness of section to be welded depending upon machine capacity.

Applications

(i) Small wires are welded to silicon semiconductor devices in the field of microelectronics.

(ii) Electrical contacts made of dissimilar metals of copper and silver.

(iii) Dissimilar metal combinations are aluminium to copper, aluminium to stainless steel and silver to aluminium etc.

(v) Thermoplastic can also be welded by using suitable ultrasonic plastic welders. Figure 3.5 shows ultrasonic welding equipment.

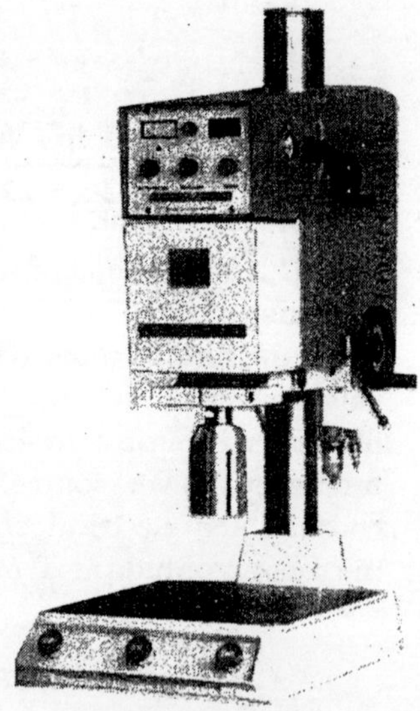

Fig. 3.5 Ultrasonic welding equipment.

Weld Quality

The welds which are produced are similar to cold pressure welds, that is, they are formed by breaking up the oxide film and pressing the freshly exposed metal surfaces together.

3.3 DIFFUSION BONDING

Process Fundamentals

Diffusion bonding is a solid phase welding process. The joining of the surfaces is produced by the applications of pressure and elevated temperature. In bonding is obtained by atomic migration as a result of the combined application of pressure and heat. The temperature required to produce joints by diffusion bonding is usually between 0.5 T_m and 0.75 T_m where T_m is the melting point of the material being bonded, in degrees absolute. Although the controlling mechanism of the process is complex, diffusion plays a major role once intimate contact between the two parts is established.

A variety of equipment can be used for diffusion bonding. Vacuum shielding is often used, and on a small scale the work-pieces may be pressed between platens within a vacuum furnace, the heating being accomplished by induction. Refer Fig. 3.6.

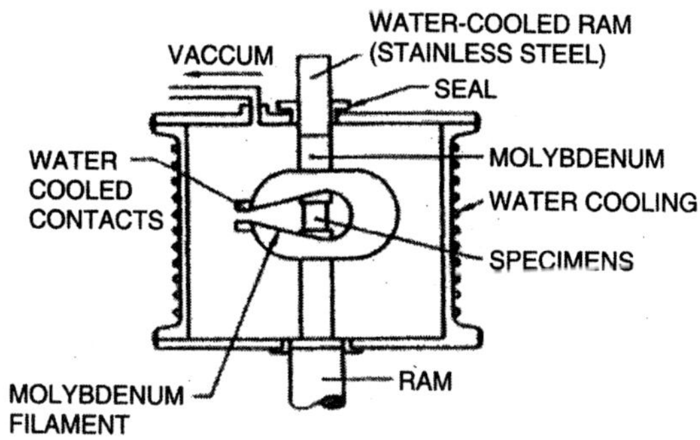

Fig. 3.6 Radiant resistance heating in a vacuum autoclave.

Process Parameters

The main practical process parameters that control the bonding operation are :

(i) Pressure
(ii) Temperature
(iii) Time
(iv) Atmosphere (vacuum).

Advantages

(i) Diffusion bonding has been used for joining many high temperature metals and alloys and dissimilar metal combination etc.
(ii) There is a great scope of producing complex hollow and multiple joints in steel and other materials.
(iii) Non-metals such as ceramics as well as powder metal products can be joined to metals in a way not possible by conventional fusion welding processes.

(iv) Diffusion bonding produces no distortion.

(v) Diffusion bonding is a process of high productivity to challenging application.

Applications

(i) Atomic energy industries : fabrication of reactor components.

(ii) Aerospace industries : titanium components.

(iii) Cutting tool manufacturing : carbide tipped cutting tools

(iv) Dissimilar metal combination for cryogenic applications.

4

CHAPTER

RESISTANCE WELDING PROCESSES

❑ INSIDE THIS CHAPTER

4.1 Spot welding, 4.2 Projection welding, 4.3 Seam welding, 4.4 Flash butt welding, 4.5 Resistance welding machines.

Resistenace welding process is widely used for sheet metal joining in industries such as antomobiles, house-hold appliances, aircraft etc. All resistance welding techniques are based on the heat produced between the aboutting metal surface due to resistance offered to the flow of electric current. The amount of heat generated basically depends on the strength of the current flow, the duration of the current flow and the resistance of the metal surfaces at the joint.

Heat generatred $H = I^2 Rt$

where, I = current

R = contact resistance of the mating surfaces.

t = time of application of the current

when the mating metal suraces reach a molten state in the localised area to an appreciable amount, pressure is applied to consolidate the joint between the two metal surfaces.

Classification

Resistance welding processes can be classified, based on the type of welds (lap or butt) produced, as shown in Fig. 4.1. Spot, seam and projection welding produce lap joints while flat and upset welding produce butt joints. In spot, seam and projection welding processes, melting occurs at the mating metal surfaces during welding. Force is always applied before, during and after the application of current to prevent arcing at the mating surfaces, and in some applications, to forge the weld metal. Flash and upset welding processes produce butt joints by heating

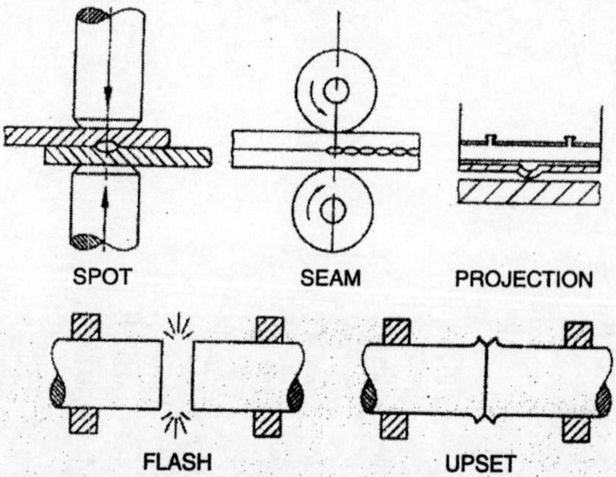

Fig. 4.1 Different resistance welding processes.

the entire cross-section. An upsetting force is applied ad some points before or after the heating cycle to bring the two mating surfaces into intimate contact.

4.1 SPOT WELDING

4.1.1 Process Fundamentals

In the resistance spot welding process, the sheets to be welded are kept between the water-cooled copper alloy electrodes, and under application of pressure, the current is passed through the mating surfaces. At the contact surface between the

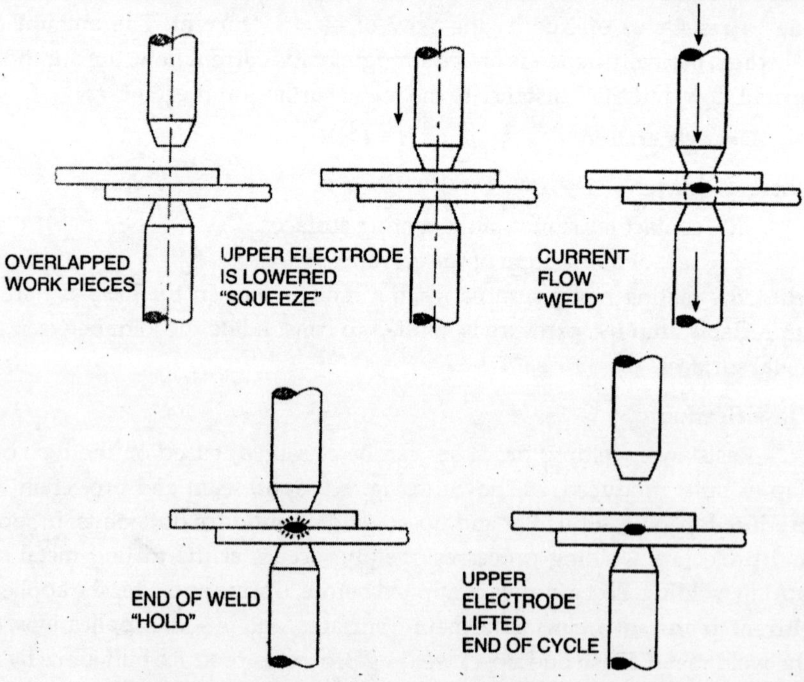

Fig. 4.2 The various stages of a spot welding - Operation.

sheets, due to high resistance to current flow, localised heating takes place and a molten welt nugget is formed. The stages involved in spot welding are illustrated in Fig. 4.2 they are as follows :

(i) The first stage involves keeping the two sheets to be spot-welded between the top and bottonm electrodes and allowing the top electrode to come down and applying pressure to develops intimate contact prior to current flows. The cluration of this stage is called squeeze time.

(ii) The second stage involves passing of current to form a matter nugget at the interface of the two sheets. This duration is called weld time.

(iii) The third stage involves applying pressure alone, without currant flow, to facilitate cooling of the nugget, and this duration is known as hold time.

(iv) The fourth stage involes parting of the electrodes to facilitate termination of one spot weld cycle and the time taken for this stage is called the off-time.

4.1.2 Characteristics of Spot-weld Nugget

With cost metals, the nugget of a spot weld has a cast structure. It is surrounded by the metal modified in structure and properties by the welding heat. The zone thus modified is called the heat affected zone. The nugget diameter in a spot weld is an important parameter which will indicate the extent of welding or quality of welding. Penetration of a nugget is also an important variable. This is expressed as the ratio of the distance from the line of joint to the boundary of the nugget '*a*' on the macro-section, to the part thickness δ. The penetration '*A*' is found for each part in a joint as follows:

Percentage $$A = \frac{a}{\delta} \times 100$$

for most metals '*A*' should be somewhere between 20 to 90%. Normally in selection of the welding parameters such as current, time etc. for a good spot weld, it is better to have penetration $A = 40 \text{ to } 70\%$ and the nugget diameter $(d) = 5$ where δ is the thickness of the sheet welded. When joining the parts of different thicknesses, the penetration of each part should be atleast 20% of the thickness of the thinner part. The nugget diameter in such cases should be based only on the thickness of the thinnest part.

Indentation is the depression created on the surfaces of the spot welded parts. Normally this depression should not exceed 10% of the thickness of the sheet.

4.1.3 Welding Techniques

Several varieties of spot welding techniques are used. They all differ in how the current is applied to the parts to be welded. Generally, two electrodes are used and the current is applied to the work form two sides. Sometimes, one of the electrodes is flat backing plate or a bar convenient for tack welding when, the parts are fitted up in a fig or fixture. For this purpose, copper backing plates are installed at the points where tack welds are to be made. It one of the two parts is three or more times thicker than the other one, the current may be applied to that part. In such a case, part should be strong enough not to deform under the electrode at the spot

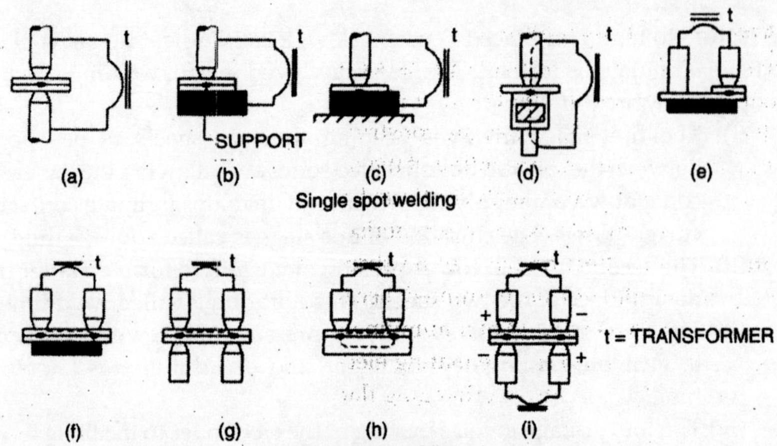

Fig. 4.3 Spot weld techniques.

weld. When one of the parts in thin-walled and hollow cylinder, a current conducting insert is used to weld the pieces together. To reduce the indentation and give a good appearance, it is preferable to use a flat electrode for one side. As an alternative, both electrodes can be on one side of the material, and the method of operation is known as twin spot-welding. Among the advantages of this method are an increased welding production rate become two spot welds can be made at a time, the equipment may have a low power rating and the welding operation can be mechanised move readily than when current is applied from two sides. In most cases, twin spot-welds are made using a current- conducting backing plate or a short-circuited bottom electrode. A current conducting backing plate may be dispensed with, if the part facing the electrodes is considerably thinner them the other. The characteristic feature of twin spot welding except the technique shown in Fig. 4.3(e) is that, part of the current conducted by the electrodes does not flow through the welding zone, but has its path completed through the upper work-piece.

The current flowing through the weld zone, its path is completed through the upper part, heats it and wears down the electrodes more rapidly. Some spot welding machines use direct twin spot-welding technique with the electrodes on both sides of the material. Figure 4.3 explains all above as explained.

4.1.4 Weld Design

There is a direct relationship between the weld strength of a given joint and its design. The factors that must be taken into consideration for a given material are the amount of overlap, the spot spacing, the spot size and the accessibility. The amount of overlap is determined by the spot size and it should be at least $2d + 3$ where 'd' is spot diameter in nm. When the spot spacing is too close, the current will shunt through the previously made spot weld, and therefore, this spacing should be at least $16t$ where 't' is the thickness of the sheet to be welded.

4.1.5 Electrode Tip Selection

Selection of a suitable electrode tip is essential. Figure 4.4 shows the typical electrode tip being used. The welding electrodes are made of copper alloys, and when the duty cycle is high, they are constructed to provide for water cooling. This prevents spreading or mushrooming of the tips. Tapered tips are generally used as they mushroom uniformly with continued wear. Domed tips, used for welding many non-ferrous materials, are characterised by their ability to withstand heavy pressure and severe heating without mushrooming. Filing the electrodes while on the machine should not be permitted because it is too difficult to restore the original contour. When a weld of minimum indentation is desired a flat tip is used, a domed tip is then used for the mating electrode. Offset electrodes are made for corner welds, for parts with overhanging flanges and for the less accessible areas.

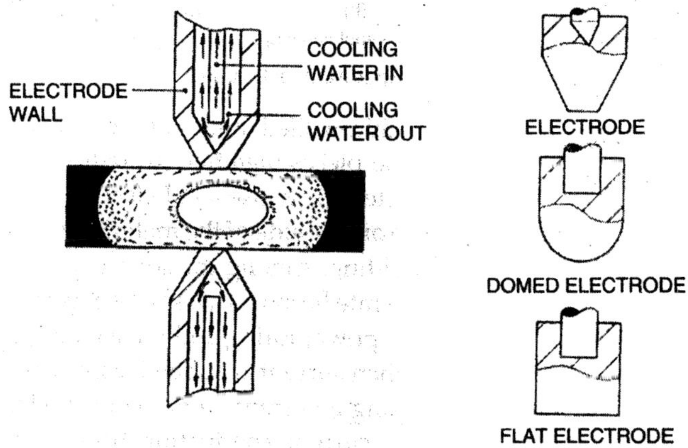

Fig. 4.4 Electrode tips.

4.1.6 Surface Prepration

Pre-weld cleaning is necessary to remove two basic types of contaminants:

- Foreign matter such as grease, oil, dust, dirt, paint, films etc.
- Oxide films. Rusty metal should not be spot welded without cleaning. Oil will not affect weld quality as much, but particles of dirt in the oil may cause erratic welding. Oil and oxides in contact with the electrode tips cause carbonisation and reduce tip life.

4.1.7 Guide Parameters

The guide parameters for spot welding of steels and non-ferrous metals are given in Table 4.1(a), 4.1(b), and 4.1(c).

Table 4.1(a) : Weld Parameters for the Spot Welding of Low Carbons Steels

Part thickness, mm	I_w rms' kA	t_w's	F_w's kgf
High-rate operations			
0.3	5–5.5	0.06–0.08	150–200
0.5	5.5–6	0.08–0.1	120–180
0.8	6.5–7	0.1–0.14	200–250
1	7.5–8	0.12–0.16	250–300

1.2	9–10	0.14–0.18	300–400
1.5	10.5–11.5	0.16–0.22	400–500
2	13–15	0.18–0.24	600–700
3	18–20	0.24–0.3	900–1000
3.5	20–22	0.3–0.4	1100–1200
4	23–20	0.4–0.56	1300–1500
Low-rate operations			
0.6	3.5	0.2	80
0.8	4	0.3	120
1	5	0.4	150
1.2	5.5	0.5	200
1.5	6.5	0.6	220
1.8	7	0.7	300
2	7.5	0.8	350
2.5	9.5	1	400
2.8	12	1.3	500

Table 4.1(b) : For Spot Welding of Stainless Steels

Part thickness, mm	I_w rms' kA	t_w's	F_w's kgf
0.3	5–5.5	0.06–0.08	150–200
0.5	4.5–5	0.08–0.12	250–300
0.8	4.5–5	0.12–0.16	300–400
1	5–5.5	0.16–0.18	350–450
1.2	5.5–6	0.18–0.2	450–550
1.5	6–7.5	0.2–0.24	500–650
2	7.5–8.5	0.24–0.3	800–900
2.5	9–10	0.3–0.34	1000–1100
3	10–11	0.34–0.38	1200–1400

Table 4.1(c) : Suggested Parametes for the Spot Welding of Non-ferrous Alloys

Part thickness in mm	Electrode force Kgf		forge period t_t	Current characteristics			
	Welding F_w	Forging F_t		I_w' rms' kA	t_r sec	t_w sec	t_f sec
	Aluminium alloys						
0.5	220	—	—	23	—	0.08	0.
	200	—	—	24	0.04	0.08	0.08
0.8	350	—	—	27	—	0.1	—
	300	—	—	26	0.04	0.1	0.1
1	450	—	—	28	—	0.12	—

	350	—	—	27	0.04	0.12	0.12
1.5	650	—	—	34	—	0.16	—
	550	—	—	33	0.06	0.16	0.16
	450	1000	0.24	31	0.06	0.16	0.16
	450	1200	0.18	32	—	0.16	—
Aluminium alloys							
0.5	130	—	—	22	—	0.04	—
0.8	190	—	—	26	—	0.04	—
1	250	—	—	30	—	0.06	—
1.5	350	—	—	34	—	0.08	—
2	500	—	—	38	—	0.1	—
Magnesium alloys							
0.8	300	—	—	20	—	0.1	—
1	400	—	—	24	—	0.12	—
1.5	500	—	—	27	—	0.14	—
2	700	—	—	30	—	0.18	—
Copper alloys							
0.5	120	—	—	15	—	0.1	—
0.8	170	—	—	17	—	0.14	—
1	200	—	—	19	—	0.16	—
1.5	300	—	—	24	—	0.2	—
2	400	—	—	30	—	0.24	—

4.2 PROJECTION WELDING

4.2.1 Process Fundamentals

Projection welding is similar to spot welding, in that the parts are welded at several projected points at a time. The projections are produced either by machining or pressing of one or both of the work pieces. In projection welding, current concentration is achieved by the low initial contact due to projection, whereas in spot welding this is achieved by proper electrode tip. Figure 4.5 illustrates the projection welding process. Under resistance heating and pressure, the projections collapse and the parts to be joined are brought in close contact.

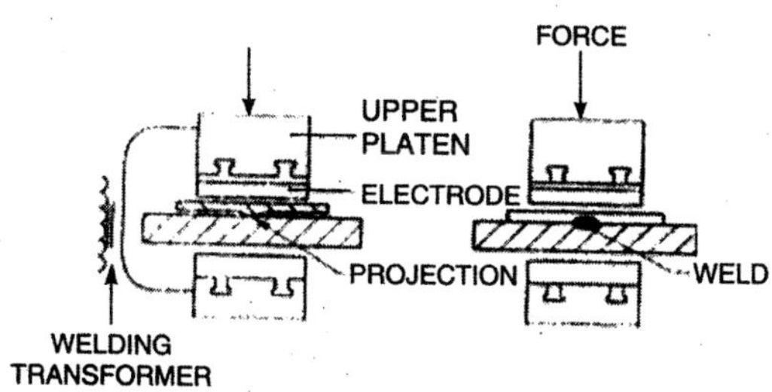

Fig. 4.5 Projection welding.

4.2.2 Welding Techniques

In addition to achieving a spot weld at a particular position different configurations can be welded as shown in Fig. 4.6. In T-projection welding, a part with a smaller cross-section is welded to a part with a longer surface area like sheet. At the joint, the smaller part is given an appropriate shape (a sphere, a cone, a closed or open projection). Cross-welding of rods, tubes and wire to a sheet may be again classified under projection welding.

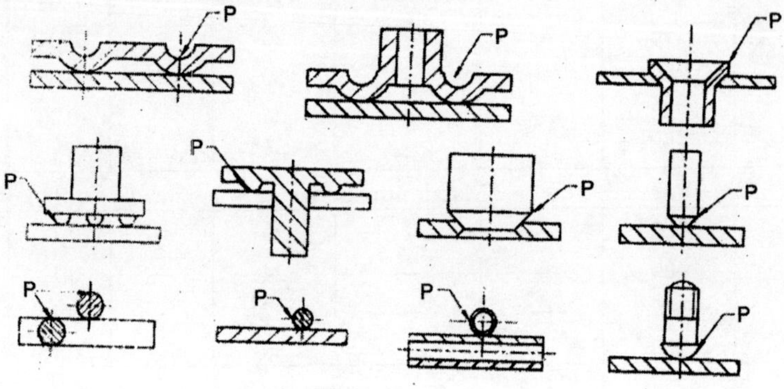

Fig. 4.6 Projection welding techniques.

Projection welding can be made on spot welding machine with a proper electrode. The electrode life is long since there is no direct contact with the hot metal. Outer or top surfaces can be produced without electrode marks, making it possible to point or plate without grinding or polishing. Projection welding can be only applied to materials such as steels or refractory metals. Non- ferrous metals such an copper, aluminium are difficult to weld. Some typical applications of projection welding are attachment of small fasteners, nuts, special bolts, studs and similar parts to larger components.

4.2.3 Guide Parameters

The recommended dimensions for design of projection welds are summarised in Table 4.2. The guide welding parameters for projection welding of low carbon steels is given in Table 4.3.

Table 4.2 : Pecommended Dimensions for Projection Weld (mm)

Part thickness	Nugget diameter	Minimum lap	Projection diameter	Projection height	Minimum spot spacing
0.3	2–2.5	3.5	1.4		6
0.5	2.5–3	4	1.7	0.4	7
0.8	3–3.5	5.5	2.4	0.45	9
1.0	4–5	7	3	0.55	10
1.2	5–6	8	3	0.7	12
1.5	6–7	10	4	0.9	15
2.0	7–9.5	12	4.8	1.1	18
2.5	8–9.5	14	5.6	1.2	23

3.0	9–10.5	17	6.8	1.4	27
3.5	10–12	20	8	1.6	30
4.0	11–13	22	8.7	1.8	34
4.5	12–14	24	9.5	2.0	38
5.0	13–15	26	10.6	2.2	42

Table 4.3 : Variables for Projection Welding of Low-carbon Steel

Part thickness, mm	I_w, rms' kA	t_w's	F_w' kgf
0.5	5.0	0.08	80
0.8	6.5	0.10	100
1.0	7.5	0.16	150
1.5	10.5	2.26	250
2.0	12.0	0.40	350
2.5	14.0	0.54	450
3.0	16.0	0.74	600
4.0	20.0	1.2	950

4.3 SEAM WELDING

This process involves making a number of spot welds (nuggets) which may or may not overlap one another. If the spot welds overlap a tight seam is produces. If the spots do not overlap, they are named roll-spot welding.The seam welding is made by means of two (or one) rotating wheel electrodes which squeeze together and roll on the parts to be welded. The welding current conveyed to the wheel electrodes heats and melts the metal at the joint. Refer Fig. 4.7.

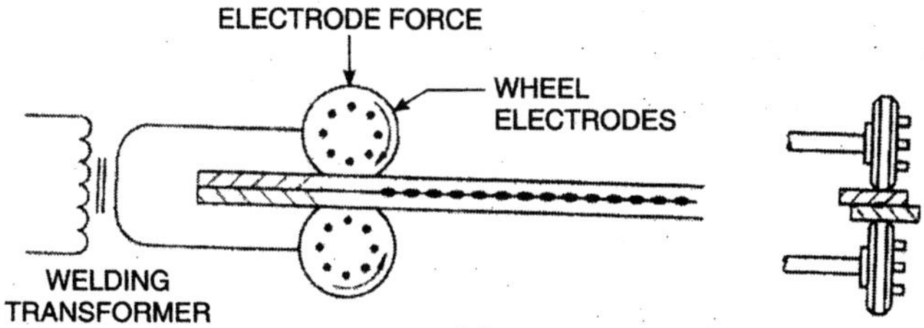

(a)

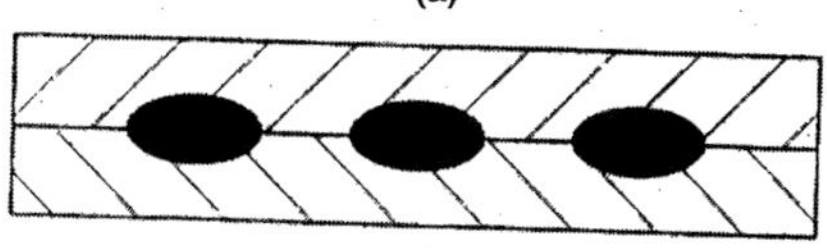

(b)

Fig. 4.7 Seam and roll spot welding.

When the sheets are moving between the electrodes and the welding currant is continuously flowing, the continuous seam welding is achieved. This process is seldom used became it is excessively overheated. Most frequently, intermittent seam welding is used, in which the work-pieces move continuously and the current is interrupted at regular intervals so that a single spot weld (nugget) is formed during each 'on' period. An overlap of the spot welds is necessary to make a pressure tight seam weld. This is controlled by the speed of the wheel electrode and the rate of the current interruption. Another variation in seam welding is stitch welding in which the work-pieces are moved intermittently and the welding current is turned on only when thy are stationary. In this case the metal in contact with the wheel electrode is cooled better compared to seam welding using external water cooling to reduce the overheating of the outer metal layers.

4.3.1 Welding Techniques

The various forms of seam welding used in practice mainly differ in the manner in which the welding current is conveyed to the parts, from one or two sides, and the position of the wheel electrodes relative to work-pieces. Refer Fig. 4.8. Two-side seam welding is similar to direct spot welding. One of the wheel electrodes may be replaced with a mandrel which is in intinate contact with the inner part. To make a circular seam weld on a stationary parts, the top wheel electrode is made to rotate about its axis and revolve around the joint axis. The bottom part is in contact with a cup-shaped electrode. Sometimes, the parts to be welded are installed on a copper backing plate, and the current can be conducted to the parts form one or two sides. In the former case backing plate may be moveable or it may be stationary an two current carrying wheel electrodes rotate about their axis and roll on parts.

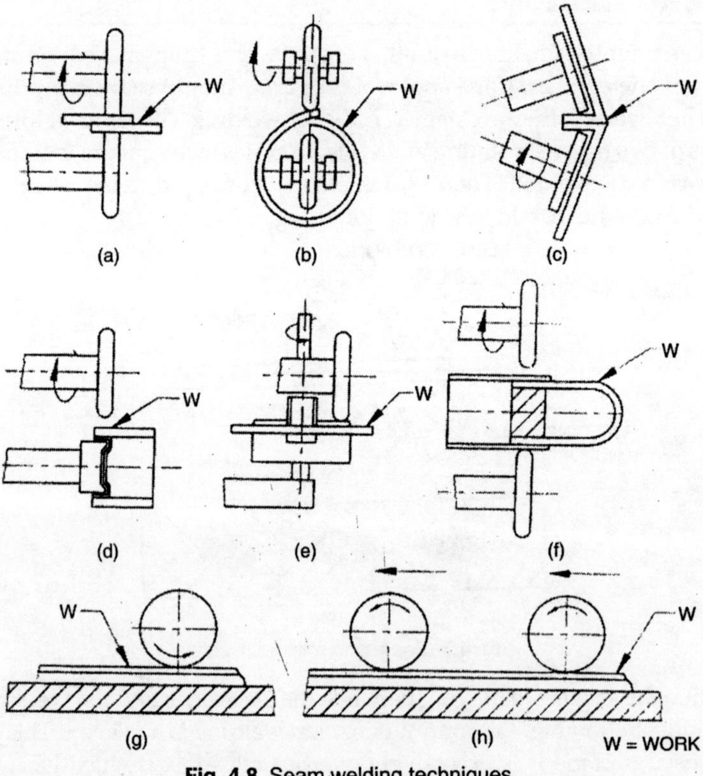

Fig. 4.8 Seam welding techniques.

4.3.2 Guide Parameters

For seam welding of low carbon steel, stainless steel and non-ferrous metals, the guide parameters are given in Table 4.4 and 4.5.

Table 4.4 : Weld Parameters for Seam Welding of Low-carbon Steels

Part thickness mm	I_w, rms' kA	Duration, s		F_w kgf	Welding speed m/min*
		t_w	t_{off}		
0.5	7–8	0.02–0.04	0.04–0.06	150–200	1.2–1.6
0.8	8.5–10	0.04–0.06	0.02–0.08	200–300	1–1.4
1	10.5–12	0.05–0.08	0.08–0.1	300–400	0.8–0.9
1.2	12–13	0.08–0.1	0.1–0.12	400–500	0.7–0.9
1.5	13–14.5	0.12–0.14	0.12–0.14	500–600	0.6–0.7
2	15.5–17	0.16–0.18	0.18–0.22	700–800	0.5–0.6
3	18–22	0.24–0.32	0.28–0.34	1000–1100	0.4–0.5

Table 4.5 : Variables for Seam Welding of Non-ferrous Metals and Alloys

Part thickness mm	Wheel pressure F_w, kgf	Welding current I_w, rms'kA	Duration	s	Welding speed m/min
			t_w	t_{off}	
Aluminium alloys of AM and AM type					
0.5	250	24	0.04	0.10	0.70
0.8	300	29	0.04	0.14	0.60
1.0	350	33	0.06	0.18	0.50
1.2	400	35	0.08	0.20	0.45
1.5	450	38	0.10	0.24	0.40
Copper alloys of π 62 type					
0.5	200	19	0.06	0.06	0.75
0.8	300	23	0.08	0.08	0.70
1.0	400	27	0.08	0.10	0.70
1.2	450	31	0.10	0.12	0.65
1.5	550	36	0.12	0.14	0.60

4.4 FLASH BUTT WELDING

4.4.1 Process Fundamentals

In butt welding, the welding faces of the parts are first brought together by a small pressure, heated by the welding current until they are at a fusion condition, and then the parts clamped in the jaws are moved towards each other at a constant or accelerated rate until they come into contact. Because the initial area of contact is small, an intense current is generated between the parts. The metal is instantly heated to the boiling point and vapourises, which is accompanied by tiny explosions of contact bridges. As a result of the explosions, some metal is expelled from the weld as spark and splashes. This is followed by upsetting by which the parts are brought together. Refer Fig. 4.9.

This upset action causes the molten and oxidised over-heated metal to be expelled from the parts and to form a welded joint. At the same time, the adjacent metal is deformed, and the outer metal fibers take a characteristic bend. In the course of flashing and upset, the initial extension of the parts is reduce by welding allowance.

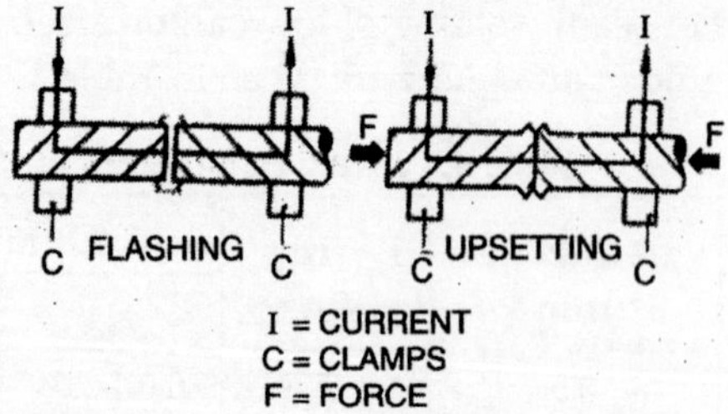

Fig. 4.9 The flash - Butt welding process.

4.4.2 Techniques

Preheating in flash butt welding is employed where large sections are to be joined. In this process, the welding faces of the work pieces are preheated as in butt welding process. The parts are intermittently brought together, heated by the current, and moved apart. After the welding faces are heated up to a proper temperature, they are flashed and upset. Flash butt welding is one of the most commonly used methods. It is used to join part having both compact and extended section such as rounds and squares in the former case, and various shapes, thin-walled tubes, thin and wide sheets in the later case. Steels and non-ferrous alloys can be flash butt welded.

Its application also include the manufacture of cutting tools, various annular blanks for flanges, wheel rims, large pipe lines and welding of hot billets.

4.4.3 Guide Parameters

The guidelines for the determination of parameters for flash butt welding of law carbon steels in respect of cross-sectional area of welding involved are given in Table 4.6.

Table 4.6 : Variables for the Flash Butt Welding of Low-carbon Steel

Cross sectional area, mm^2	Allowance in the case of continuous burn off, mm		Allowance in the case of burn off and preheat mm		
	for burn off	for upset	for preheat	for burn off	for upset
75	5–5.5	2.0–3.0	2.5	4.5	3.5
100	5.5–6.5	2.4–3.0	3	5	3.5
200	7–7.5	3.0–3.5	4	5.5	3.5
300	9–10	3.5–4.0	4	7	4
600	11–12.5	4.0–4.5	4	7	4
1000	14–17.5	4.5–5.5	4	10	4
1250	17–20	5–5.5	4	11	4
1500	18–22	5.5–6.5	4.5	11.5	4.5
2000	20–25	6–6.5	5	12	5

4.5 RESISTANCE WELDING MACHINES

The two main functions that any resistance welding machine performa are to press together and heat the parts to be joined. A seam welding machine will additionally advance the parts, and a butt welding machine will clamp the parts in the jaws (electrodes). The design of a welding machine is adapted to the welding techniques examined above. All welding machines can be divided into two groups essentially differing in design, namely machines for spot seams and projection welding and machines for upset and flash butt welding. A basic resistance welding machine can be considered to have a mechanical and an electrical sections.

4.5.1 Resistance, Spot, Seam and Projection Welding Machines

The main part of the mechanical section in a resistance spot welding machines is a frame, on which are mounted an upper arm and a lower horn, which carries an electrode in an electrode holder. The lower arm is usually made free to move up and down so that the electrode clearance can be adjusted to suit the shape and size of the components being welded. The upper arm mounts a penumatic drive unit, providing the necessary electrode force. It is attached to an upper horn carrying an electrode holder. The force applied by a pneumatic drive is taken up by the frame, the upper and lower arms and horns, so that they should have a stiffness to minimize the deformations due to this force. Refer Fig. 4.10.

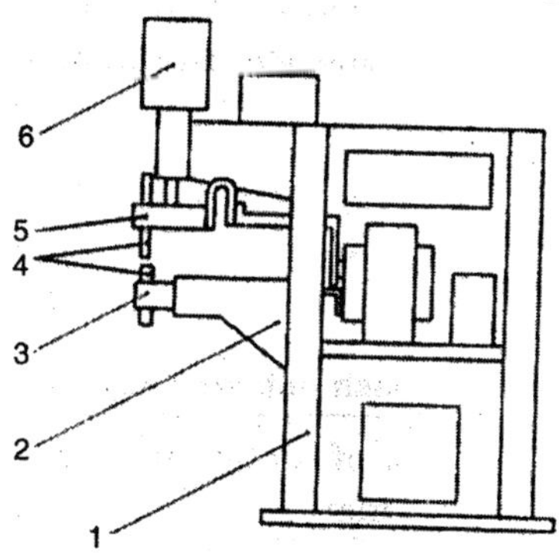

1. FRAME
2. LOWER ARM
3. ELECTRODE HOLDERS
4. ELECTRODE
5. ELECTRODE HOLDERS
6. PNEUMATIC CYLINDER

Fig. 4.10 Schematic diagram of resistance spot welding machine.

The electrical section of the machine consists of a welding transformer, a tap-change switch, a contactor and a control unit as shown in Fig. 4.11. Often, the control gear is mounted in a separate cabinet. The contactor connects and disconnects the welding transform to and from the supply mains. The electrical section includes the secondary circuit or loop, formed by the conductors, conveying current from the transformer to the components being welded. From the, transformer, the current is applied via stiff and flexible buses to the upper horn and lower horn, carrying electrode holders. It is easy to see that the horns and the

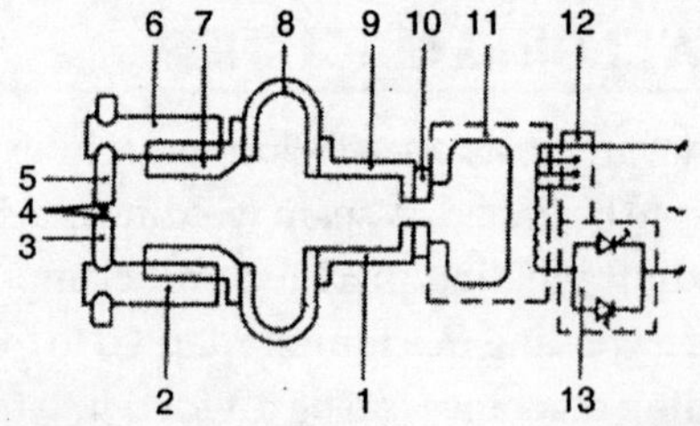

1. BUS BAR
2. LOWER HORN
3. LOWER ELECTRODE HOLDER
4. ELECTRODE TIP
5. UPPER ELECTRODE HOLDER
6. UPPER HORN
7, 8, 9, 10. BUS BAR
11. TRANSFORMER
12. TAP SETTING
13. THYRISTOR CONTROL

Fig. 4.11 Power circuit of a resistance spot welding machine.

electrodes convey both current and force, so thėy belong to both the electrical and mechanical sections of the machine. All components of the secondary circuit are made of copper or copper alloys having a high electrical conductivity. Most of the secondary circuit components, the welding transformers and the contactors have an internal water cooling system. Projection welding machines use backing plates instead of electrodes and seam welding machines used electrode heads with wheels. Seam welders have a wheel drive.

4.5.2 Butt Welding Machine

The butt welding machine has been showin in Fig. 4.12. The mecahnical section of a butt welder consists essentially of a frame with guides, a stationary platen, and a moving platen, refer Fig. 4.13. Each platen gives suppprt to clamping drives, with jaws. The moving platen is connected to a clamp and upset drive. The frame should be able to stand up to the large clamping and upset forces and hold the workpieces precisely aligned without any deformation during the welding cycle.

Fig. 4.12 Butt welding machine.

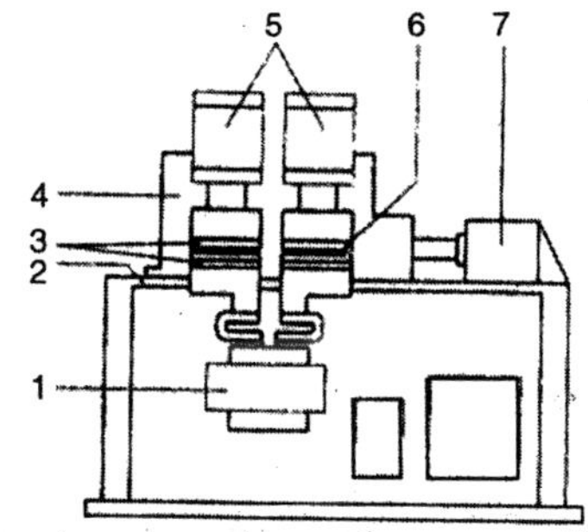

1. TRANSFORMER
2. MACHINE FRAME
3. CLAMPING JAWS
4. FIXED PLATEN
5. HYDRAULIC CYLINDER
6. MOVABLE PLATEN
7. HYDRAULIC CYLINDER

Fig. 4.13 Mechanical section of a butt welding machine.

The electrical section of a butt welder consists of a welding transformer, a tap-changer also called regulator, a contactor, and a control unit. The secondary circuit of the butt welder is simpler than the other types of machines. It consists of a flexible conductors and transformer terminal blocks connected to the jaws. In most of the machines, the upper jaws are moved vertically to clamp the parts to be welded.

❑❑❑

CHAPTER 5

SPECIAL WELDING PROCESSES

❑ **INSIDE THIS CHAPTER**

5.1 Underwater welding, 5.2 Electron beam and laser beam welding.

5.1 UNDERWATER WELDING

In underwater welding the welding can be carried out by welders under wet environment or dry environment, using habitats or hyper baric chambers. Underwater welding had been tried during 1900s for patch up work on ships. During 1960s varnish coated or wax coated, water proof electrodes were extensively used for underwater welding.

In 1962, there were only 62 off-shore drilling platform (rigs) and during 1978, there were 470 platforms. There are many types of off-shore drilling rigs such as jack up type, semi-submersible type, ship/barge type and the submersible type.. Around 50% of the rigs are in 120 meters deep in water. Other than oil rigs we also have ocean bottom mining, ocean exploration, ocean thermal energy conversion, salvaging etc. Hence, under water welding is required for:

(a) Repair of existing old off-shore structures.

(b) Erection of new off-shore structures.

(c) Extending underwater facilities to deep waters.

(d) Repair of ships while they are afloat.

(e) Repair of nuclear reactor components underwater.

5.1.1 Wet Welding Techniques

By wet welding it is meant that underwater welding is done without specially developed enclosures around the welding spot or joint. Hence, the electrodes or welding gums have additional waterproofing arrangements, like wax coating in electro due or gas screens in torch welding.

Underwater wet welding is restored to for a fast, temporary and inexpensive repair. Today, the off-shore industry needs quality wet welding repairs. Wet welding is a fast method of repair, requiring less support equipments, Manual metal arc welding and MIG/CO_2 welding are the processes used for wet welding.

(a) Gravity Welding

This make use of a simple apparatus having an electrode holder fixed to a sliding bar. When the electrode is fixed and the arc is initialed at the tip of the electrode. The electrode burns and come down due to gravity, thereby welding is achieved. This apparatus is simple, cheap and employed successfully at low water pressure. Refer Fig. 5.1.

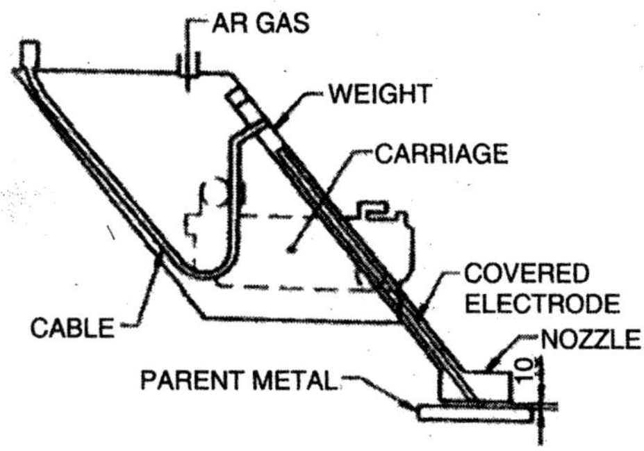

Fig. 5.1 Schematic of gas - Shielded gravity welding set.

(b) Shielded Metal Arc Welding

The welder can reach portions of offshore stimctunes that cannot be welded using other methods. No enclosures are required and no time in lost in building or rigging them. Standard welding machines are used. It has more freedom for repair and easier fit up of the repair sections. The major problem is in hydrogen pick-up and high cooling rates causing hydrogen induced cracking.

(c) Gas Metal Arc Welding (MIG/CO_2)

A water certain type CO_2/MIG welding has been developed in Japan. This user dual nozzles which have the shielding gas flow from an inner nozzle and a concentric flown of water from an ovter nozzle. With this type of torch even overhead welding and be done easily. The same principle is successfully used for plasma welding which is popular now-a-days in achieving good welds.

(d) Local Cavity Welding

The process developed by the Japanese, makes use of special torch where the gas is fed into the torch to create a dry zone in the region surrounded by water. A water jet acts as a shutoff wall to the exterior water. The obtained cavity is stable and durable to the external and internal perturbations and to torch travel. The stability of the arc and the bead formation is excellent. Refer Fig. 5.2.

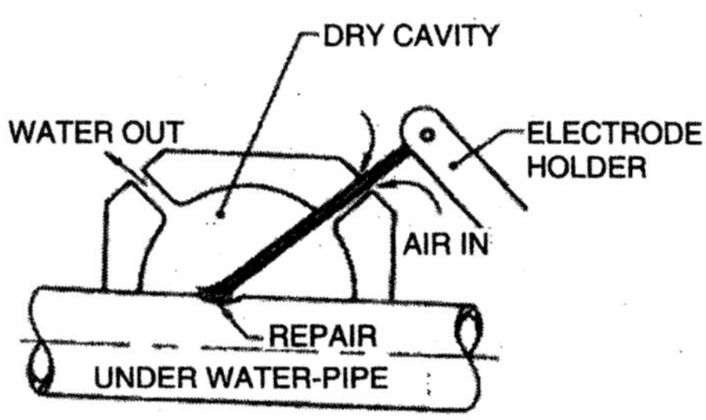

Fig. 5.2 Local cavity welding - Principle.

5.1.2 Habitat Welding

Welding is performed at ambient pressure in a large chamber form which water has been displaced. The welder/diver does not work in the diving equipment. The habitats may be large habitats or mini habitats. In the case of mini habitats welding is performed in a simple, open bottom-dry chamber which accomdates the head and shoulders of the welder/diver who is in full diving suit. Refer Fig. 5.3, which shows a habitat scheme for under water welding.

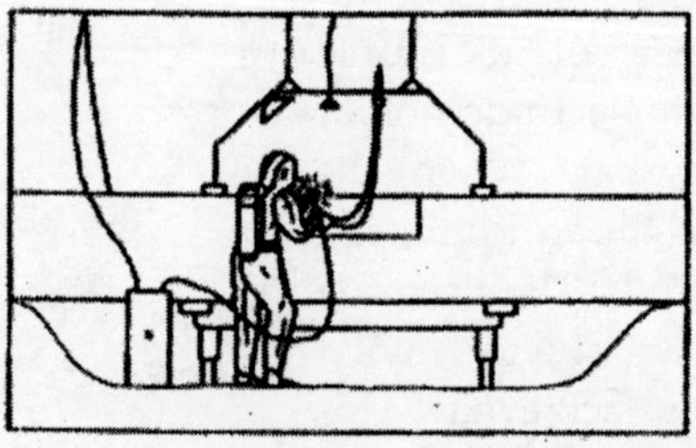

Fig. 5.3 Patch welding in a simple habitat.

Welding done at extreme depths, will have very high ambient pressures, and hence, very special system have been prescribed at ambient conditions for pipe line repairs. This is achieved by using a split pressure vessel (Fig. 5.4) that closes over the weld area. This is also known as hyperbaric chamber.

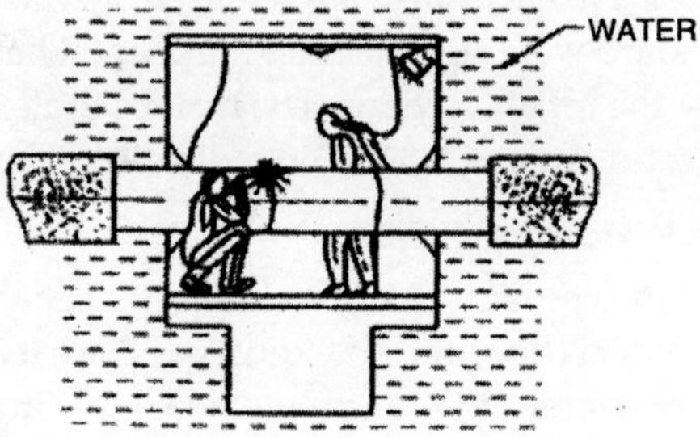

Fig. 5.4 Sketch of welding operation in a hyperbaric dry habitat.

The vessel is evacuated to atmospheric pressure for welding. After the completion of the weld and non-destructive testing, the vessel is flooded with water, opened and lifted to clear of the pipe.

Shielded metal arc welding with special underwater welding electrodes, TIG welding, MIG welding, flux-cored arc welding, plasma welding with special torches are used in a hyper baric chamber. Sometimes welding Robots are also employed within the hyperbaric chambers, which are immune to high pressures.

5.1.3 Underwater Weld Quality

AWS specification based on quality requirements of underwater welding may be referred.

- Type A are intended for structural applications and are made in accordance with a qualified welding procedure.

- Type B are intended for limited structural applications and are made in accordance with a qualified welding procedure.
- Type C are crack-free welds made in accordance with a qualified welding procedure where structural quality is not critical.
- Type O These are of quality equivalent to that produced on the surface in compliance with a code applicable to the particular construction involved. Type O welds have metallurgical and mechanical properties which are essentially the same as those of welds made in air.

It is clear from above that welding producer needs to essentially developed in all the cases.

Generally the quality of welds can be classified as :

(a) High quality repairs of underwater constructions.
(b) Permanent repairs of the underwater structures.
(c) Emergency repairs of underwater structures.

5.1.4 Safety Aspects

Conventional diving, rather surface diving is suited for short time missions in shallow waters. Decompression is required after only a few minutes of working below depth of 100 ft. When air is used for breathing, the safe depth limit is above 60 meters, with the helium/oxygen mixture it is less than 120 meters. Using saturation diving techniques, divers can be kept in a chamber even at 300 meters equivalent pressure, and gradually transported to the job-site for a work shift period of 8 hours. Several commercial companies use ambient pressure chambers. In these cambers, water is replaced by pressurized gas. The divers enter from the bottom and work inside on a platform. A helium- oxygen mixture with an oxygen partial pressure up to 0.5 kg/cm^2 is used. The welder breaths through a mask using a separate system of gases. The atmosphere inside the chamber is humid and therefore special equipment is used to reduce the humidity.

A submerged chamber with a pressure of approximated 1.0 atmosphere is preferable. In this a personal transfer capsule moves the welders to the chamber, and hence. There are not exposed to pressure or other diving hazards.

Submersibles of a variety of designs and capabilities to woke upto 20000 ft are available commercially. Manned submersibles or remote controlled underwater vehicles are popular now-a-days. There is a danger of fire when welding in compressed air environment. compressed air is not a safe environment to work in. Heliox (Helium-oxygen) atmosphere is sufficient to sustain life, also it is impossible for heliox to maintain flame propagation. Hence it is safe.

Generally Argon, Arganox (argon -oxygen), helium of heliox gases are used as a chamber gas. Even though nitrogen is a good alternative to avoid fire risk, it is not preferred since it affect the weld quality.

5.2 ELECTRON BEAM AND LASER BEAM WELDING

Electron Beam Welding (EBW) and Laser Beam Welding (LBW) are modern fusion welding processes. In conventional arc welding, the anc initiated from the electrode or wire diverges and takes a conical shape. Due to this diverging nature of the arc, the heat is dissipated on a large surface area of the metals being welded. This results in wider weld bead and less weld penetration associated with wide heat affected zone (HAZ). To achieve the required, full penetration of the joint, a wide groove edge preparation and multi pass welding is necessary.

These problems can be overcome by adopting EBW and LBW processor. The beam is focussed as a tiny spot on the weld joint, resulting in increased power density (watts/cm^2) compared to arc welding processes. The heat energy applied is highly localised due to the converging nature of the focussed beam.

The main features of EBM and LBW processes are:

(i) Single pass narrow deep penetration welds.
(ii) Low thermal distorting due to less heat input.
(iii) Narrow heat affected zone.
(iv) Faster welding speeds.
(v) No necessity for edge prepration.
(vi) No filler metal required normally.
(vii) High productivity processes.
(viii) Consistent weld quality.
(ix) Can be easily automated.
(x) Possibility of welding inaccessible joints.

The drawbacks of EBW and LBW processes are :

(i) High equipment cost.
(ii) High cost of maintenance of there machines.
(iii) Critical joints fit up requirements.
(iv) Not suited for site applications.

5.2.1 Electron Beam Welding (EBW)

(a) Process Fundamental

Electron Beam welding is a process in which the welding heat required is obtained from the impact of a concentrate beam of high velocity stream of electrons on the work piece. These negatively charged electrons bombard the metal surface, causing such intense local heat that it almost instantly vapourises a hole through the entire thickness of the weld joint.

The walls of this hole are molten and as the electron beam moves across the joint, the metal flows into the joint producing as extremely deep and narrow weld. This method of welding is known as key-holing (Fig. 5.5) EB welds may have a depth to width ratio of more than 10:1 due to extremely high heat concentration.

EBW is mostly done in high or medium vacuum environment, though out of vacuum EBW is also possible. The vacuum is a necessity became the electrons will

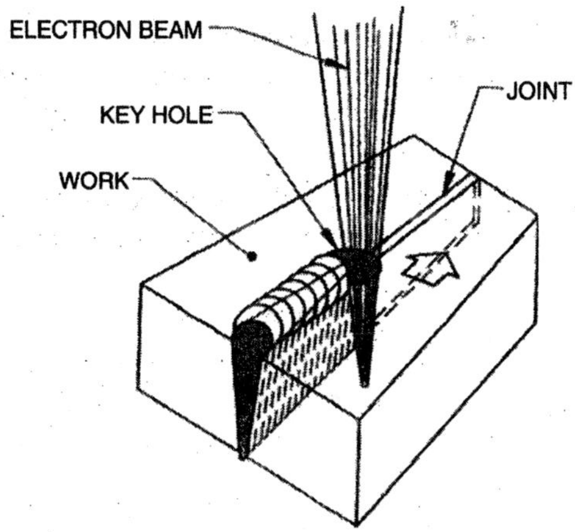

Fig. 5.5 Key holing.

interact with the gases in the atmosphere and diffuse the beam. So, a certain amount os energy and concentration of the beam will be lost. Lim effect will deprive the basic advantages of EBW.

(b) EBW Equipments

The major parts of EBW equipments are :

(a) Electron gun.
(b) Vacuum chamber.
(c) High voltage transformer.
(d) Electric/electronic controls cubicle.
(e) Chamber vacuum pumps.
(f) Guns vacuum pumps.

The required electron beam is generated, controlled and accelerated inside the electron gun. The working principle of a electron gun is similar to that of a vacuum triode tube. The cathode is made of tantanum or tungsten and heated to

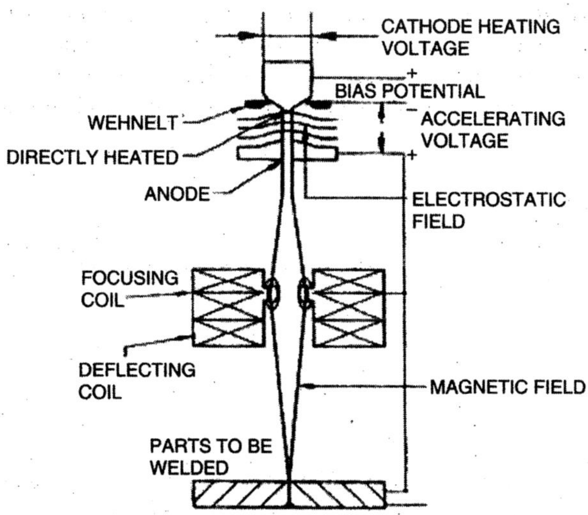

Fig. 5.6 Schematic of an electron gun.

high temperature. An electron cloud is generated due to thermionic effect. The electric field due to high voltage between the cathode and anode accelerates the electrons and sets them free with a considerable amount of energy. The electrons pass though the hole at the anode and flow towards the past to be welded. The wenhelt (grid) is a metallic shield around the electron cloud, which is biased negatively in relation to the cathode. Its function is to regulate and control the beam current. Figure 5.6 explains the system.

Focusing and deflection coils are also mounted inside the gun. Focus coils act as magnetic lens, which focusses the beam on the work-piece. Deflecting coil is used to deflect the beam to follow a particular path, if required. The EB gun can be either movable or fixed type. The welding chamber is provided with rotary and linear motion capabilities depending on the job requirements. High vacuum EBW uses a vacuum work chamber, where air is pumped out of the chamber. This accomplishes two things. First, it provides a clear path for the electron beam without collision air molecules. Second, the vacuum provides a clean atmosphere for the molten weld metal, which is most suited for welding. Harmful atmospheric contaiminants such as oxygen and nitrogen are removed in a vacuum environment. Vacuum chambers for EBW have been made in may shapes and sizes depending upon the job requirements. Technicians performing EBW should be aware of the production of X-rays while welding some metals. Most electron beam vacuum chamber have lead glass windows, and/ or closed circuit TVs for viewing the weld from a safe of shielded area.

(c) Welding Parameters and Their influence in EBW

Accelerating Voltage

The importance of the accelerating voltage lies in its ability of control the spot size of the focussed beam. For a particular beam power (beam current multiplied by accelerating voltage), if a high accelerating voltage in elected, the beam current can be reduced. This, in turn reduces the number of electrons in the beam. With fewer electrons The beam gets narrower and the spot size is reduced as determined by the relation:

$$\text{Spot diameter} = (\text{Beam current}/\text{Accelerating voltage})^{3/8}$$

This accelerating voltage is normally limited to 60 to 150 KV.

Beam Current

The depth of weld penetration attainable depends on the beam power, which is directly proportional to the beam current. This current range varies form 50 milliamperes to 1 ampere.

Focus Current

The weld quality in EBW, especially penetration, to a large extent depends on the position of the focul point of the beam in relation to the work-piece The beam diameter at the work-piece is dependant on the position of the focal point and thus the power density is varied, resulting a different bead profiles. As a consequence, certain weld defects like porosity etc. can be eliminated by adjusting the focus current/focal position. The beam spot size at the work-pices is normally adjusted to be about 0.5 to 0.8 mm in diameter.

Welding Speed

Like in arc welding the welding speed directly controls the heat input. The correct choice of the welding speed depends on the depth width of the weld desired and also on the physical properties of the materials being welded like thermal conductivity etc.

(d) Weld Defects and Control Methods in EBW

(i) **Porosity :** This occurs due to the evolution of the trapped gain from the liquid metal. Porosity may also arise from surface contamination of grease of oxide containing moisture. The gas contents in the beam metal should be at the minimum possible level which has to be taken care of during the procurement of the raw material.

(ii) **Cracking :** The chance of cracking is reduced with EBW, in comparison with conventional fusion welding processes, although cracking still presents a problem for many materials and joint designs.

(iii) **Solidification cracking :** This occurs due to the presence of low melting point compounds formed by impurities like sulphur and phosphorus in steel. This cracking can be reduced by the use of transverse beam oscillation.

(iv) **Necklace cracking :** This is a peculiar defect in EBW and has been found in a variety of materials including titanium alloys, stainless steels, nickel alloys and carbon steels. It occurs mainly in blind and very narrow fully penetrating welds. This defect can be eliminated by widening the weld bead intentionally.

(e) Application

EBW has unquestionably provided a facility to weld materials which are difficult to be welded by conventional welding techniques. EBW has been successfully applied in following fields and components.

Automobiles

(a) Diesel truck piston.
(b) Shock absorbers.
(c) Gears and gearboxes.

Aircraft and Aerospace

(a) Rotor assemblies.
(b) Aircraft landing gear and mats.
(c) Repair of blades and vanes.
(d) Butt welding of blades.
(e) Conc. for satellite.

Nuclear

(a) Fuel rods.
(b) Fuel casings.
(c) Centre wing box.

Mechanical and Miscellaneous Industries

(a) Steam- tunbine vanes.
(b) Steam boiler tubes.
(c) Tube to header plate assembly.
(d) Bi-metallic band saw blades.
(e) Diaphragms.

An electron beam welding equipment has been shown in Fig. 5.7.

Fig. 5.7 Electron beam welding equipment.

5.2.2 Laser Beam Welding (LBW)

Laser is an acronym for light amplification by stimulated emission of radiation. Under normal conditions, most atoms or molecules remain at their lowest energy level or ground state. But if these particles are excited into higher energy states by

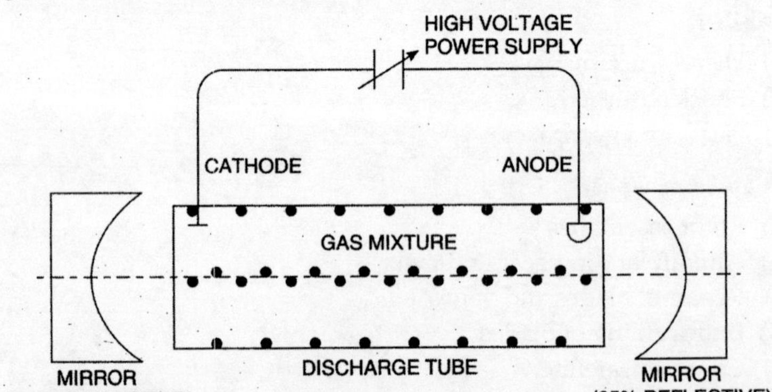

Fig. 5.8 Laser configuration.

an intense flash of light, by an electric charge, or by any other means, they will emit incoherent light, when dropping back to the normal ground state. In a laser cavity, such emitted photons (Particles of light) are trapped between highly polished and parallel mirrors, forcing them to bounce back and forth in the cavity (Fig. 5.8)

Whenever a photon passes close to another excited particle of the same wave length, the second particle will also be stimulated to emit a photon that in identical in wave length, phase and spatial coherence to the first. Both photons are now capable of stimulating the emission of more photons like themselves. This results in a growing wave between the parallel mirrors (Fig. 5.9). It one of the mirrors is partially transparent, a highly disciplined, intense and now coherent laser beam is emitted.

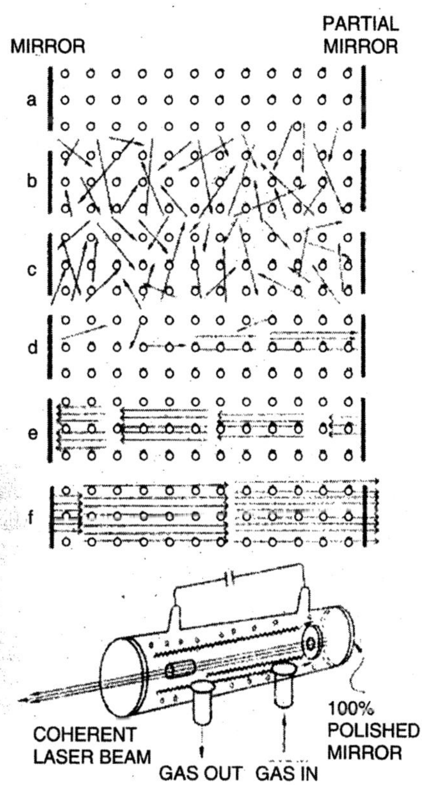

Fig. 5.9 Events in a laser resonator.

Not all materials can qualify as laser media since. The successful laser medium must be excitable enough to achieve a condition known as "population inversion". That is, more photons must be generated by the medium than are being absorbed. In a CO_2 laser equipment, high voltage electric discharge is used to excite

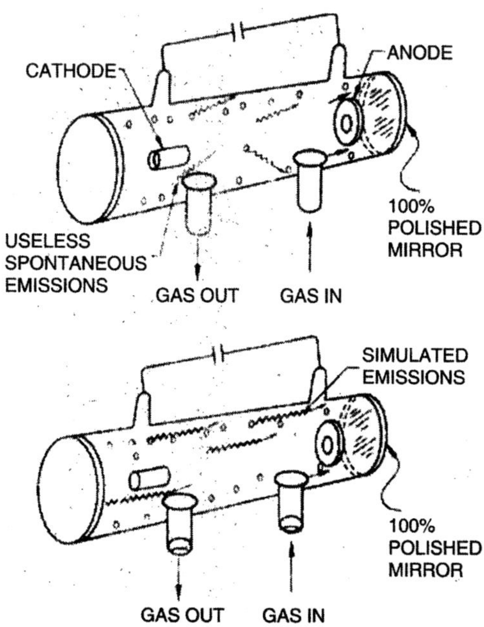

Fig. 5.10 Conceptual view of lasing process.

CO_2 environment, each level receives a share of the molecules, but the lower energy states gain the most, preventing the formation of population inversion. To make a population inversion possible, the CO_2 medium must be modified in some way. This is done by mixing nitrogen (N_2) gas molecules with carbon dioxide. Helium (He) is also included in the mixture, but its purpose is to increase thermal conductivity. Figure 5.10 shows a conceptual view of the laser process, from initiation to stable beam output.

In contrast to the fluorescent lamp light, the laser beam has a single wave length (Monochromatic), parallel rays with low divergence and coherence. A laser beam can be focused by a lens to a small spot, the size of which is theoretically limited only by beam divergence and the focal length of the lens.

Type of Lasers

Many kinds of lasers are now commercially available. Some of the important industrial lasers are given below in table. However, there are many other types are also available such as scientific, medical, communication and of military importance. Among the gas leasers, CO_2 is the most commonly used type in industrial application. 25 kW (maximum capacity) CO_2 lasers are available commercially. Among the solid state lasers, Nd : YAG and Nd : Glass types are the most commonly employed in industrial applications. Both are pumped by intense light from flash lamps.

Table 5.1 : Types of Lasers

Gas	Solid-state
Argon	Synthetic ruby.
Carbon dioxide	Gallium-Aluminium-Arsenide.
Helium-cadmium	Gallium-Arsenide.
Helium-Neon	Nd : Glass (Neodymium : Glass) Nd : YAG (Neodymium : Yttrium-Aluminium-Garnet)

Capabilities of Laser Beam

A laser beam is an unique tool in an industry which has got the capability of the following material processing operations:

(i) Welding
(ii) Brazing and soldering
(iii) Cutting
(iv) Drilling
(v) Machining of trimming
(vi) Engraving of scribing
(vii) Surface hardening
(viii) Surface cladding or alloying

Hence, a laser is an excellent took for processing metal as well as non-metals. The process parameters which influence the results of the process viz, penetration depth and quality of welds are:

(i) Beam power
(ii) Welding speed
(iii) Focussing conditions
(iv) Shielding gas and flow rates.

Laser beam welding is most suited for thin job applications and especially for thermal sensitive jobs which are difficult to weld by conventional are welding processes. LBW is the best choice for special material like titanium or titanium alloys. Laser beam welding of mass produced automobile parts using power up to SKW is on increase. The applications range from precision transmission parts to engine parts, body panels, laminated cores, transformer core, battery can, exhaust gas, oxygen sensor etc. LBW is also used in nuclear field for various applications. LBW has been already applied in Aerospace applications viz gas turbines and aircraft panels switch welding. Battery cams have been hermetically seal welded by lasers. In the electrical and electronics field, LBW has been applied for micro-motor coil, lamp and relays manufacture. In the manufacture of heart pace-makers also, laser welding is employed.

❑❑❑

6

CHAPTER

ARC THEORY

❑ INSIDE THIS CHAPTER

6.1 INTRODUCTION

Matter in nature exists in three forms, namely solid, liquid and gas. Matter is made of atoms. In the case of solids, the atoms of molecules are closely packed and held together by bonding forces. In liquids and gases the atoms are loosely packed. The outer electron in atoms or molecules in metallic solids are shared by many atoms. Hence, the electrons of metals are somewhat loosely held and they become easily free when electromotive force or emf is applied.

The solids are classified according to their electrical conductivity as insulators, conductors or semi-conductors. In the case of the energy to become free from the bond and to flow as a current of electricity. Hence there materials will naturally have high resistance to flow of electricity at normal temperature and normal electromotive force 'emf' or voltage.

Matter in liquid form can also conduct electricity. The charge carriers may only be electrons as in the case of mercury. The charge carriers may be ions of both positive and negative charges as in the case of electrolytes. When current passes through mercury, only electrons move whereas in the case of electrolytes (*e.g.*, NaCl in water) the ions Na^+, Cl^- move in opposite directions to produce electrolysis.

6.1.1 Conduction of Electricity in Gases

In the case of gases, the atoms or molecules are not packed and they are free to move. Infact if we confine a gas in a container, the molecules of gas move in free and randonm fashion. They collide with each other. They impinge on the walls of the container. The pressure of the gas is the momentum that is imparted on the wall of the container by the impinging atoms. If the electrons are made to come end of the binding of atoms, that is, if they are made 'free' then we can have charge carriers to carry electricity. The process of 'freeing' the electrons from the atoms is called the ionization of gas. The electrons are bound to the atom, therefore, to free the electrons we have to supply energy to the electron. The energy which is supplied to, remove the electron from the atom is called the 'ionization energy'. When we ionize the gas, the gas becomes conductive becasues now the charge carriers are available. If we now apply a potential in the ionized medium, the charges will start to move and we will have current flowing through the ionized medium. The ions which have positive charge will move towords the cathode (–) and electrons will move towards the anode (+).

6.1.2 Electric Discharge Phenomena

The flow of electricity through metallic solids or thorough liquid metals is called conduction. The flow of electricity through ionic electrolyte solution such as NaCl gives rise to the pheonomena called 'electrolysis'. The flow of electricity through

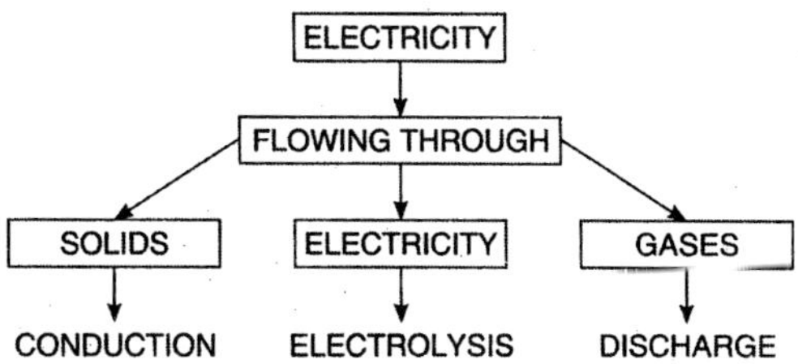

Fig. 6.1 Types of electric flow.

gas (ionized) is called electric discharge. Figure 6.1. furnished the types of electric flow in substances.

There are many types of electric discharges. For example, there is an electric discharge that is taking place inside a flourescant tube, neon signs etc. There is another type of discharge occuring in the spark plug of a scooter and then we have the eletric arc. The discharge in a tube light is called glow discharges. In welding the arc discharge is occuring with low voltage but high current. The types of eletric discharge are shown in Fig. 6.2. When we talk about a discharge we can take tht pressure (voltage) and flow (current) of discharge as important parameters. Electric discharges are also classified according to their voltage and current.

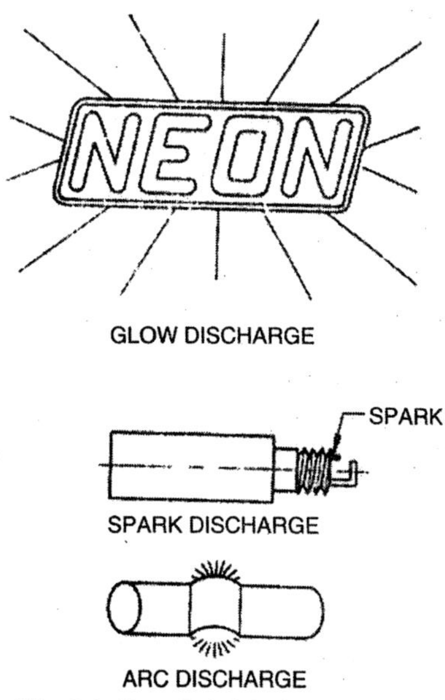

Fig. 6.2 Types of electric discharges.

6.2 THE ARC DISCHARGE AND ITS CHARACTERISTICS

The arc discharge takes place after the current exceeds a limit (threshold value) and the ionization increases drastically, in that the gas becomes highly conductive. Hence, the arc discharge can be designated as a high current low voltage discharge. Figure 6.3 shows a typical gas discharge characteristics and the parameters of, the different types of discharges. The first type is called 'townsend' discharge after the scientist who had studied the discharge behaviour. In this region there is no light and hence, it can also be called as "dark discharge". The second type is the "glow discharge" where light is produced. The third type of arc discharge produces light as well as heat. The "electric arc" was named after its very shape when it was observed first as a discharge between two carbon rods (Fig. 6.4). The arc discharge covers a very wide range of current from as low as 10 A to as high as 1000 A. The volt ampere characteristics of the arc discharge is shown in Fig. 6.5.

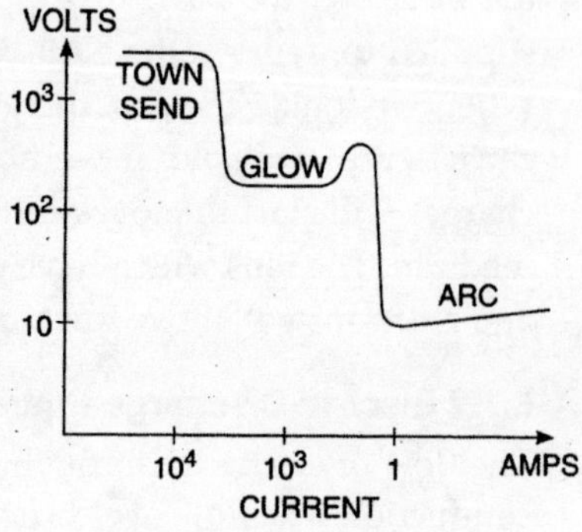

Fig. 6.3 Discharge characteristics in gas.

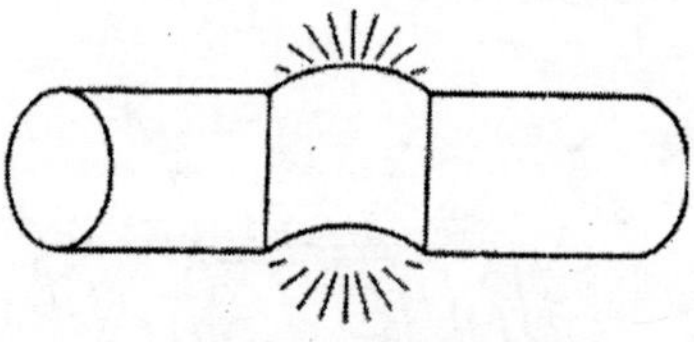

Fig. 6.4 Carbon arc.

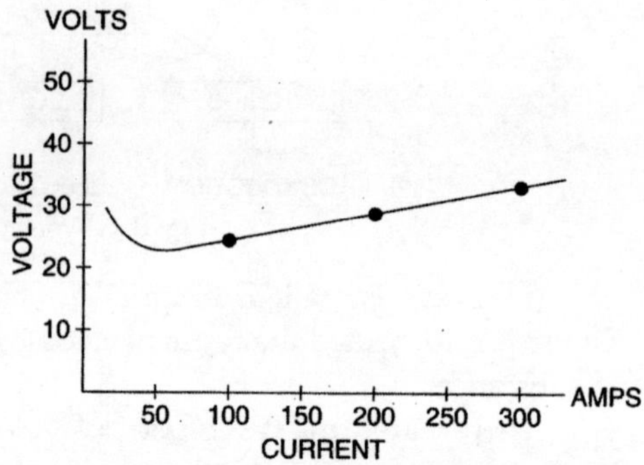

Fig. 6.5 Typical arc characteristics.

It has been found that, as the length of arc discharge increases the chartacteristics curve shifts upwards in the graph, and when the length decreases, the characteristic curve shift downwards (Fig. 6.6). For the same current, the arc discharge requires increased emf for increased length of dischrage. To put it in simple terms : longer the arc higher is the arc voltage. Comparing with the current voltage characteristics of a simple resistor governed by Ohm's law we find that the

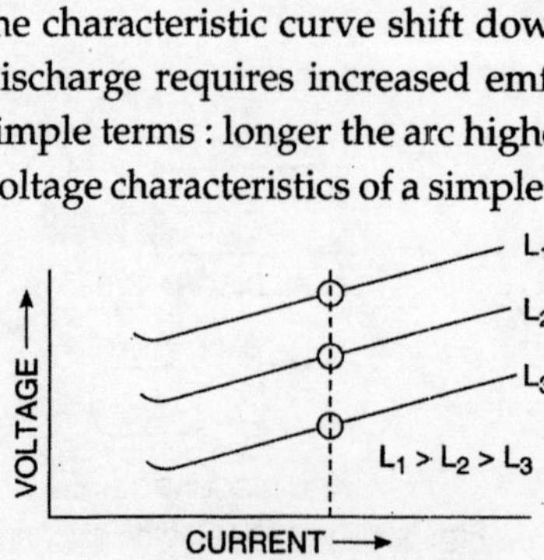

Fig. 6.6 Variation of arc length.

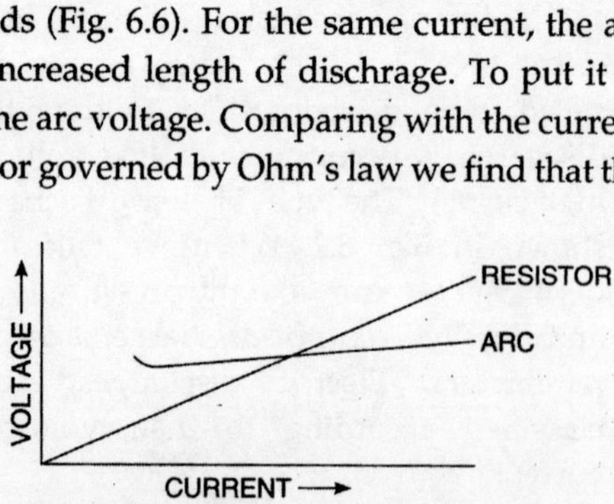

Fig. 6.7 Comparison of resistor and arc.

arc discharge does not follow Ohm's law fully (Fig. 6.7). One more difference to be observed is that the resistor cannot have any arbitrary voltage and current. The current and voltage are restricted by Ohm's law as $V = IR$ where R is the resistance value, but in the case of an arc discharge, the arc discharge once formed can have different arc voltages for the same current with different lengtsh of arc discharges. Current-voltage characteristic of an electric arc not only depends on the length of the discharge but also on the type of gas used. For example, each gas will produce a different characteristic curve. A gas with a lower ionization potential will require less emf for the same arc length (Fig. 6.8). Gas A requires higher voltage than gas B for the same length of arc and current because the ionization potential of gas A is higher than that of gas B.

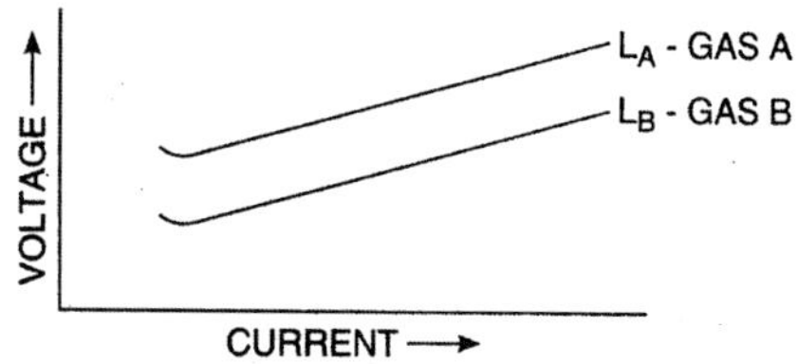

Fig. 6.8 Arc characteristics for different gases.

The potential distribution along the length of the arc is shown in Fig. 6.9. The potential drops heavily both near the anode and the cathode. The names of the regions where there is a large voltage drop are anode and the cathode regions. It is also surprising to note that the width of the anode and cathode regions are very small. Since the same current is flowing across the arc, the power that is developed is concentrated near the anode and the cathode. This concentrated power or heat is responsible for heating and subsequent melting of the electrodes.

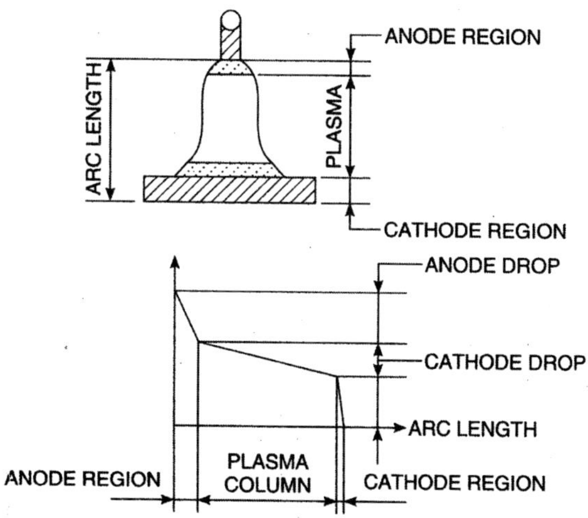

Fig. 6.9 Potential distribution in electric arc.

6.3 TEMPERATURE DISTRIBUTION

As there is a voltage gradient along a arc, there is also thermal gradient. The thermal gradient is not only along the length it is also there across the cross-section of the arc column since the arc is exchanging energy with the surrounding. The heat flow towards the anode and cathode is mostly conduction whereas the heat flow to the ambient is mostly convection and partially radiation. A typical temperature distribution of an arc column is shown in Fig. 6.10.

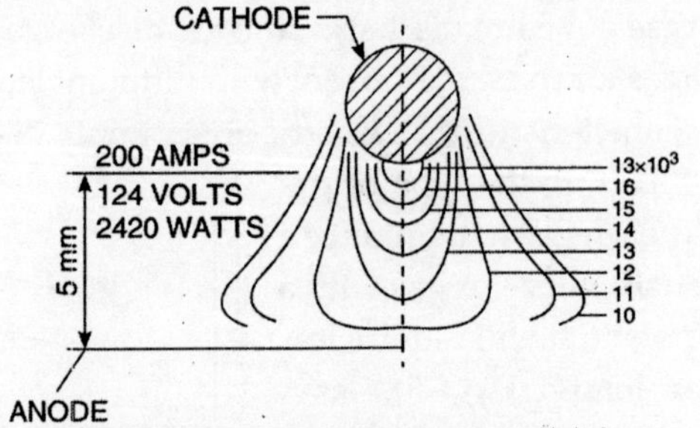

Fig. 6.10 Temperature distribution in arc.

6.4 THE WELDING ARC

The heat that is developed at the cathode and anode of an arc discharge is able to melt most of the metals and alloys. Hence the arc is used as an intense heat source for joining of metals. For welding applications the arc may be structure between a non-consumable electrode and a metal that has to be welded. Example are TIG welding and Plasma welding. The arc can also be struck between a consumable electrode and a base plate, as in flux and gas shielded welding processes. In atomic hydrogen welding the arc is struck between two tungsten (non-consumable) electrodes. The electric arc can operate at a current as low as 1 amp and even less in case of microplasma welding. It can also operate at currents of the order of 1500 amps in case of submerged arc welding. This wide range of current for arc discharge has made it possible to be used for variety applications.

The media through which the arc discharge takes place in the case of welding applications are also many. The arc discharge may happen in pure Argon, Hydrogen and Helium. It can happen in pure CO_2 gas. Gas mixtures like Argon plus Oxygen, CO_2 plus argaon can also be the medium. The gas produced by vapourisation of certain chemicals as in the case of stick electrodes, flux-cored wire and submerged arc welding can be the medium of the arc.

The welding arc causes melting of metals hence, the molten metal has to be shielded from the atmospheric air until the liquid cools down to room temperature to prevent oxidation.

6.5 METAL TRANSFER

The welding arc is struck between the consumable electrode and the metal to be welded. Arc heat melts the electrode and the base plate. The molten metal is transferred from the electrode to the base plate across the arc. These are various forces which are aiding and also opposing the transfer of molten metal across the arc column.

(i) The gravitational force aids the transfer during down-hand welding but it becomes an opposing force while overheat welding is done.

(ii) The surface tension of the molten liquid makes the drop to stick to the electrode thus opposing the transfer of metal. In the case of overhead wleing, the same surface tension force prevents the molten pool at the base plate from falling down.

(iii) The electronmagnetic forces developed due to the flow of current play a major role in the dropet transfer.

6.6 THE PINCH EFFECT

The effect of magnetic field created by the current, flowing in a conductor is to square or apply a radial force (Fig. 6.11) towards the centre of the conductor. When the metal at the tip of the electrode melts, the surface tension forces make the molten metal to grow in the form of a globule. The globule is attached to the tip of the electrode by adhesive forces. The current entering into the globule from the tip produces a squeezing at the neck where the globule is attached to the tip. This produces what is called "the pinching effect" which detached the metal droplet from the tip.

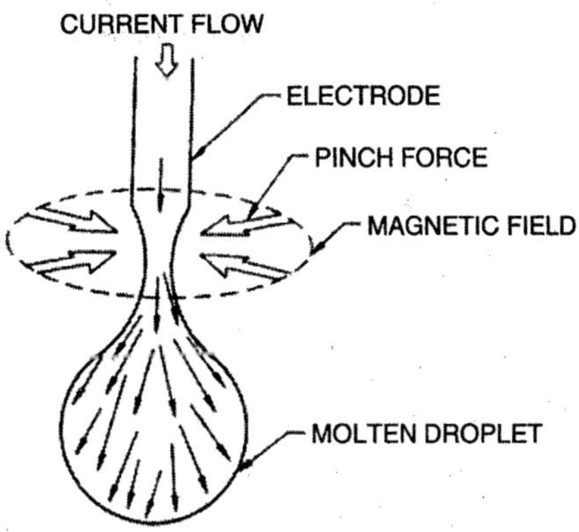

Fig. 6.11 Pinch effect.

6.7 IGNITING THE WELDING ARC

The welding arc is started by any one of the following methods :

(i) Touch the retract.

(ii) High frequency starting

(iii) Pilot arc starting.

(iv) Short-circuiting and fusing.

(i) In the touch and retract method, which is used in MMAW (stick electrode welding), the electrode is just touched and immediately withdrawn from the plate. For a short time the distance of the electrode from the plate is so small and the electric field between the electrode and plate is so intense that it is able to break the gas molecule, ionise it and start the discharge. When the arc gap increases the arc is sustained due to the potential difference.

(ii) In TIG welding if touch and retract method is used, the possibility of tungsten electrode melting and tungsten inclusion is more. As this is not acceptable, the method is generally not preformed. A high frequency, high voltage unit is used alongwith the main welding power supply to initiate a spark discharge, and thereby, ionizing the column to start the arc discharge. The high voltage is given at a high frequency, keeping in mind the safely factor.

(iii) In plasma welding, the main arc is struck after a pilot arc, a low current is first initiated and it makes available ionized column for the main arc to start easily. The pilot arc itself may be started by either the touch and retract method or by high frequency starting.

(iv) MIG/CO_2 welding systems and some-types of submerged arc welding systems employ a different method for starting the arc. The electrode wire which is normally fed by a wire feeding system through the welding torch is made to contact the plate. The power supply delivers a heavy short-circuit current and the length of the wire between the torch and the plate is blown like an electric fuse. The fusing action always proceeds with an initiation of arc discharge some where along the length where the material got ruptured.

6.8 ARC BLOW

In certain cases, depending upon where the return plate (earth) connections are given for the welding current, the arc may be deflected towards a direction of weaker electromagnetic field. This phenomenon may become a nuisance with high current D.C. arc. The arc will be deflected away from the weld zone if proper earth connections are not given. Figure 6.12 shows the electromagnetic forces involved in arc blow. Following guidelines are useful to avoid this problem :

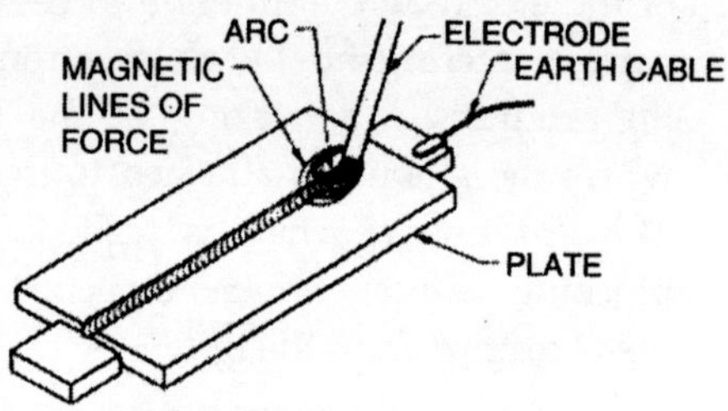

Fig. 6.12 Electromagnetic forces involved in arc blow.

(i) Attach the return connections on the ends of the weld, not in the corner of the plates.

(ii) Start welding from return cable connections.

(iii) It is preferable to have both ends of a weld seam connected to the return cable for long welds.

(iv) For circumferential seam welding, jobs connected to return terminal on both the ends will reduce the arc below.

❑❑❑

7

CHAPTER

THERMAL CUTTING PROCESSES

❑ INSIDE THIS CHAPTER

7.1 Oxy-acetylene cutting, 7.2 Plasma arc cutting, 7.3 Laser cutting, 7.4 Metal gouging.

7.1 OXY-ACETYLENE CUTTING

Oxy-acetylene cutting or gas cutting process in which a small area of metal is first preheated to its ignition temperature and a strem of high purity oxygen is impinged on the heated area to scoop the metal out. Before the oxyacetylene cutting was invented, shearing was done to cut the plate. To cut large holes and thicker plates, drills and chipping hammers were used.

Early cutting torches comprised a welding torch and with diameter of copper tube about 6 mm with a small brasstip attached to it. This torch requires two oxygen cylinders, one to feed the preheat flame and the second for cutting. In 1910, it was cutting tip with that a cutting fuel at the centre, encircles by the preheat flame was developed. The preheat flame raises the work-piece to ignition temperature, adds heat to the work during cutting to cutting reaction forms a protective shield between the cutting oxygen stream and the atmosphere and removes rust, scale and point from the top surface of the work.

Once the preheating flame raises the cutting area of the work-piece to ignition temperature (above 870°C for ferrous alloys), the high velocity stream of oxygen flows and cut the preheated metal. When cutting oxygen flows to the heated work, an exothermic reaction produces carbondioxide and water and releases intense heat. The oxygen jet blows the metal oxides out of the materials to form the cut.

7.1.1 Equipment

The equipment setup for oxyacetylene cutting is shown in Fig. 7.1. It includes a torch assembly, a hose and connections, cylinders containing oxygen and acetylene

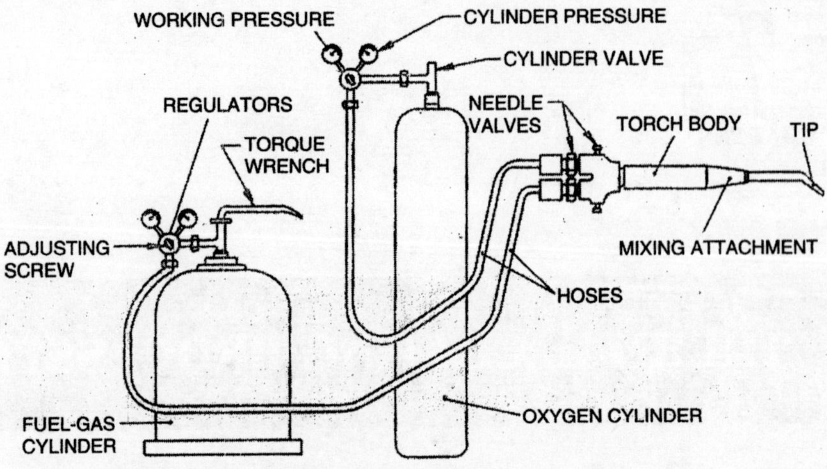

Fig. 7.1 Equipment setup for OFC and OFW.

gas, regulators that set and maintain the gas pressure from the cylinder to the cutting torch, check valves to allow the gas flow only from the cylinders to the torch and a flame striker.

7.1.2 Torch

The oxyacetylene cutting torch comprises a handle, a mixer and a cutting tip. The cutting attachment is shown in Fig. 7.2. The cutting attachment comprises the tubes

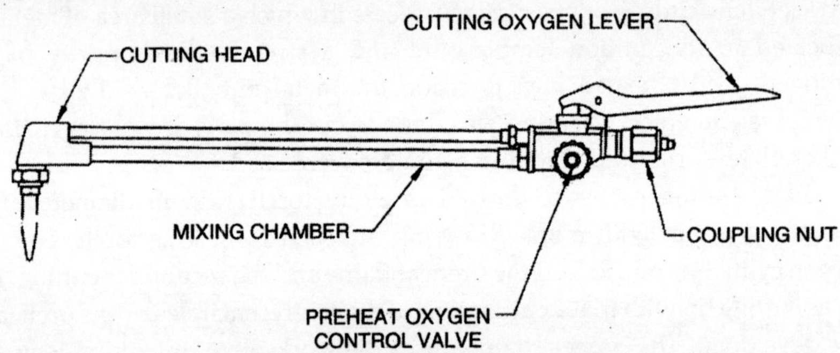

Fig. 7.2 Oxyfuel cutting attachment.

that carry the preheat gases and cutting oxygen which are joined to the torch handle. Valves on the torch handle control the preheat gas flow. A spring-loaded valve on the cutting attachment regulates the cutting oxygen flow. The fuel gas and oxygen mix in a chamber betweenthe torch handle and cutting head. The mixing chamber blends oxygen and the fuel gas well for efficient combustion. After preheating, to start cutting action, the operator depresses the cutting oxygen lever, for cutting, the oxygen valve on the torch handle remaining wide open.

7.1.3 Cutting Tips

Cutting tips are designed with a centre orifice for cutting oxygen, surrounded by the preheat flame ports (Fig. 7.3). Standard tips carry straight bore centre holes.

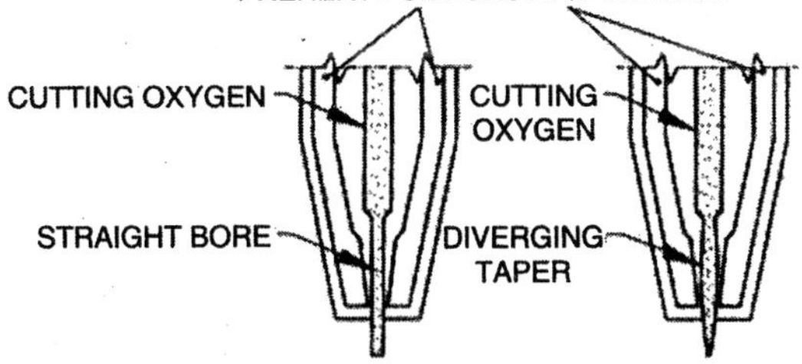

Fig. 7.3 Tapered tips for high-speed cutting.

They are employed for manual and for mechanised cutting at oxygen pressures 2 to 4 kgf/cm^2. The orifice in a high speed tip has a diverging taper which allows the use of oxygen at 4 to 7 kgf/cm^2 pressure. This will increase cutting speed by 25%. The selection of the tip hole size for given thickness of the material to be cut is shown in Table 7.1. This can be considered as a guide for determining more precise settings for a particular job.

Table 7.1

Hose dia., in.	Tip drill size	Base material thickiness, in.	Fuel-gas pressure, kgf/cm^2		Oxygen kgf/cm^2
			Acetylene	LP gas, natural gas	
6.3	68	3.15	0.14 - 0.14		2.46
6.3	62	4.7 - 9.5	0.21 - 0.35	0.21 - 0.35	1.75 - 2.24
6.3	56	12.7 - 22	0.21 - 0.35	0.21 - 0.42	2.1 - 3.51
6.3	53	25.4 - 24.4 - 12.7	0.21 - 0.42	0.28 - 0.56	2.46 - 3.51
6.3	51	50.8	0.35	0.56	3.16
19	46	76.2	0.42	0.56	2.81
19	42	101.6 - 152.4	0.42 - 0.56	0.42 - 0.63	2.81 - 3.86
19	35	177.8 - 203	0.42 - 0.56	0.42 - 0.63	3.51 - 3.86
19	30	228.6 - 304.8	0.56 - 0.70	0.49 - 0.70	3.86 - 4.92
19	25	330.2 - 406.4	0.70 - 0.84	0.49 - 0.70	5.62 - 6.32

7.1.4 Regulators

Gas regulators reduce the cylinder gas from the storage pressure to a lower working pressure at the torch. They maintain the working pressure during cutting. Regulators reduce oxygen pressure from as high as 150 kgf/cm^2 to a working pressure 1 to 2 kgf/cm^2 and the fuel gas pressure 18 kgf/cm^2 to the working pressure of 0.4 to 0.9 kgf/cm^2.

7.1.5 Gases

(a) Oxygen

Oxygen for cutting should be atleast 99.5% pure. Lower purity reduces the cutting speed. A 1% decrease in purity reduces the cutting speed by 15% and increases oxygen consumption for cutting by 25%.

(b) Fuel Gases

The fuel gases most commonly used for preheating in oxy-fuel cutting are acetylene, natural gas and propane.

(c) Acetylene

Acetylene gas is widely used as a fuel gas for oxygen cutting. Its advantages are :

(i) Easy availability and
(ii) High flame temperature.

(d) Natural Gas and Propane

They are used as fuel gas for oxygen cutting because of availability and low cost. The same cutting tips are generally used for both natural gas and propane.

7.1.6 Oxy-acetylene Cutting Techniques

The neutral preheat flame setting is the most common flame setting for oxy-fuel cutting. The neutral flame will neither add carbon to steel nor oxidise it. It protect the molten steel from oxidation. The sequence of cutting is indicated in Fig. 7.4.

(a) Manual Cutting

There are three known methods to start a cut. They are edge start, piercing start and sheet metal cutting.

(i) Edge Start

In this method the torch tip is held half way over the edge of the steel, keeping the inner preheat flame cone about 1.6 to 3.2 mm above the surface of the material to be cut. When the top corner of the plate reaches a reddish yellow colour, the cutting oxygen lever is slowly depressed. A smooth stream of sparks indicates that cutting oxygen has penetrated the plate. The torch is moved at the appropriate cutting speed suitable for the fuel and the section thickness while the preheat flame is held 1.6 to 3.2 mm above the steel. This method produces a precision drop cut. Alternatively, the tip is kept entirely over the edge of material to be cut. The preheat flame is moved back forth along the line of cut for a distance equal to the thickness of the plate. The preheat flame cone should be 1.6 to 3.2 mm above the surface to be cut. When the edge reaches a reddish yellow colour, the tip is brough just off the edge and the cutting oxygen lever is depressed. The torch is moved into the steel to begin the cut. This method produces sharper corners at the beginning of the cut. Though it takes a little longer to start the cut, this is the best method to use when the edge of the part is also the edge of the steel plate to be cut.

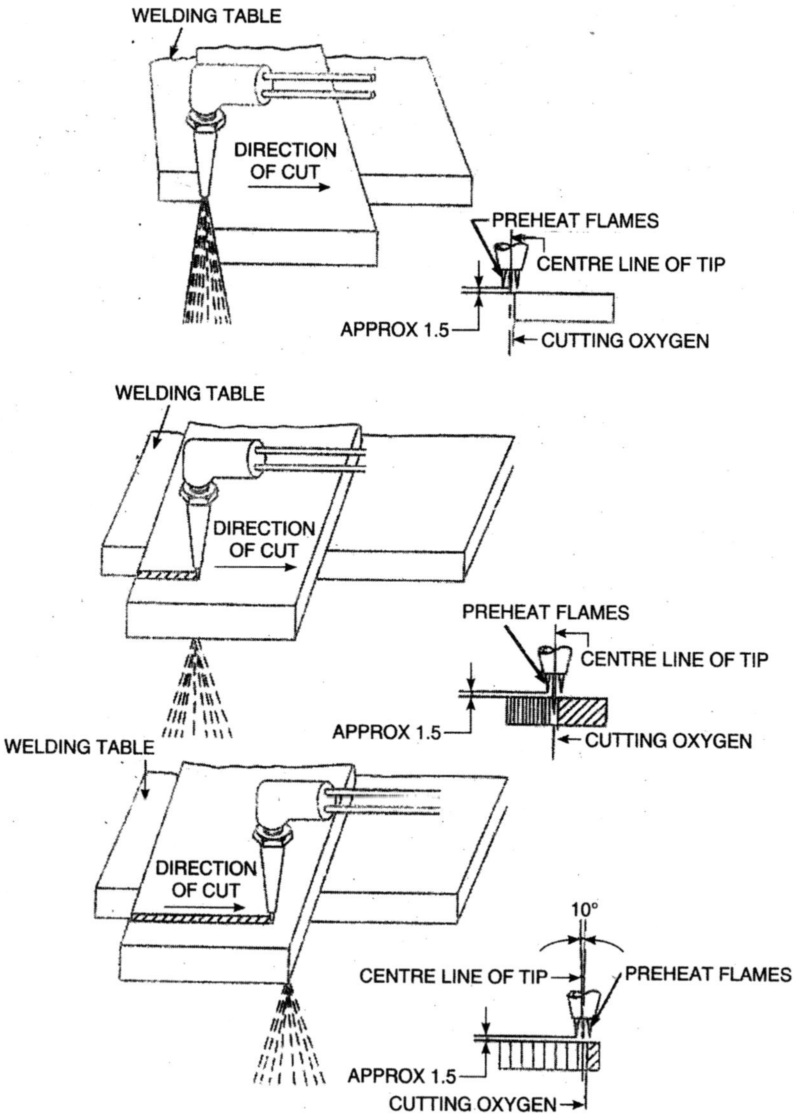

Fig. 7.4 How to make an oxyfuel cut.

(ii) Piercing Start

To cut steels of higher thickness about 500 mm, somewhere beside the edge, a hole is to be drilled to start cutting. This starting hole can be produced faster with a torch. This technique is called piercing. Piercing a thicker plate requires larger preheat flames than are used for edge start. The preheat flame also should be slightly oxidising to create more heat. Locate the spot in the scrap area outside the part to be cut, if possible. The preheat flame is allowed to touch the steel. This will rapidly bring the material to a reddish yellow colour. Slight up and down motions of the torch tip will indicate the proper distance to be kept between the torch and the surface. The torch tip is moved slowly to bring the entire area of piercing to reddish colour. The cutting oxygen lever is squeezed steadily and the tip simultaneously lifted from the plate to keep the slag from being blown into the torch tip. As seen on the hole is completely pierced and breaks through the bottom side, the cutting oxygen lever is turned on all the way and cutting started.

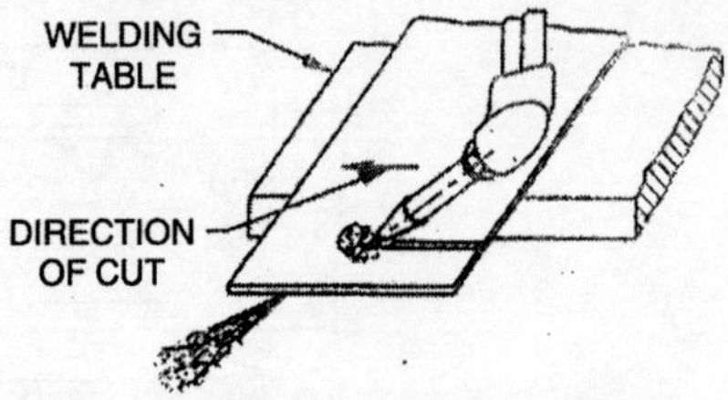

(iii) Cutting Sheet Metal

Sheet metal cutting is shown in Fig. 7.5. To cut material having thickness less than 6 mm, the tip end is turned in the direction of the cut. The cutting oxygen will blow the molten slag away from the cut and will prevent the cut edges from fusing.

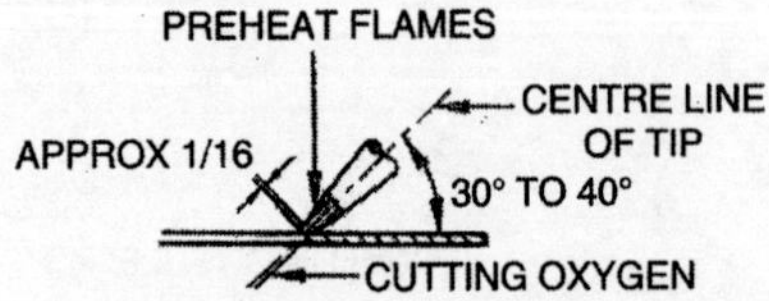

Fig. 7.5 Cutting sheet metal.

7.1.7 Cutting Quality

The factors to be controlled to get a high quality cut are :

(i) Coupling distance
(ii) Cutting speed
(iii) Torch movement
(iv) Tip angle

(i) Coupling Distance

Coupling distance or the stand-off distance is the distance from the inner preheat flame cone tip to the surface of the plate. It is not the distance from the torch tip to the plate surface. Coupling distance is such that it holds the preheat flame with the highest amount of heat close to the surface of the plate. Acetylene flame has more heat in the outer flame cone. The correct coupling distance depends on the fuel gas used. To cut plates of 6 to 36 mm thickness, a coupling distance of 1.6 to 3.2 mm is recommended. To pierce or to make a very fast start, the preheat cone is allowed to just touch the surface.

(ii) Cutting Speed

The cutting speed must be just fast enough so that the cutting oxygen jet passes completely through the plate thickness to make a clean cut on the top and bottom of the steel plate. If the speed is too slow, it will create a lot of oxide slag. This slag will be thrown out or hang from the bottom. The kerf surface will be rough and irregular. If the speed is too fast, then the full thickness of the plate may not get cut and we have to restart again. Fig. 7.6 shows the effect of cutting speeds.

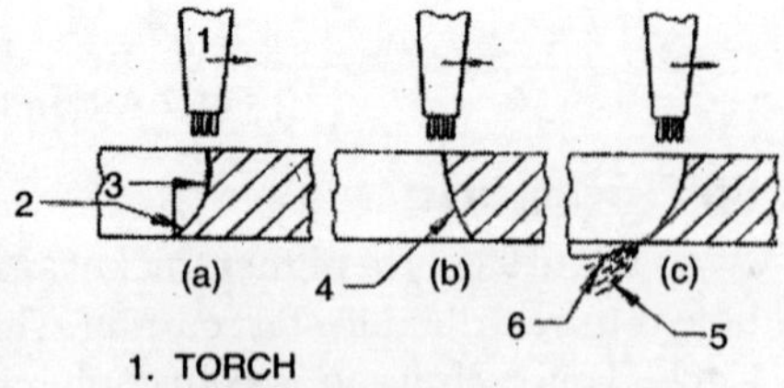

1. TORCH
2, 4. LINES OF CUT
3. DRAG
5. SPRAY
6. GLOBULE OF MOLTEN METAL

Fig. 7.6 Torch speed and its effect on cutting.

(iii) Torch Movement

The movement of the torch depends on the type of metal to be cut (Fig. 7.7). Carbon steel requires less preheat than cast iron and so the torch is moved straight ahead a steady, slow speed. Cast iron needs more preheat than carbon steel, so the torch is moved in an arc that keeps the preheat flame in one area for a longer time.

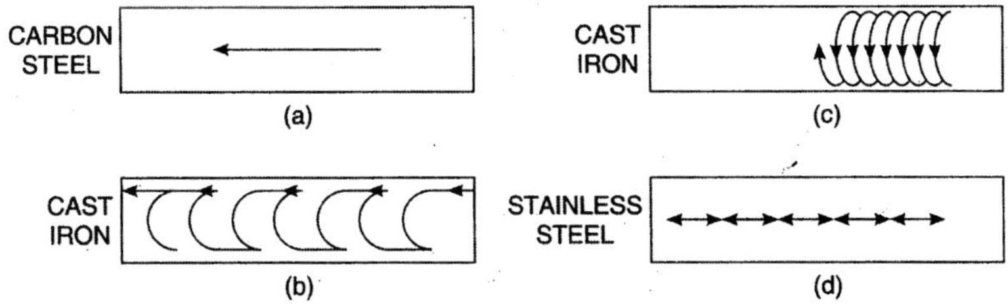

Fig. 7.7 Torch movement for different materials.

(iv) Tip Angle

The torch head and the cutting oxygen jet must be in line with the cut. The torch head can be held at an angle of 60° with the horizontal to cut thin plates. It will minimise melting of the edges, prevent overheating of base metal and there will be lesser distortion of the work. For cutting thicker plates of more than 12 mm. Thickness, hold the torch tip straight up and down, perpendicular to the face of the horizontal surface.

A good quality cut produces no slag at the bottom of the cut, and the cut edge surface is smooth and square. Drag lines slope slightly. A bit of the scale may form at the top of the cut caused by the preheat flame. Slag forms on the cut surface if the cutting tip is dirty. If pressure marks appears on the cut surface, use a smaller tip, reduced oxygen pressure or increased travel speed will be helpful. Grooves in the cut surface result from holding the tip too close to the work or part of the preheat flame then burning inside the kerf and the cutting oxygen stream get deflected out of the cut.

7.1.8 Kerf

Kerf is the gap created by the removal of material by the cutting jet. The kerf width depends on :

(i) The size of the nozzle.
(ii) The type of nozzle used.
(iii) The speed of cutting.
(iv) The rate of the flow of cutting oxygen and preheat gases.

Refer Fig. 7.8.

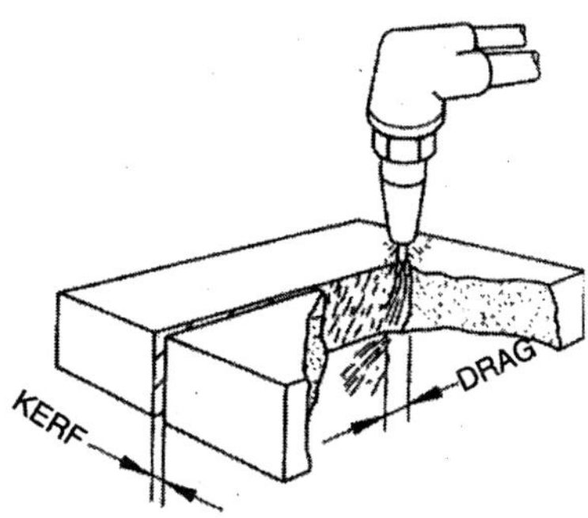

Fig. 7.8 Kerf and drag.

7.1.9 Drag

Drag is the measure of the deviation of the oxygen stream at its exit from the work (Fig. 7.8) measured along the length of cut. A 10% drag would mean that the far side of the cut is behind the near side of the cut by an amount equal to 10% of the thickness of the material being cut. If the cut has zero drag, then it is called a drop cut. If the cutting speed is increased or if the cutting oxygen flow is less than the recommended value the drag will increase. The problems normally encountered in flame cutting are :

(i) Crack in the edges.

(ii) The cut edges will be hard and would be difficult to machine.

If the carbon content of the work is less than about 0.25% the above problems are rare.

To cut steels with higher carbon content and alloy content, preheating and post-heating have to be carried out to prevent hardening and cracking of the steel along the cut surfaces.

Applications : Manual oxy-fuel cutting is used in steel fabricating industries. Steel mills and foundaries use oxy-fuel cutting to remove caps, gates and risers from billets and castings. It is used to cut plates to the required shapes. Oxy-fuel cutting gauges remove old welds, cut grooves, rough-machines and stack-cut plates. Machines that carry oxy-fuel torches are used to size plates and to make contour parts from plates, prepare work-piece edges (square or bevelling).

7.2 PLASMA ARC CUTTING

Plasma arc cutting process was introduced in the industries in the mid of 1950s. The process is used to cut all metals and non-metals. The common oxy-fuel cutting process (based on a chemical process) is suitable for cutting carbon steel and low alloy steel cutting only. Materials such as copper, aluminium and stainless steel were earlier separated by sawing, drilling or sometimes by powder flame cutting. These materials are now cut using a plasma torch at faster rates and more economically. The plasma cutting process is basically a thermal cutting process, free of any chemical reaction, that means without oxidation. In plasma arc cutting an extremely high temperature and high velocity constricted arc is used.

7.2.1 Principle of Operation

Plasma arc cutting is a process resulting from ionising a column of gas (argon, nitrogen, helium, air, hydrogen or their mixtures) with extreme heat of an electric arc. The ionised gas alongwith the arc is forced through a very small nozzle orifice, resulting into plasma stream of high velocity (speed upto 600 m/sec) and high temperature (upto 20000°K). When this high speed is reached, high temperature plasma stream and electric arc strike the work-piece and ions in the plasma recombine into gas atoms and liberate a great amount of latent heat. This heat melts the work-piece, vapourises part of the material and the balance is blasted away in the form of molten metal through the heat.

7.2.2 Plasma Cutting System

Plasma cutting requires a cutting torch, a control unit a power supply, one or more cutting gases and a supply of clean cooling water (in case water-cooled torch is used). Equipment is available for both manual or mechanical cutting. A basic plasma arc cutting circuit is shown in Fig. 7.9. It employs direct current straight polarity (DCEN). The nozzle surrounding the electrode is connected to the work-piece (positive) through a current limiting resistor and a pilot arc relay contact. The pilot arc between the electrode and nozzle is initiated by a high frequency generator connected between the electrode and nozzle. The orifice gas ionised by the pilot arc is blown through the constricted nozzle orifice and forms a low resistance path to ignite the main transferred arc between the electrode and the work-piece when the ON/OFF switch is closed. The pilot arc relay may be opened automatically when the main arc ignites, to avoid unnecessary heating of the constricting nozzle. The constricting nozzle is of copper and normally water cooled to withstand the high plasma flame temperature (about 20000°K) and to have longer life.

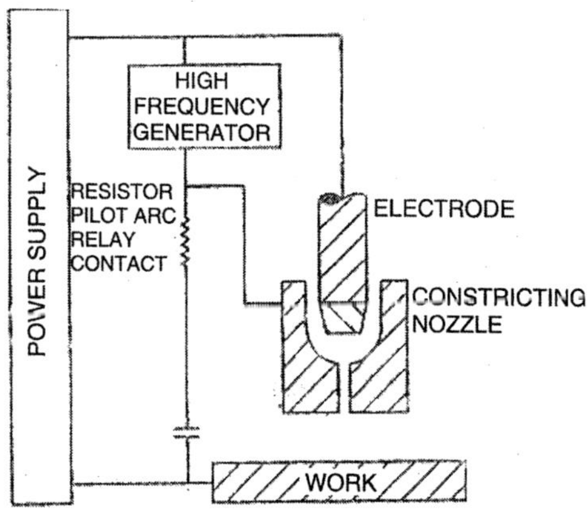

Fig. 7.9 Basic plasma arc cutting circuitry

In conventional gas plasma cutting, discussed above, the cutting gas can be argon, nitrogen, (argon + hydrogen), or compressed air. For all the cutting gases other than compressed air, the non-consumable electrode material is 2% thoriated tungsten. In air-plasma cutting (Fig. 7.10) where dry, clean compressed air is used as the cutting gas, the electrode of hafnium or zirconium is used because tungsten is rapidly eroded in air. Wet and dirty compressed air reduces the useful life of consumable parts and produces poor quality. Several process variations are used to improve the cut quality for particular applications. Auxiliary shielding in

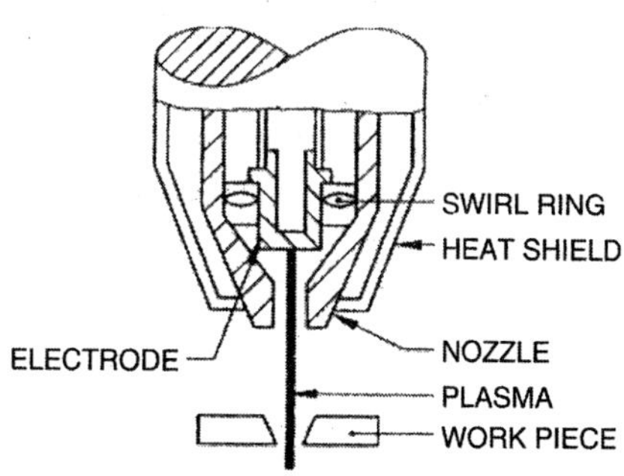

Fig. 7.10 Components of an air-pac torch.

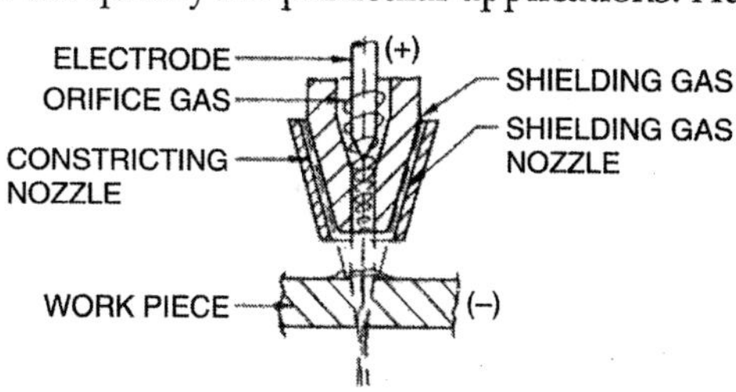

Fig. 7.11 Dual flow plasma arc cutting.

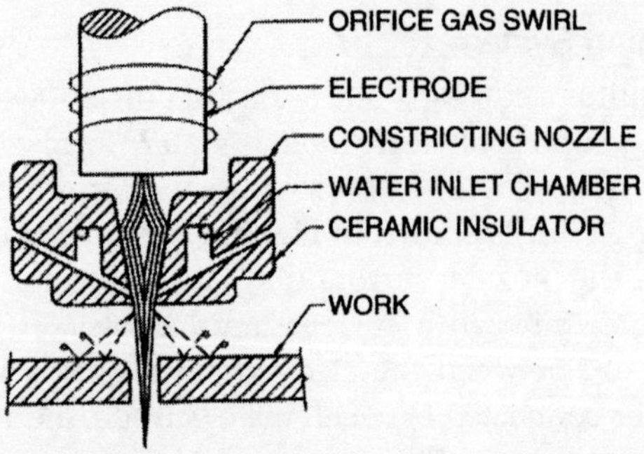

Fig. 7.12 Water injection plasma arc cutting arrangement.

the form of gas or water is used. Figure 7.11 to improve the cut quality and to improve the nozzle life. Water injection plasma cutting (Fig. 7.12) uses a symmetrical impinging water jet near the constricting nozzle orifice to further constrict the plasma flame and to increase the nozzle life. Good quality cut with sharp and clear edges with little or no dross is possible in water injection plasma cutting.

7.2.3 Process Variables

(i) Torch design-constricting nozzle shape and size.
(ii) Process variation-dual gas flow, water injection, air-plasma.
(iii) Cutting gas type and its flow rate.
(iv) Distance between nozzle and job.
(v) Cutting speed.
(vi) Plasma cutting current.
(vii) Power used during cutting.
(viii) Manual/machine cutting.
(ix) Material to be cut and its thickness.
(x) Quality of cut required-rough or smooth.
(xi) The bevel angle and round off corner etc.

Refer Fig. 7.13 and 7.14.

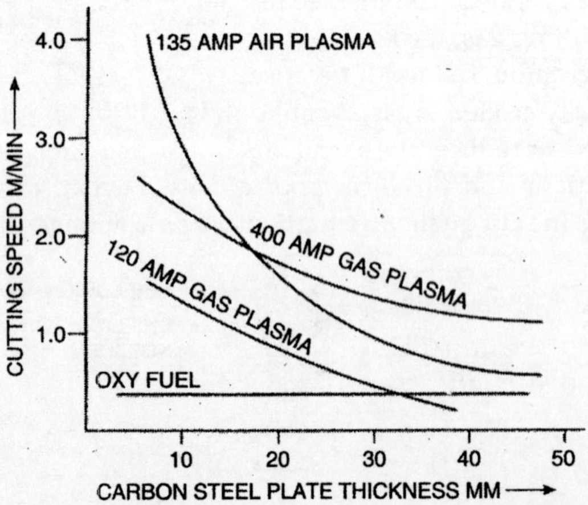

Fig. 7.13 Comparison of oxy fuel and plasma cutting processes.

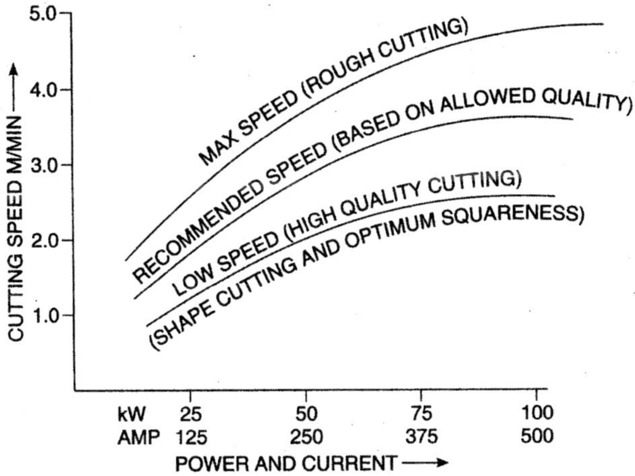

Fig. 7.14 Relationship between cutting speed and power.

Advantages of Plasma Cutting

(i) All metals and non-metals can becut due to high temperature and high velocity plasma flame.

(ii) Cuts are of very clear form with little or no dross.

(iii) High speed piercing is achieved.

(iv) Cutting of piled plates is possible even with different materials.

(v) Cutting cost is quite low as compared to other processes especially for stainless steel.

(vi) Cutting speed is high.

(vii) Cutting is possible in all positions and locations (under water also).

Applications

(i) Straight and sharp cutting of all metals and non-metals.

(ii) Cutting of risers and gates of forgings and castings.

(iii) Stack cutting of several sheets of 1.5 to 6 mm thickiness.

(iv) For making holes in thick sheets (by piercing operation).

(v) For gouging, rough-machining etc.

(vi) For sizing the scrap.

7.2.4 Safety Precautions in Plasma Cutting

The operators and persons in the vicinity of plasma cutting operation must be protected from :

(i) Arc radiation and spatter — protect body and eyes.

(ii) Metal fumes and gases — use breathing mask, proper ventilation.

(iii) Noise - upto 115 dB — use ear plugs.

(iv) Electrical shocks — high operating voltage (180-400 V) and both anode and cathode in torch, input supply is to be switched off before attending to the torch etc.

7.3 LASER CUTTING

The laser beam is able to cut through most materials by evaporating the material and creating a hole and traversing it. With materials that melt, a coaxial gas jet is used to blow the molten product away (Fig. 7.15). Failure to do this would produce a weld. This gas jet can also add heat in the cutting zone, if the material reacts exothermically with it, *e.g.*, steel and O_2. The heat affected zone is in any case, narrow due to the high processing speeds and highly localised nature of energy. The main features of laser cutting are as follows :

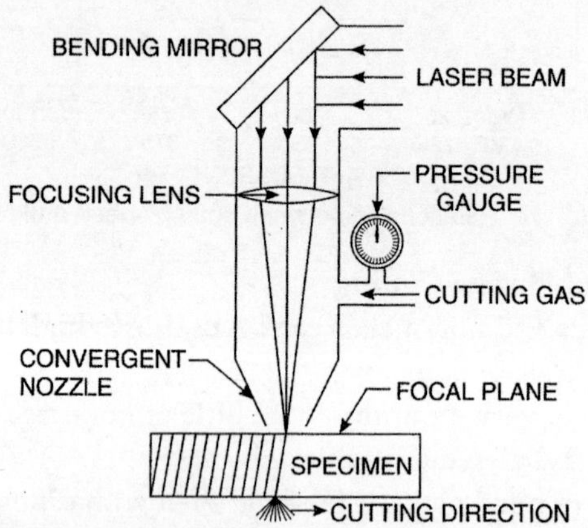

Fig. 7.15 Schematic diagram of laser gas cutting head.

(i) The cut has a very narrow kerf thus effecting saving in material.

(ii) The cut has a very narrow heat-affected zone (HAZ) and therefore gives low thermal distortion.

(iii) It gives an almost machine smooth cut surface reducing the need for subsequent trimming.

(iv) The cut has square edges.

(v) There is no tool wear.

(vi) The work-piece does not need holding against tool drag and work holding demands are reduced to merely a locating role.

(vii) It is possible to cut through friable, brittle, soft or hard materials with almost equal ease.

(viii) It is a non-contact method with no cutting forces and is ideal for cutting flexible materials such as cloth, rubber, corrugated cardboard etc.

(ix) It is non-directional, therefore, it cuts equally well in all directions and it can be started anywhere.

(x) Blind cuts can be made.

(xi) The cutting speed is higher as compared to other methods.

(xii) Multi-station operations from one laser are possible.

(xiii) It is easily automated by numerically controlled.

(xiv) Possible to weld the cut edge directly.

There are five different ways in which a laser can be used to cut different materials :

(i) **Vapourisation :** The beam energy heats and substrate to above its boiling point and the material leaves as vapour and gets ejected.

(ii) Melting and blowing the beam energy melts and substrate and a jet of inert gas blows the melt out of the cut region.

(iii) **Burning in reactive gas :** (i) the beam energy heats materials to the kindling temperature which then burns in a reactive gas jet, as in (ii) the jet also clears the dross away.

(iv) **Thermal stress cracking or controlled fracturing :** The beam energy sets up a thermal field in a brittle material, *e.g.*, glass such that it can guide a crack in any direction.

(v) **Scribing :** A blind cut is used as a stress raiser allowing mechanical snapping along the scribed lines.

What distinguishes these five mechanisms is how the heat is subsequently coupled and its effects on the work-piece.

Lasers have been particularly successful in cutting titanium for the aircraft industry. Before lasers were used, this difficult materials was cut with plasma arc. This produced a wide kerf and a wide HAZ, which had to be machined off. The narrow HAZ from laser cutting simply requires a minimal post machining step to clean the edge surface. A comparison of metal cutting techniques is given in Table 7.2. As can be seen, laser cutting is far more efficient than either plasma arc or oxy-acetylene cutting.

Table 7.2 : Comparison of Various Cutting Processes

	Laser	Oxy-acetylene	Plasma-arc
Heat affected zone	0.25 mm to 0.76 mm	6.35 mm	2.54 mm
Kerf width	0.89 to 1.4 mm	2.03 mm	7.92 mm
Edge machining allowance	1.27 mm	9.52 mm	5.08 mm

A reacting cutting process is used in the laser cutting of metals. In this process oxygen is supplied by a coaxial gas jet. It is this oxygen that actually makes the cut,

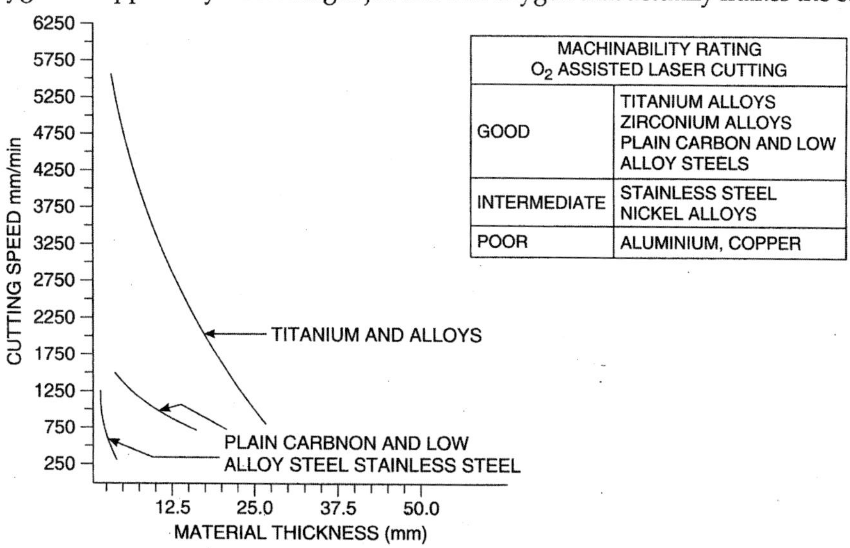

Fig. 7.16 Laser cutting rates for metals.

while the laser sustains the cutting reaction. The variation in laser cutting rates for different metals are shown in Fig. 7.16. Among many applications of laser cutting, both for metals and non-metals, only a few examples are given here. One example is that in the cutting of stainless steel helicopter blades, the laser took 1.6 minutes compared to the previous method which took 35 minutes in addition to six special cutters. Another example is the manufacturing of air-conditioning units. Using a numerically controlled laser obviates the need to mark out the work while ensuring the cut is distortion-free through galvinised or plastic coated steel sheets. This application gave a saving of 50,000 US dollars in one year in 1976, approximately the cost of laser.

7.4 METAL GOUGING

In welding technology, gouging means removal or washing away of metal. This removal or washing away of metals is accomplished with a flame (usually oxy-fuel flame) or by the air-carbon are process. Gouging is used to remove weld defects, tack welds, temporary supports, attachments, casting defects and also in the preparation of edges for welding.

7.4.1 Flame Gouging

The main principle behind flame gouging is the oxidation of metals. At elevated temperatures the oxidation process in enhanced and with a stream of pure oxygen directed against the white hot spot, the metal burns rapidly and the jet of gases removes the oxidised and burn metal readily. The heat is generated due to the chemical action of fuel gas used in the process with the mixture of oxygen. The fuel gases, commonly used, are acetylene, natural gas, propane, butane etc. Selection of these gases depends mostly on their cost and their availability. Among all those fuel gases, acetylene gives the hottest flame around 3120°C.

Applications

Flame gouging is widely used where oxidation of metal is possible like steel and wrought iron. Flame gouging is generally not used for non-ferrous metals. Flame gouging is used for quick and accurate removal of a narrow strip of surface metal from a steel plate, forgings and castings without penetrating through the entire thickness of metal. It is also used for removal of weld and casting defects, tack welds, metal from the back side of weld seams etc. When used for removing defects the process is called spot-gouging edge preparations are also frequently done by the flame gouging process.

Equipments

The same gas cutting torch is used with a specially designed gouging nozzle for flame gouging. The equipment used in the flame gouging system consists of fuel and oxygen lines, cylinders, gas regulators, adequate hosing and a torch body. The flame cutting or gouging torch has a high pressure valve that is activated by a

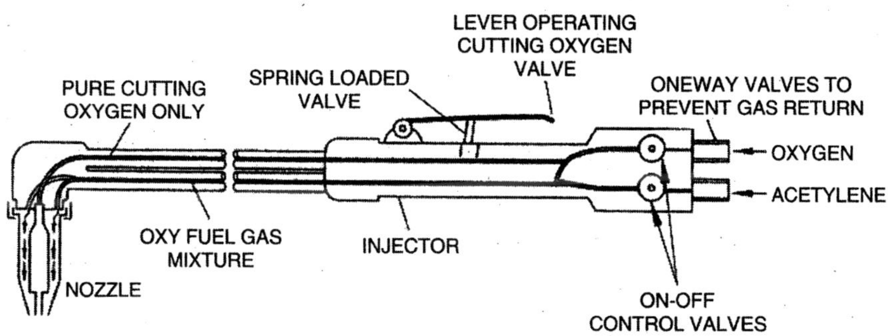

Fig. 7.17 Flame gouging blow pipe.

lever, trigger or button. In addition to the oxygen or fuel gas valves, the high pressure oxygen valve when pressed releases a jet of oxygen through the tip. The nozzle tips have preheat orifices and one central orifice that directs the flow of oxygen for the gouging operation. A flame gouging flow pipe and a nozzle are shown in Figs. 7.17 and 7.18 respectively.

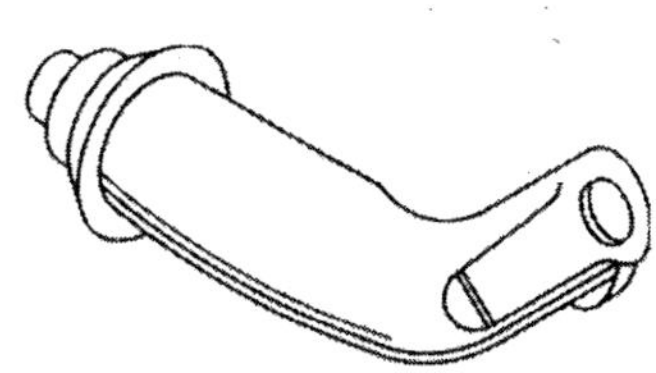

Fig. 7.18 Oxy-acetylene gouging nozzle.

Technique

There are two basic gouging techniques :

(i) Progressive (Fig. 7.19)

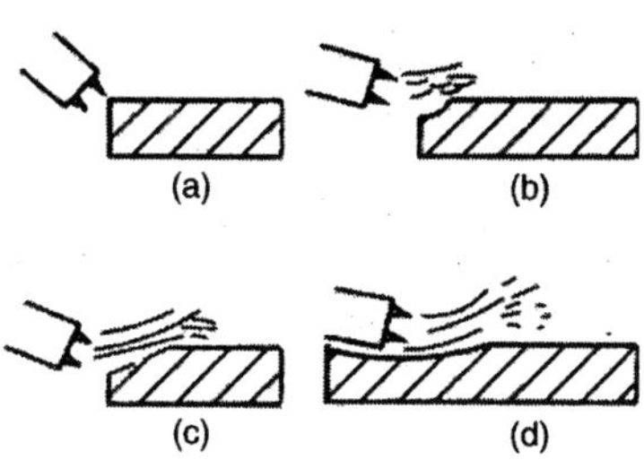

Fig. 7.19 Progressive gouging.

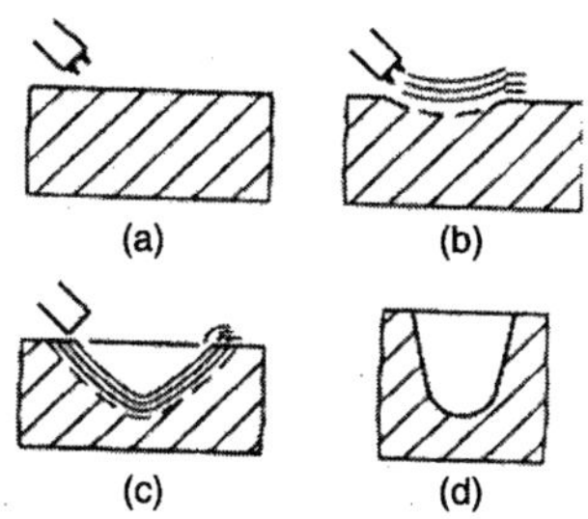

Fig. 7.20 Spot gouging.

(ii) Spot (Fig. 7.20)

(i) Prior to the start of gouging at the edge of a plate, the plate edge is preheated to a bright red colour. As the nozzle of the flame torch is reduced to the operating angle, the cutting oxygen is turned on to make a gradual starting cut. When the desired depths of groove are reached, the gouging nozzle is moved along the line of cut.

(ii) In spot gouging, small areas are removed sometimes to quite a depth. The cross-sectional view is also shown in figure.

In certain applications preheating of the jobs prior to flame gouging may be needed as in the case of welding.

7.4.2 Air-carbon Arc Gouging

Principle

The air-carbon arc gouging process consists of 'melting' of metal with the heat of a carbon arc and blowing away the molten metal with a parallel jet of high pressure compressed air. This process does not require oxidation of metals as in flame gouging and hence both ferrous and non-ferrous metals may be gouged easily.

Applications

The air-carbon arc gouging process is widely used to gouge or even to sever rolled steel, to remove flows and cracks from castings and welds, for back gouging of weld seams, to prepare root of welds and weld edge preparations, to remove tack welds, temporary attachments, sharp edges etc.

Equipments

The equipments consists of a special insulated torch or electrode holder with a dual purpose cable which carries the current from the power source and a supply of compressed air at approximately 6 to 7 kgf/cm^2 pressure (Fig. 7.21). Mechanised torches are also available.

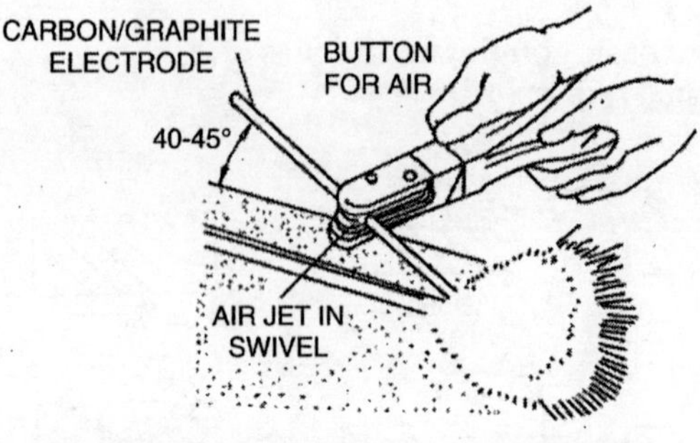

Fig. 7.21 Electrode holder — Arc air gouging.

Power Sources

Any arc welding source (A.C./D.C.) of sufficient capacity may be used for the air-carbon arc gouging process. However, the open circuit voltage (OCV) should be sufficiently high and atleast 60 V. The arc voltage ranges from a low 35 V to a high 56 V and governed by the size of carbon electrodes used. With a D.C. power source, the electrode is connected to the positive pole (DCEP).

Consumables

There are three basic types of electrodes

(i) D.C. coated electrodes

(ii) D.C. plain electrodes

(iii) A.C. coated electrodes

The D.C. coated electrodes are more popularly used. These electrodes are called carbon-arc gouging electrodes because they are made from a special mixture of carbon and graphite. The coated electrodes are enveloped in a thin layer of high purity copper or aluminium. No such coatings are given on plain carbon electrodes. The carbon in the electrode provides resistance to erosion; the graphite provides a lower speed of oxidation and better electrical and thermal conductivity.

The copper or aluminium sheath over the carbon electrode prevents the electrode tip from tapering excessively, allows higher currents and restricts the radiation of heat as well as giving protection to the carbon/graphite electrode. The plain electrodes erode more rapidly than the coated electrodes. The electrodes are available in the ranges of diameter 4 mm to 19 mm and also is different shapes like flat, half-round and rectangular for specific applications. The A.C. coated electrodes have some rare earth metals incorporated with carbon and graphite mixture to provide better arc stability with an alternating current.

Technique

The carbon electrode is inserted in the holder at approximately half its length (150-155 mm maximum), with the air jets positioned underneath the electrode to enable the molten metal to flow away. For aluminium alloys the electrode extension should be reduced to 100-105 mm. The gouging angle is approximately 45° from the work but may be reduced to 20° for shallow gouging or washing. For certain applications preheating of the jobs may be necessary prior to gouging as in the case of welding. Based on the size of the electrode proper and suitable current values are chosen. A short arc should be maintained by progressing the direction of the gouging, fast enough to keep up with the metal removal. The gouging can be done in all positions, *viz.*, flat, horizontal, vertical and overhead. The smoothness of the gouged surface depends upon the skill of the person gouging, the steadiness of progression and the speed of gouging. The suggested current ranges for the commonly used electrode types and sizes are given in Table 7.3. Recommended preheating temperature depends on material composition, job thickness etc. as applicable in case of welding.

Table 7.3 : Suggested Current Ranges for the Commonly Used Electrode Types and Sizes

Electrode diameter	D.C. electrode with dcrp[a]		A.C. electrode with A.C.		D.C. electrode with dcsp[b]	
mm	Min A	Max A	Min A	Max A	Min A	Max A
3.2	30	60	40	90	150	
6.4	200	400	200	300	200	250
9.5	350	600	350	500	300	400
12.7	600	1000	400	600	400	500
15.9	800	1250				
19.1	1200	1600				

Advantages of Air-Carbon Arc Gouging Process :

(i) Fast. Five times faster than chipping or grinding. Gouges a groove of 9.5 mm deep at over 600 mm per minute. More ideal for back gouging of welds and removing weld defects.

(ii) Easily controllable; removes defects with precision.

(iii) With a good gouging, the surfaces require little or no cleaning and there is lack of dross.

(iv) Low equipment cost and economical to operate. No oxygen or fuel gas required. No cylinders or regulators are necessary.

(v) Compact and versatile.

(vi) Can be used for both ferrous and non-ferrous metals.

(vii) Even 'U'-groove edge preparations for welding can be accomplished using an automatic machine.

❑❑❑

8

CHAPTER

SURFACING

❑ **INSIDE THIS CHAPTER**

8.1 SURFACING PROCESSES

The service conditions of components include heat, stress corrosion impact, wear, fatigue etc. or their combination in varying degrees. Under these conditions the components are required to give satisfactory performance. Depending upon their severity, these environment factors make the service conditions complicated and demanding. The component surfaces which are exposed to such severe conditions are generally coated with a suitable alloy by welding or allied processes. Such a coating enables the component to perform satisfactorily. The process providing such a coating is called surfacing. Surfacing is done for the following purposes :

(i) To build up the worn out component to its original dimensions.

(ii) To import special properties to the surface of the component. Coating of the given component with a material, having special properties which the base material does not possess, is done to achieve a satisfactory or enhanced performance in actual service. The special properties of the coating material are, resistance to various types of wear, corrosion and high temperature seizure. A stainless coating on the interior of low alloy steel chemical plant vessel for improved corrosion resistance; hard facing the mining and earth moving equipment components and surfacing valve seat rings are some of the typical industrial applications of weld surfacing.

8.1.1 Wear

Wear is defined as the removal of material from a surface due to the mechanical properties present in the service environment. The major types of wear are as follows :

(i) Abrasive wear which is caused by the sliding of an abrasive particle on a surface. Depending upon the load or force applied on the abrasive particle the severity of the wear increases.
(ii) Impact wear occurs when the material is subjected to impact loading, when the impact load is applied through the abrading material itself.
(iii) Erosive wear is the removal of material due to impingement of solid particles as in a fluid flow.
(iv) Cavitation, which is another form of material removal, occurring of turbulant flowing fluid medium.

8.1.2 Corrosion

Corrosion can be defined as the degradation of the metal due to the destructive attack by a chemical or electrochemical reaction. The major corrosion types are described below :

(a) Uniform Attack

The corrosive medium uniformly attacks the exposed surface of the product. Examples are rusting of iron, high temperature, oxidation of metals.

(b) Galvanic Corrosion

If two dissimilar metals are in electrical contact in the same corrosive medium then the anode of the galvanic couple is corroded.

(c) Pitting Corrosion

It is the result of localised corrosion occurring at the surface of the metal, where the rate of corrosion is greater than in other areas. Examples of the attack are the corrosion of iron buried in soil and stainless steel submerged in sea water.

(d) Intergranular Corrosion

It is the selective corrosion at the grain boundaries of a metal resulting in loss of strength and ductility. It occurs when the chemistry of the grain boundary is different from that of the bulk of the alloy. This type of corrosion is found in stainless steels and aluminium alloys.

(e) High Temperature Corrosion

Effects of wear, corrosion and stress are enhanced by elevated temperature thus accelerating the failure of the component in service. High temperatures oxidize the materials which reduces the strength and other mechanical properties. It is essential that the material of the component for operation in environments having wear corrosion and high temperature should be carefully selected. Surfacing is an effective method of improving life of such components in service.

8.1.3 Surfacing

Surfacing can be done by any one of the several welding or spraying techniques. The selection of a special technique depends upon several factors like the base material characteristics, component size, shape and weight, number of components to be surfaced, cost of surfacing, consumable availability, thickness of the deposit required, type of bonding between the deposited material and the substrate etc. The protective material can be deposited by either welding or spraying.

(a) Welding

Surfacing can be carried out by many welding process. The processes that are employed for this purpose are :

(i) Gas welding

(ii) Arc welding :

(a) Manual stick electrode welding.

(b) Gas metal arc welding.

(c) Plasma welding.

(d) Submerged arc welding.

(e) Strip cladding.

(f) Electroslag welding.

(iii) Laser surfacing.

(iv) Explosive welding.

The welding process is employed when thick metal deposits with very good bonding between weld deposit and the substrate are required. Welding produces a permanent metallurgical bond which is strongest bonding in material. Also the deposition rates are much faster in welding when compared to spraying. Further the material to be deposited is available in the form of wires, rods, strips and powder. In Fig. 8.1, strip cladding method of surfacing is illustrated. Wires both solid and flux or metal powder-cored wires are available.

Fig. 8.1 Strip cladding.

The exact form of these consumables depends upon the process and its versatility :

(i) Oxy-acetylene Welding : Surfacing can be done by the oxy-acetylene welding process. This is a process in which the base material is melted. The dilution rates are lower than in arc welding processes. The deposition rates are also lower than those in arc welding processes due to low dilution. Oxy-acetylene weld surfacing generally produces high quality deposits.

(ii) Arc Welding Processes : Arc welding processes are used for surface application. Arc welding generally produces higher dilution which can reach upto 60%. Processes like plasma arc welding, MIG welding and submerged arc welding can give high deposition rates. Processes like TIG welding and manual stick electrode welding give lower deposition rates. There is no limitation on the upper thickness of the weld deposit and on the lower side. The minimum thickness of the welded deposit will be 1.5 mm in surfacing using arc welding processes.

(iii) Laser Surfacing : A laser beam is used to obtain the controlled melting of the work-piece surface to the desired depth. At the same time, an alloy powder to get the required suitable weld deposit properties is fed to the metal surface. The powder melts and forms the weld deposit with the required characteristics. Some of the main advantages of laser surfacing are :

(a) Minimum heat input in the base material,

(b) Access to the surface to be coated is better as compared to other arc or flame based processes, and

(c) Deposition of high melting point alloy on low melting component surfaces is possible.

(iv) Explosive Cladding : It is a process in which the cladding material is separated from the base plate by a predetermined distance. The cladding is covered with suitable explosives. When the explosives are detonated, the two plates progressively slam against each other. In the process unwanted surface films are expelled and a wave like surface bond at the interface is produced. This technique is more suitable for cladding steels with materials like aluminium and titanium.

For selecting a suitable process for weld hard facing, the following factors should be carefully considered :

(i) Size, shape and weight of the component.

(ii) Base material composition, its melting range, thermal expansion and contraction characteristics.

(iii) Form (rod, wire, powder, strips etc.) and availability of welding consumables.

(iv) Welder's skill.

(v) Location of surfacing whether in a shop or at site.

(vi) Accessibility for surfacing.

(vii) Welding position for surfacing.

(viii) Number of components to be surfaced.

(ix) Quality requirements of the weld deposits to be made.

(x) Service conditions.

(xi) Availability of the suitable welding equipments and power sources.

(xii) Cost of surfacing.

8.1.4 Spraying

Apart from welding, surface coating can be done by various spraying techniques. Both metallic and ceramic coatings can be applied using spraying techniques. The spraying techniques can be classified broadly as follows :

(i) Flame spraying

(ii) Arc spraying.

Spraying techniques produce a mechanical bonding of the coating material with the substrate. Coating thickness per pass of 0.12 mm to 0.75 mm can be achieved by using proper technique.

Selection of Materials for Surfacing

Materials for surfacing can be classified broadly as those for build up (reclamation to original size) and for hard facing. Materials for reclamation are mainly governed by the chemical composition of the base material. For hard facing applications, a variety of materials are available. It is normal practice to choose two or three materials for specific application and then decide the best out of the selected consumables, depending upon various factors like the cost, amount of work, field performance etc. The major classification of the welding filler materials and their area of application are indicated in Table 8.1.

Table 8.1 : Criteria of Selection of Weld-hardfacing Materials

	Low ←	Abrasion Resistance		→ High	
Low ↑	14% Manganese steels	Martensitic alloy steels	Arc deposited carbide	Flame deposited carbide	Low ↑
Cost	Martensitic stainless steel				Heat and corrosion resistance
↓	18 8 Cr-Ni Stainless steel	High speed steel		Austenitic and martensitic steel	↓
High	Cobalt and nickel based alloys				High
	High ←	Impact Resistance		→ Low	

Advantages of Surfacing

(i) Increased service life of the component.

(ii) Higher productivity through less downtime for maintainence.

(iii) Higher operating efficiency by accomodating increased/higher service operating parameters like load, stress, impact etc.

(iv) Reduced maintainence cost due to economical reclamation of worn out parts.

(v) Optimum balance of deposit properties (between wear and toughness) through the use of suitable hard facing materials.

8.2 THERMAL SPRAYING

Thermal spraying is a coating process where the material to be sprayed is heated substantially to its molten stage and the molten particles are atomised to fine particles and projected on the base material (job) in the form of jet (Fig. 8.2). These projected particles adhere to the base material (the substrate) mechanically,

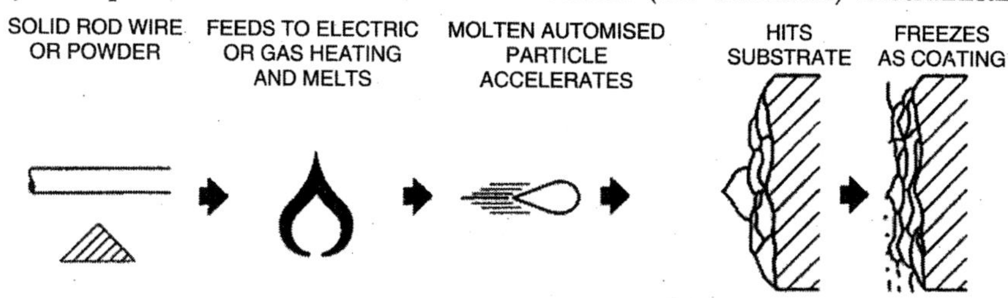

Fig. 8.2 Principle of thermal spray coating.

metallurgically or chemically to give the required coating. The thermal spraying method is also called metallizing or spraying. The temperature of the particles, job temperature, particle speed and size and job roughness influence the bond strength. Users want good coating to job adherence, high coating density, uniform deposition thickness and smooth consistent surface appearance. Thermal spraying is used for a large number of applications in various fields and in different industries. It is often the most economical and most effective to increase service life of the new or wornout mechanical parts. Thermal spraying is used effectively for producing heat and wear resistant coatings, corrosion resistant coatings, electrically conducting and insulating coatings and surface build up coatings for various industrial applications.

8.2.1 Types of Thermal Spraying Processes

Flame, electric arc, plasma and detonation spraying are the commonly used processes. Now-a-days a high speed oxy-fuel process is also developed to produce the high coating quality. Each process has special features and advantages as given in Table 8.2. The final choice can be governed by factors such as the material to be sprayed, cost aspect, deposition rates, availability of the system and the quality of the sprayed deposits (*i.e.*, bond strength and percentage porosity) required to meet the service conditions. These aspects give a general comparison of these thermal spraying processes.

Table 8.2 : Features of Thermal Spraying Processes

	Flame	Arc	Plasma	Detonation
Impingement speed m/sec	100	150	300	800
Approx. temp °C	3000	5000	20000	4500
Typical deposit porosity %	10 - 15	5 - 15	1 - 10	1 - 2
Typical adherence MN/m^2	7	10	30	60
Advantages	Cheap	Cheap	Low porosity	Very low porosity
	High deposition rates	Low contamination	Good adhesion	Very good adhesion
		Cool substrate	Cool substrate	Cool substrate
			Low contamination	
Limitations	Normally high porosity	Only electrical conductors	Fairly expensive	Very expensive
	Inferior adhesion	Normally high porosity	-	Low efficiency
	Heated workpiece	-	-	-

(a) Flame Spraying

Flame spraying, the simplest and the most versatile of all thermal spraying processes, sprays material supplied as wire, rod or powder. Fuel gases like acetylene, propane, natural gas or hydrogen mixed with oxygen heat up to melt spray material. In the simplest form of powder flame spraying, oxygen aspirates the powder from a container attached to the spray gun and injects it into an oxy-fuel flame (Fig. 8.3). In some powder flame spray guns, the pressurised inert

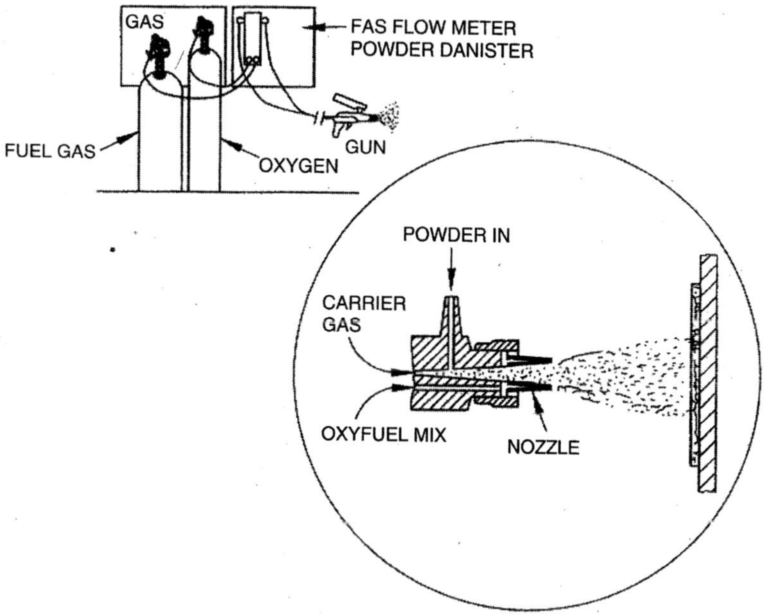

Fig. 8.3 Powder flame spraying.

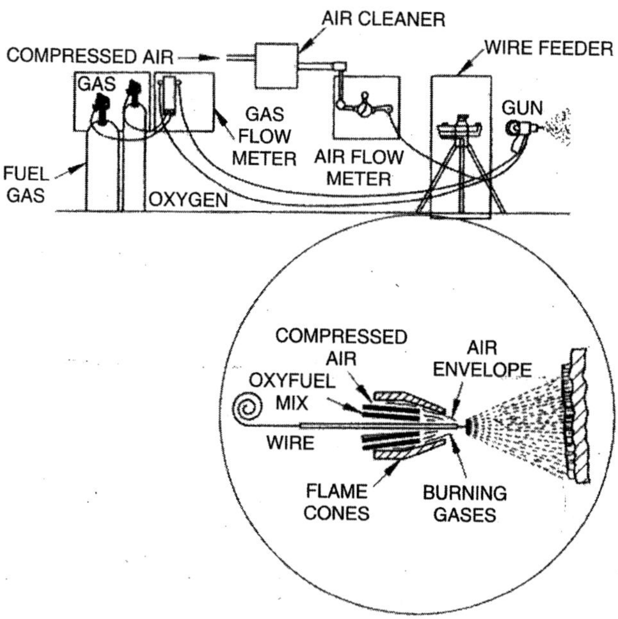

Fig. 8.4 Wire of rod flame spraying.

gas from remote powder feeder carries powder into the flame. When the wire or rod is used as the spray material, air-motor driven gears draw the material into the gun to melt into the flame (Fig. 8.4). In the gun compressed air or inert gas is used to atomise the molten particle and to propel them on to the job surface for spraying. The inert gas helps in another way to reduce oxidation of the particles and the substrate. Flame spray guns are inexpensive, of light weight and compact. But compared to other spraying processes, particle speeds and temperatures are low, producing more porous, low density coating of low bond strength. Due to the limited heat of oxy-fuel flame the deposit rates are low. Flame spray is used to coat any material that melts below the flame temperature, *i.e.*, 2000°C-2500°C. The wire flame spray uses usually metals that can be drawn into a wire. Ceramic can be sprayed with guns for wire spraying either in the form of rods or powder filled plastic cords.

(b) Electric Arc Spraying

Electric arc spraying can be used for the electrically conductive materials which can be drawn in a continuous wire form. In electric arc spraying, two consumables (material to be sprayed) wires are continuously fed at the same speed through on air or electric motor (Fig. 8.5). An electric arc is created between the tips of these wires through a constant current D.C. power supply (open circuit voltage –18 V to 35 V, current – 150 to 800 Amps). The material melted through the electric arc heat is atomised and propelled on to the substrate through compressed air flowing at 50 to 70 liters per minute and 2.5 to 6 kg/cm^2 pressure. Electric arc guns (quite light weight but heavier than flame spraying guns) spray 1.5 to 50 kgs of wire per hour. The rate of deposition depends on the current and type of material sprayed. High arc temperatures upto 5000°C melts the wires fast depositing particles with high heat content and fluidity

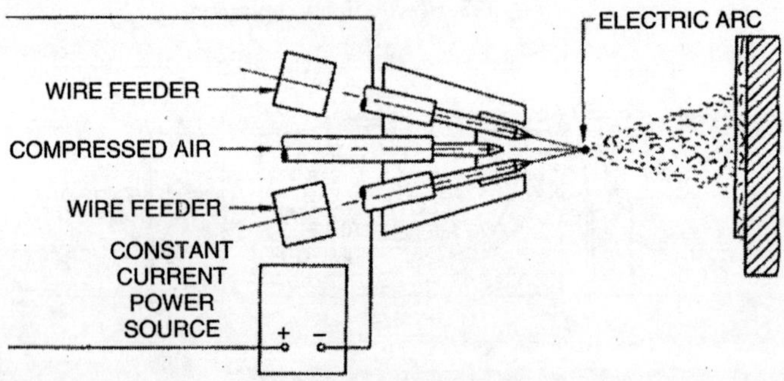

Fig. 8.5 Electric arc spray.

Due to diffusion of sprayed material into the substrate, on exceptionally strong bond strength can be achieved in electric arc spraying. The quantity of oxides in the deposit may be controlled by the spray conditions. The chemistry of deposit may vary because an arc is created between the deposit wires. So selection of material is difficult in arc spraying. Electric arc spraying is widely used economically for spraying aluminium, stainless steel, zinc, molybdenum, carbon steel etc.

(c) Plasma Spraying Process

In the the plasma spraying process (Fig. 8.6) the coating material in powder form (size 20–120 μm) carried by the carrier gas is injected into a gas plasma where it is melted, atomised and propelled on to a substrate surface at a high velocity upto 600 m/sec. To produce the shield, gas plasma, a D.C. arc is struck through a D.C. constant current power source between a non-consumable tungsten electrode and a water-cooled copper nozzle of a plasma torch in an inert gas shield, normally argon, nitrogen or argon + hydrogen etc. The plasma, thus formed, due to arc construction by the water-cooled copper nozzle, has a very high velocity (upto 3000 m/sec) and a high temperature (upto 20000°C), depending on torch design, gas mixture and other operating parameters like current, voltage etc.

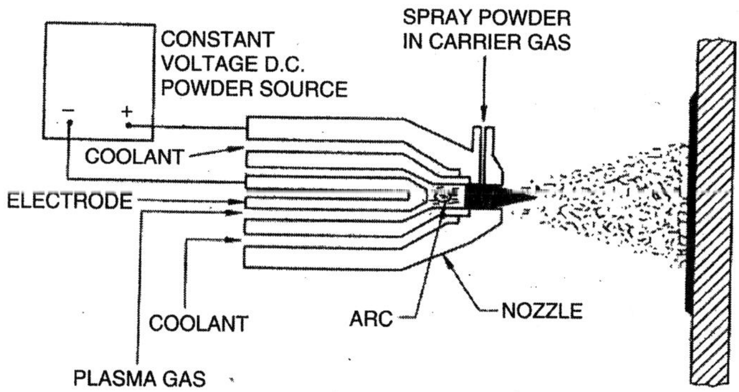

Fig. 8.6 Plasma spray.

Plasma spray, because it uses an inert gas and deposits at high temperature and particle velocity, produces coatings with mechanical and metallurgical properties generally superior to the flame spray coatings. Porosity and oxide inclusions are low, the bond strength between substrate and coating can be upto 5 kg/sq mm and the deposition rate can be from 1.0 to 15 kg/hour depending upon the material and powder used during spraying. However these equipments are complex, heavy and costly as compared to the flame and arc spraying equipments. Now-a-days in critical applications plasma spraying is carried out in a vacuum chamber to increase the particle velocity, avoid the oxide layer inclusion and to produce a better quality of coating.

(d) Detonation Spraying Process

The detonation spray process uses a gun that resembles a small caliber cannon. In it, air or oxygen and fuel gas, along with a charge of coating powder mix in a chamber (Fig. 8.7). A spark device ignites the mixture, setting of detonations

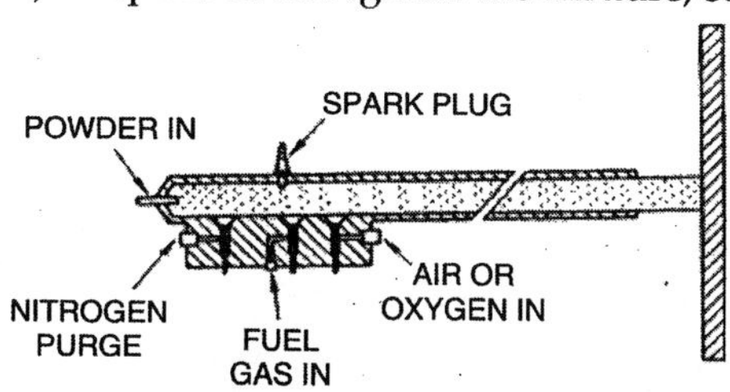

Fig. 8.7 Detonation spraying.

several times per second to propel the particles out of the barrel on to the work-piece at upto 800 m/sec. In some guns the nitrogen purges the chamber after each explosion. The temperature in the gun reaches 4500°C. A gas cooling stream keeps the job temperature below 200°C. Detonation guns produce coatings with a very high bond strength and density and very low porosity. The process is slightly risky by nature therefore should be handled by skilled operators only. The noise level during spraying is quite high.

8.2.2 Steps in Thermal Spraying Operation

Bonding relies on mechanical interlocks between the frozen particles and the substrate surface. To achieve a good quality thermal spraying coating, the following steps are to be understood.

(a) Surface Cleaning

The surface to be sprayed should be thoroughly clean and free from contamination, mill scale, oil, rust, oxide and other foreign materials. Clean, non-porous surfaces are obtained by vapour degreasing, steam or hot detergent washing or application of an industrial cleaning solvent based on the condition of the substrate. Porous materials such as castings etc. may need baking at 200-300°C to vaporise oil, greese etc. Mill scales and contaminated materials in the arc sometimes removed by abrasive blasting or machining operations.

(b) Under-cutting or Masking Operation

Machine undercutting provides for the thickness of the coating with mechanical tolerances on parts such as shafts, pistons, cylinders etc. Mask the areas surrounding those to be sprayed to protect them from blasting and spraying, using high temperature masking tapes, paints or metal sheets etc.

(c) Surface Preparation

After cleaning and masking, the surfaces to be sprayed should be roughened by blasting with clean, angular chilled iron or Al_2O_3 or silicon carbide grits (mesh size SAE Gr.16 to G.40) to achieve better mechanical bonding. Thread cutting or knurling can be done additionally on shafts and cylinders etc. in the place of blasting operation, if required. After grit blasting, spraying should be carried out within one hour to avoid oxidation or contamination on the base material surface.

(d) Preheating

Preheating the work-piece upto 150°C immediately before coating drives off moisture from the job surface. Heat also helps by expanding the work-piece so that the substrate cools with the coating, minimising the stress created in coating during cooling.

(e) Thermal Spray Operation

After the work-piece surface is properly cleaned, masked, grit-blasted and preheated, spraying has to be carried out using the predetermined operating parameters for the material to be sprayed. Spray with a smooth and uniform relative motion of the spray gun with respect to the job, holding the spraying stream perpendicular to the substrate surface to apply the spray material in thickness of 0.1 mm maximum-each layer. The gun to work-piece distance should be 50 mm to 250 mm, depending on the spraying material and spraying process. Correct setting of feed rate and the gun to work-piece distance improve the deposition efficiency. During spraying, the relative motion of the gun to work-piece should be such that the job temperature should not exceed 250°C to get a good bonding. Sometimes to have better bond strength between the job and the coated material, a deposit of 0.5 mm to 0.1 mm of bond coating material is applied as a first coat prior to final coating of the material. The bond coating materials react exothermically and metallurgically as they strike the cool surface of the substrate during spraying and the spray particles get locally welded into place.

(f) Post Spraying Operation

After spraying, the surface can be buffed, machined or ground based on requirements. Sometimes, to reduce the surface porosity, fusing or sintering of the components at high temperature or painting with special paints is also carried out to suit specific service conditions.

8.2.3 Safety and Health Hazards in Thermal Spraying

Hazards in thermal spraying processes like those of welding are inevitable. They include electric shock, arc radiation, fire, dust, noise, fumes, gases or even explosion. The operator needs protection from all these. All thermal spraying processes are noisy. The higher the temperature and velocity, the better the coating but the noiser the process. Occupational Safety and Health Administration—USA (OSHA) recommends permissible limits for noise exposure in each thermal process are given in Table 8.3. For plugs are normally enough for flame and arc spraying but better ear muffs or isolation eyes should be well protected against radiation (Table 8.4). Special care is required against fumes and metal dust to avoid inhailing by the operator. The best method is to use supplied air respiration system or atleast breathing masks with filters. Adequate ventilation should be provided at the spray station.

Table 8.3 : Noise Levels - Thermal Spray Processes

Permissioble noise levels per Osha regulations	
Duration per day, h	**Sound level, dBA continuous**
8	90
6	92
4	95
3	97
2	100
1.5	102

1	105
0.5	110
0.25 or less	115
Noise of thermal spray coating processes	
Process	**Sound level dBA**
Flame spray	95 - 105
Electric arc spray	105
Plasma spray	95 - 120
Detonation spray	Above 115

Table 8.4 : Recommended Eye Filter Plates for TSC

Operation	Filter shade number
Wire flame spray (Except molybdenum)	2 - 4
Wire flame spray of molybdenum	3 - 4
Powder flame spray metals	3 - 6
Powder flame spray exothermics, ceramics	4 - 8
Electric arc spray	3 - 6
Plasma spray	9 - 12
Fusing	4 - 6

8.3 ARC SPRAYING

Spraying is the process by which a material in the form of wire or powder is melted by suitable means and the particles projected on to a surface suitably prepared for the purpose, so as to obtain a build up of a coating. This coating performs many functions. When an electric arc is used to melt and fuse the particles of the coating material, the process is called arc spraying.

8.3.1 Principle of the Arc Spraying

In electric arc spraying, the arc is struck between two wires of the metal to be coated/sprayed to produce the heat required for melting and fusion. Generally a stream of compressed gas, usually air, projects the metal droplets on to the surface to be coated. This principle of arc spraying is explained in the schematic, (Fig. 8.8). As in the case of other spraying processes, treatment of the surface to be sprayed is indispensable. This is a prerequisite to achieve a good bond. In arc spraying, in addition to mechanical and physical adhesion there is evidence of the metallurgical bond. Welding at localised spots has also been found to exist.

The adhesion strength under comparable conditions is higher in arc spraying than in flame spraying. The velocities of particles are lower than in flame spraying, being between 100 and 180 m/sec. However, the particle sizes are greater and the temperatures are also higher. Those two factors are said to be the cause of localised welding.

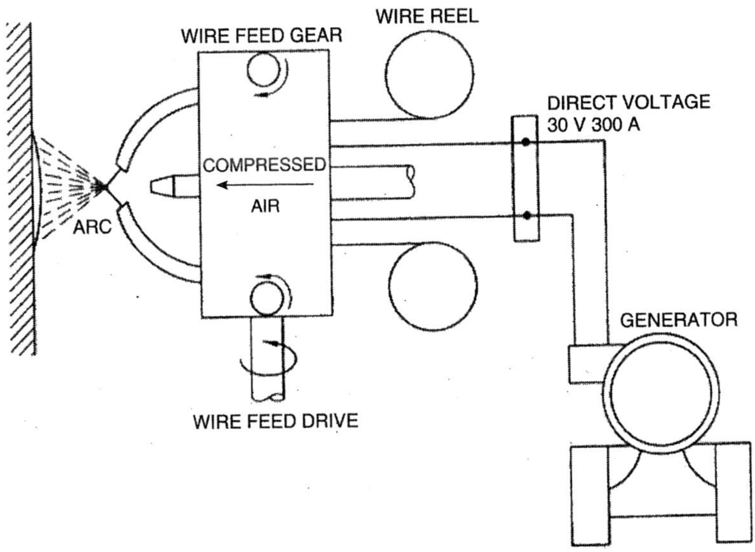

Fig. 8.8 Schematic diagram of an arc spraying equipment.

8.3.2 Process Parameters

(a) Surface Preparation

Proper preparation of the base material is the most important part of the job. Rough threading and/or grit blasting are usually done. Under optimal conditions, partial welding will be obtained between the surface and coating. To achieve this condition it is imperative that the surface must be absolutely clean, free from even the slightest trace of oil, grease or moistue. Therefore, it is recommended that the rough threading or grit blasting be performed just before the spraying is done. Cast-iron should always be grit blasted after threading is done to remove traces of carbon that may be left as residue on the surface. If grit blasting is impractical, a very fine rough threading should be considered as a satisfactory alternative.

(b) Degreasing

This is generally advocated for machine parts which have been in use for sometime. Only trichloroethylene is considered as a suitable agent for degreasing.

(c) Stoving/Furnacing

Generally parts that may be impregnated with oil, especially those of cast irons, should be furnaced or stoved upto 400°C. This operation may be replaced by heating with an oxy-fuel flame torch to burn away the oil.

(d) Undercutting

This is a machined step provided for coating purposes. The end of an undercut should have an open angle but no feather edge. The base metal on the shaft ends should be chamfered to allow the coating to get over the edge.

(e) Threading

The objective of threading is to provide for more surface area and thus to increase the area of contact. The next important function of the thread is to restrain the contraction of the coating. On cast irons 16-20 tpi is recommended.

(f) Grit Blasting

Only sharp steel grit blasting is permitted. Aluminium oxide, corundum and such other abrassive materials are unsuitable as they leave residual particles on the surface which are detrimental to spraying. Generally, surfaces of jobs to be electric arc sprayed are prepared by grit blasting only. However, on long roller surfaces and flat surfaces, making of 0.5 to 1 mm deep grooves at 300 mm intervals, with 6 mm grinding wheel is recommended. Rollers under heavy load and hydraulic rams are always surface-threaded.

(g) Masking

This operation is done to prevent the spray from affecting portions of the surface which should not be coated. For this purpose, special solutions with a carbon base should be brushed on all such surfaces. Masking tapes and asbestos bands may also be used. Oil holes are plugged suitably. Keyways are protected with dummy keys made of carbon.

(h) Preheating

Preheating generally is not recommended. However, in colder climates a preheat of 100-500°C will be helpful as the adhesion is greatly improved.

(i) Spraying Process

Depending on the thickness of coating required, spraying may be done in many passes. However, the first pass is significantly important.

(j) Spraying the First Pass

The temperature at the tip of the wire during melting may be around 4000°C. To obtain proper welding effect of the coating particles with the surface, this first pass must be as hot as possible. Therefore, the effect of the atomising air towards reducing this temperature should be borne in mind while setting the parameters. Generally, the volume of the air may be cut down and/or the voltage increased to raise the temperature. The spraying distance may also have to be kept at a bare minimum to maintain high particle temperature.

(k) Building up the Further Passes

Second and further passes will have to be made continuously at normal air pressure and lowest voltages possible. It is preferable that the work is finished in one continuous operation. The spraying distance is kept between 70 and 200 mm and the speed of rotation of the job kept high (50-100 mm/min). The traverse speed of the spray gun should not be high but 'screw' like finish of the coated surface should be avoided. Care should be exercised while surfacing with 18Cr 8Ni stainless steels. The traverse direction is only one way. This will give rise to less stress and reduce cracking in thick coatings. Thickness of 18/8 stainless steel coatings in excess of 3.5 mm should, therefore, be avoided for above reason.

Wires of diameter 1.6 mm are generally used. However, it is not uncommon to use wires of diameter 2.4 mm or even 3 mm for special applications. When spraying a non-ferrous metal directly on to a non-ferrous material, it is usual to preheat the base material to 150°C. Short spraying distances coupled with high voltages should be employed to obtain a satisfactory bond. While working inside bores and cylinders, it is recommended that the work is kept preheated throughout the spraying operation.

8.3.3 Precautions

Spraying, as a rule, should be done continuously without interruption. This is basically to ensure that the surface does not suffer damage. If in the case, the job has to be continued after a lapse of around 12 hrs, *e.g.*, continuing the work next day following precautions will have to be observed :

(i) Stainless steels and brozes : Clean brown paper to be wound tightly and fixed with a tape. Job to be passed in blankets to keep warm.

(ii) Carbon steels : A light overspray of nickel or stainless steel is recommended.

(iii) Avoid leaving the spraying unfinished in the machining zone.

(iv) Spraying is to be done immediately after surface preparation operation.

(v) As a general precaution, if oil is found to ooze, spraying should be stopped. Cleaning with trichloro ethylene should be done before restarting the spraying operation.

8.3.4 Common Problems/Defects

(i) Spitting

This may be caused by uneven wire feed, worn out nozzles or faulty wire.

(ii) Collar Build-up

Collar will build-up at the edges of the areas to be sprayed. These must be continuously chiselled or chipped off. If not done in time, they will show up as dark lines after machining.

8.3.5 Safety Precautions

Personnel employed in spraying should be protected against radiation of the arc. They should be protected with leather gloves, a welding helmet or goggles. While spraying copper, and copper and tin bearing alloys, a pressure fed mask is necessary. Special light goggles are a must while spraying zinc and aluminium to counter radiation hazard to the eyes. Spraying releases oxides in the form of fine dust in the spray. This will have to be effectively extracted by using fume/dust extraction systems. Apart from local extraction the shop floor should be well ventilated.

8.3.6 Equipment for Arc Spraying

In general, the equipment for arc spraying constitutes from four main parts :

(i) The power source.
(ii) The regulating or control unit.
(iii) The wire roll holding units.
(iv) The spray gun or pistol.

(i) The Power Source

The power source used for arc spraying is usually a specially wound transformer with protective devices such as overload switches etc. fitted on to it. The units are designed to operate on 3-phase, 50 c/s supply at voltages in the range of 380-440 V. The current ranges are generally above 250 A and differ with the units capacity.

(ii) The Regulating or Control Unit

This consists of regulators and pressure gauges for controlling the air being fed to the wire feed area.

(iii) The Wire Roll Holding Units

Usually, two wire roll holders built into the rear of the system, form this unit. These carry the reels for the wire and are specially designed and electrically insulated. The construction is simple so as to permit easy and quick changing of wires.

(iv) The Spray Gun or Pistol

The gun usually consists of a compressed air-wire feed mechanism and a nozzle assembly. Two copper contract tubes guide the wires into the arc zone at the top of the nozzle. The gun is so constructed that the wires will meet with an included angle often between 30° and 60° between them. The contact tubes carry the electric current. The rest of the gun is well insulated. It is made from light weight material to facilitate manual handling. The gun max also be mounted on a tool post to facilitate automatic spraying.

Applications : Spraying using the electric arc is usually handy for maintainence work. It is also used for corrosion-resistant material on other substrates. Hard facing and acid-resistant coatings are other examples of arc spraying applications. The materials used for coating vary from mild steels, austenitic stainless steels and martensitic steels to chromy manganese deposits. Nickel, nickel-copper alloys, aluminium, zinc, tin and brozes are other materials which also find wide applications with the arc spraying process.

Thus the process for spraying is very versatile and permits a wide spectrum of metals and alloys to be sprayed or coated. It also lends itself to automation. Even though the capital cost is higher than flame spraying, arc spraying ensures higher particle bond strength and other properties.

❑❑❑

CHAPTER 9

CLASSIFICATION OF STEELS AND THEIR PROPERTIES

❑ INSIDE THIS CHAPTER

9.1 Atomic bonding, 9.2 Solidification of structures, 9.3 Crystal structure, 9.4 Crystal defects, 9.5 Phase transformations, 9.6 Complete solid solubility, 9.7 Limited solid solubility, 9.8 Eutectic system and solid solutions, 9.9 Iron/carbon phase diagram, 9.10 Phase transformation, 9.11 Time temperature transformation (TTT), 9.12 Continuous cooling transformation (CCT curves), 9.13 Classification of steels.

Engineering materials can be classified broadly into three groups according to their nature :

(i) Metals and alloys.

(ii) Ceramics and glasses.

(iii) Organic polymers.

Basically, metals and alloys are capable of changing shapes. They possess good thermal and electrical conductivity and are strong and tough. Ceramics are non-metallic inorganic compounds having good insulating properties but they are brittle. Organic polymers are light having high degree of viscoelasticity and plasticity. A schematic explaining the classification of materials is given in Fig. 9.1. Figure 9.1 also includes a new range of materials called composites where in either a metal fibre is incorporated in ceramics (RCC) Reinforced Ceramic Composites or a ceramic fibre in

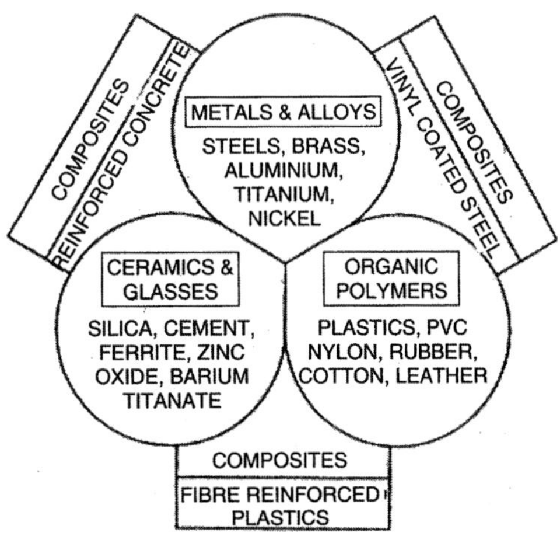

Fig. 9.1 Major classification of materials.

introduced in plastics (FRP) Fibre Reinforced Plastics. These materials derive their inherent characteristics because of the way atomic bonding takes place in them.

9.1 ATOMIC BONDING

The various types of bonding are :

(i) Electrovalent bonding.

(ii) Covalent bonding.

(iii) Metallic bonding.

(iv) Vander Waals bonding.

Infact different properties of materials arise mainly from the widely differing nature of these power types of bonding. The first three types of bonds are called primary bonds and are strong. The Vander Waals bond is a weak secondary bond often present between molecules or between layers of atoms.

(i) Ionic Bonding (Eectrovalent Bonding)

It is perhaps the easiest type of bond to understand. It arises out of an electrostatic force of attraction between the positive and negative ions in solid crystals. A typical example is in sodium chloride (NaCl) with positie ions of sodium (Na) and negative ions of chlorine (Cl). Refer Fig. 9.2. In the NaCl crystal, the Na ions is surrounded by 6Cl ions and vice-versa. Ionic bonding can also be found in ceramic materials which contain compounds of metallic and non-metallic elements.

Ceramic range from very simple type like MgO (with only two kinds of atoms) to very complex type like Mica. Ionic bonding can be found in ceramic materials which contain compounds of metallic and non-metallic elements.

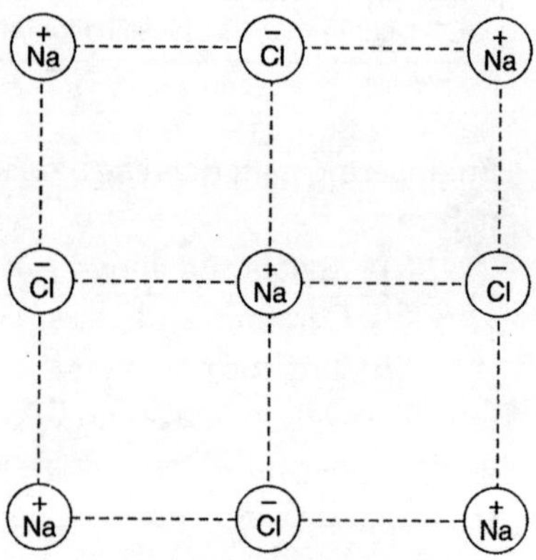

Fig. 9.2 NaCl structure.

(ii) Covalent Bonding

This type of bonding results from the sharing of electrons between atoms to achieve a stable structure and for each one covalent bond one pair of electrons in involved. A typical solid which consists entirely of covalent bond is diamond (Fig. 9.3). These bonds are localised and their directional nature leads to a tetrahedral structure. Because of strong C-C bonds, diamond is very hard and stable upto 3000°C. Most organic materials are polymers based on covalent bond formed by carbon. The' electrons are bonded strongly and hence polymers possess good thermal as well as electrical conductivity.

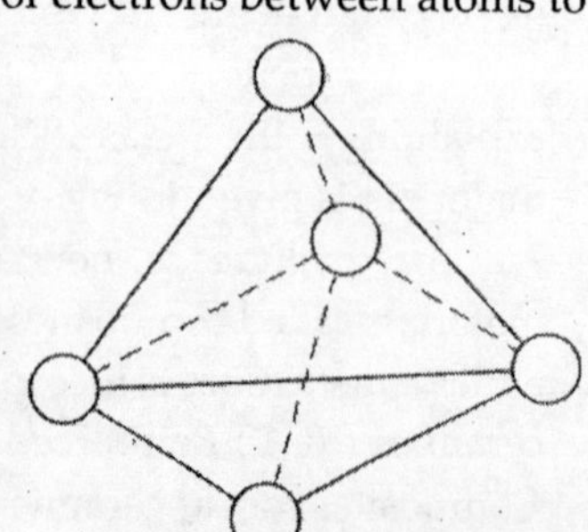
Fig. 9.3 Carbon tetrahedron in diamond.

(iii) Metallic Bonding

This type of bond occurs in all the metals. The characteristic feature is de-localised nature of bonds (Fig. 9.4). The metallic crystals consist of positively charged ions with de-localised valence or conducting electrons moving about freely in the crystal. This accounts for the higher electrical conductivity of the metals. Further the bonds are not directional. The atoms tend to form crystal structures which resemble close packing of spheres with the electrons forming an electron gas or cloud. Because of this, the metallic ions move easily with respect to each other, leading to high plasticity of metals.

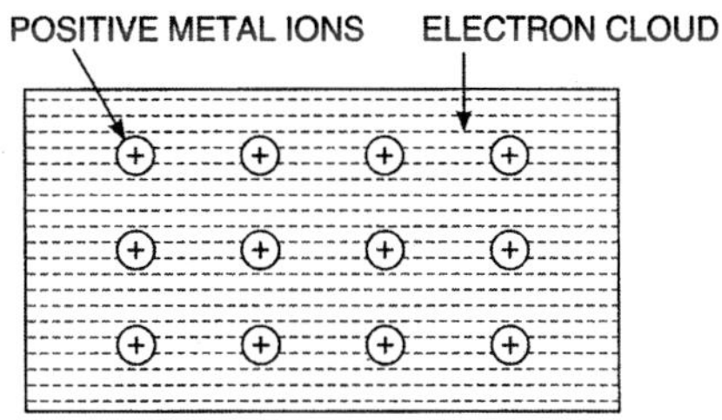

Fig. 9.4 Metallic bond.

(iv) Vander Waals Bonding

These are weak, secondary bonds responsible for attraction between the gas molecules. They are also found between convalently bonded pairs of atoms in polymers. This bond occurs as a polarization of the electric charges of the atoms. Polarization is the separation of the centres of positive and negative charges in an electrically neutral atom or molecule. Most of polymeric molecules have atomic chain with covalent bonds and the chains themselves are held by Vander Waals forces.

9.2 SOLIDIFICATION STRUCTURES

When liquid metal is allowed to cool, as in the case of a casting, small nuclei are formed and each nucleus grows into the liquid in a tree like or dentrite pattern (Fig. 9.5). A dentrite may show a long primary axis and side branches. The size of the dentrite depends on the rate of cooling. Since the impurities tend to remain in the liquid, they collect at the space between the dentrites. This is termed as interdentric segregation which is not desirable. Also, pores are formed in the inter-dentric region due to shrinkage in volume during solidification. If a pool of metal is allowed to cool rapidly in a casting, three zones are formed (Fig. 9.6).

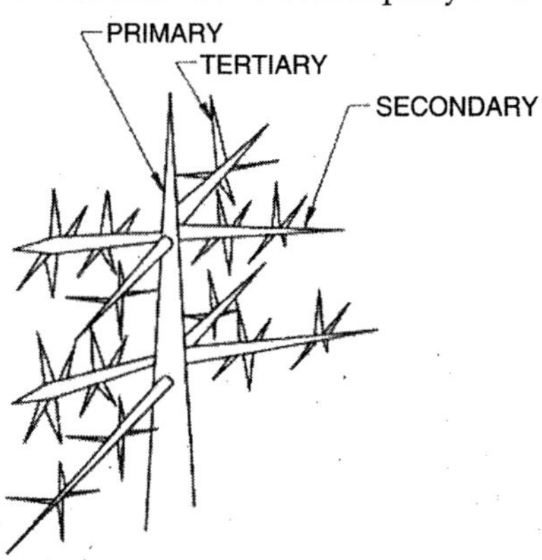

Fig. 9.5 Shape of a dendrite.

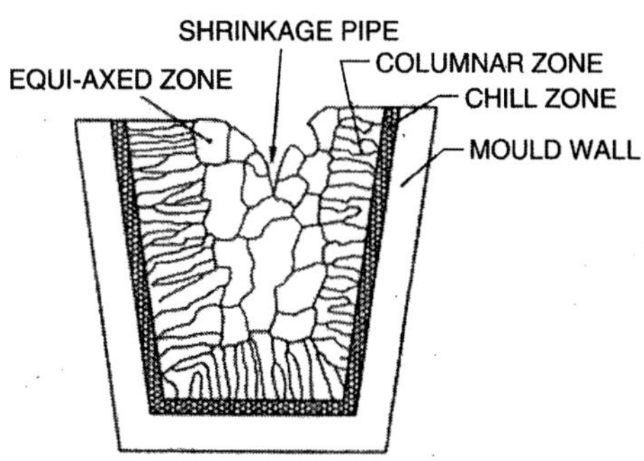

Fig. 9.6 Ingot structure.

(i) A chill zone formed very close to the mould wall having very fine crystals.
(ii) A columnar zone consisting of length in crystals growing perpendicular to the wall.
(iii) Equi-axed zone consisting of uniform large and equi-axed crystals. The heat extraction of the mould is much less compared to the side wall regions. The extent of each zone depends on the mould wall material or its thickness and the physical and metallurgical characteristics of the cooling melt.

In the case of fusion welding, the amount of liquid metal deposited is small and the heat extraction rates are very large as compared to casting. Unlike casting, nuclei are already present in the unmelted parts of the crystals left on the edges of fusion zone which is the portion of the parent metal actually melted by the weld deposit. Columnar crystals readily grow from such nuclei as shown in Fig. 9.7. It can be seen that the columnar grains of the weld metal grow along the direction guided by the coarse grains and this type of growth is termed as epit-axial growth. The thickness of the columnar grain primarily depends on the rate of heat extraction. Higher the solidification rate, finer will be the columnar grain thickiness. In the case of casting, the interdentritic structure is not desirable and is broken down by homogenization heat-treatment. Such solidification structures are sometimes found in the weld metal.

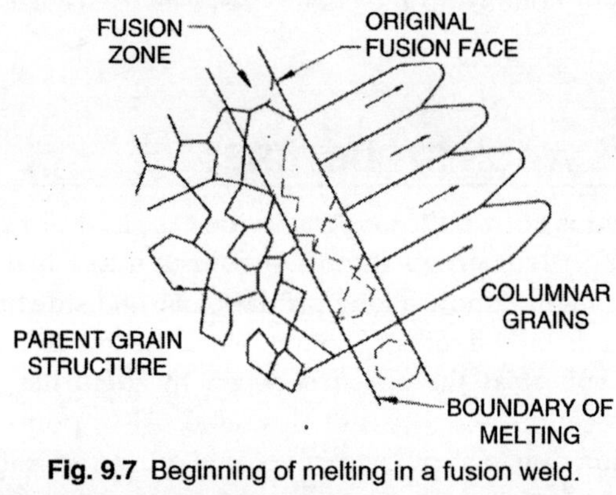

Fig. 9.7 Beginning of melting in a fusion weld.

9.3 CRYSTAL STRUCTURE

A distribution of atoms in three dimensions is said to form a space lattice if every point has identical surroundings. Because of the regularity in the distribution of the points (atoms) that compose a lattice, the essential geometry can be described by three sides (lattice vectors) a, b and c as shown in Fig. 9.8. These sides (vectors) describe the unit cell of the space lattice. The geometry of the space lattice is completely specified by the vector lengths a, b and c and the interaxial angles so that we have 7 crystal systems as summarised in Table 9.1.

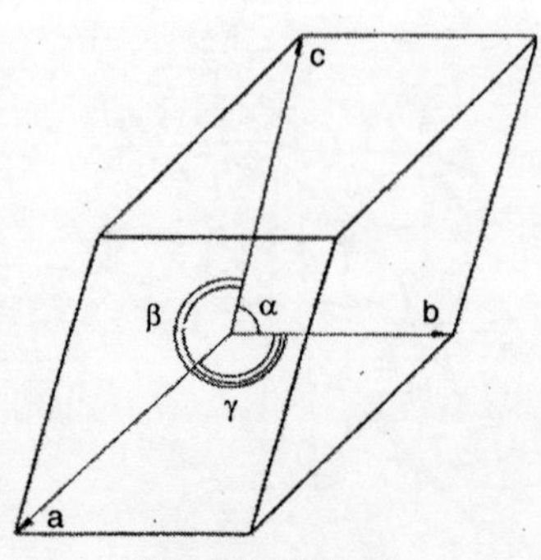

Fig. 9.8 Unit cell.

Table 9.1 : 14 Space Lattices

Crystal System	Lattice	Axial Relation	Angular Relation
Triclinic	Primitive	$a \neq b \neq c$	$\alpha \neq \beta \neq \gamma = 90°$
Mono-clinic	Primitive base centered	$a \neq b \neq c$	$\alpha \approx \beta \approx 90°$
Ortho-rhombic	Primitive FCC BCC base centered	$a \neq b \neq c$	$\alpha \approx \beta \approx \gamma \approx 90°$
Tetragonal	Primitive BCC	$a \neq b \neq c$	$\alpha \approx \beta \approx \gamma \approx 90°$
Hexagonal	Primitive	$a \neq b \neq c$	$\alpha \approx \beta \approx 90°$ $\gamma \approx 120°$
Rhombohedral	Primitive	$a \neq b \neq c$	$\alpha \approx 90°$
Cubic	Primitive BCC FCC	$a \neq b \neq c$	$\alpha \approx \beta \approx \gamma \approx 90°$

There are 14 possible arrangements of points in all, that satisfy the definition of space lattice. These 14 space lattices can be grouped according to the relative lengths of the vectors and the inter-axial angles so that we have 7 crystal systems as summarised in Table 9.1. Even though there are 14 different structures that are not possible, most of metals crystalize in these common forms : face-centered, body-centered and the hexagonal close packed structures.

9.3.1 Body-centered Cubic Structures

A crystal is defined as an orderly array of atoms in space. The unit cell is the smallest group of atoms possessing the symmetry of the crystal, which when repeated in all directions, will develop the crystal lattice as shown in Fig. 9.9. Let us consider the unit cell of the body-centered cubic lattice. The corner atom does not belong to one unit cell, but is a put of all 8 unit cells that surrounds it. Therefore, it can be said that only 1/8 of this corner atom belongs to any one unit cell. Use can be made of this fact in computing the number of atoms permit cell in a body-centered cubic crystal. Each atom lying at the corner of the unit cell contributes 1/8 of an atom located at the centre that is not shared with other unit cells. The BCC lattice thus has two atoms per unit cell. The coordination number of a crystal structure equals the number of nearest neighbours that on atom possesses in the lattice. In BCC lattice the coordination number is 8. Examples, iron in ferrite, vanadium, molybdenum, tungsten etc.

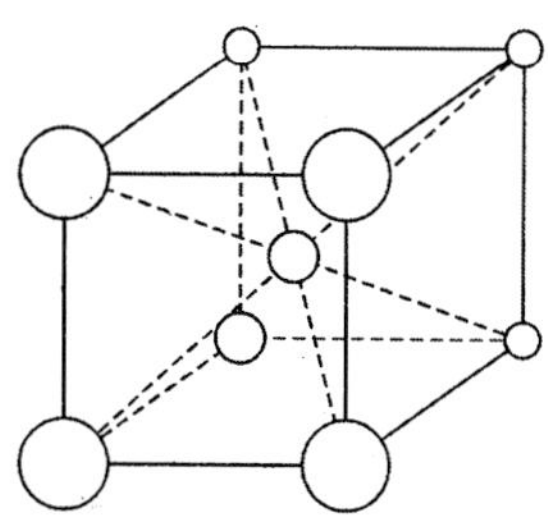

Fig. 9.9 BCC structure.

9.3.2 Face-centered Cubic Structure

In unit cell of the face-centered cubic lattice as illustrated in Fig. 9.10 has an atom located in the centre of each face of the cell. The FCC lattice has a total of 4 atoms

(1/8 × 8 + 1/2 × 6) per unit cell. The close packed directions correspond to diagonals that cross the faces of the cube. There are in all 6 close-packed directions in the FCC lattice. The coordination number of the systems of the 2 closest packing is 10, 6 neighbour atoms inthe same plane, 3 above and 3 below the plane. Examples : iron in Austenite, Nickel, Copper, Gold, Lead etc.

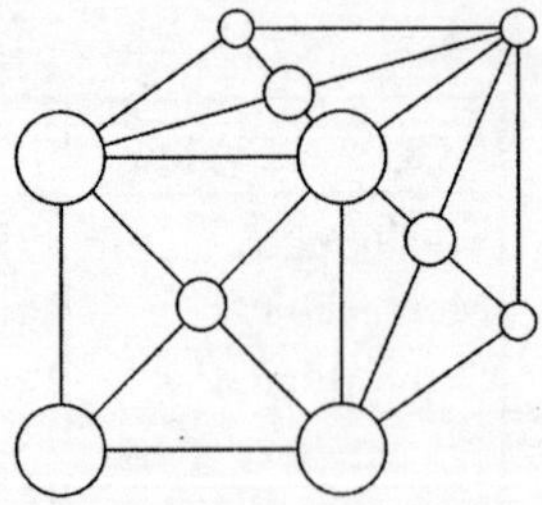

Fig. 9.10 FCC structure.

9.3.3 Hexagonal Close-packed Structure

In configuration of atoms that is more frequently used to represent the close-packed hexagonal lattice is illustrated in Fig. 9.11. It contains more than the minimum number of atoms needed to form an elementary building block for the lattice and therefore it is not a true unit cell. However, because the arrangement brings out important crystallographic features, it is commonly used as the unit cell of the close-packed hexagonal structure and the coordination number equals 12. Examples : Titanium, Cobal, Zinc, Zirconium etc.

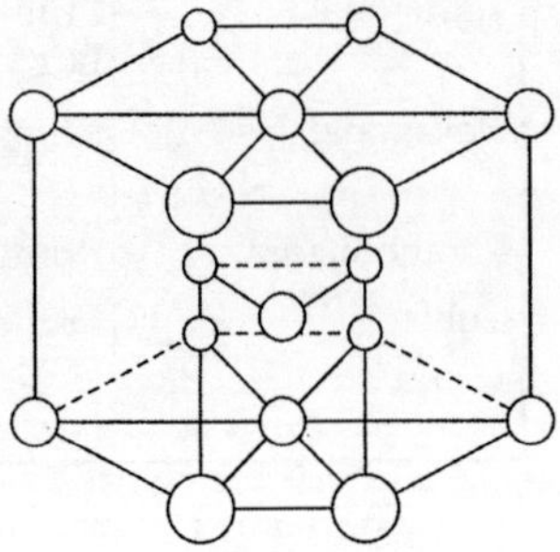

Fig. 9.11 Hexagonal closed packed structure.

9.4 CRYSTAL DEFECTS

Almost all crystals contain some defects since the atoms are not perfectly arranged. The defects influence the physical and mechanical properties. The defects can be classified as :

(i) Point defects (vacancy, interstitials)
(ii) Line defect (dislocation)
(iii) Surface defects (grain boundary, laminations)
(iv) Volume defects (voids, slag inclusions)

Point defects arise out of missing atomic lattice sites (vacancies) or additional atoms (interstitials). These defects are important in understanding the movement of atoms or diffusion. The number of vacancies is usually very small which

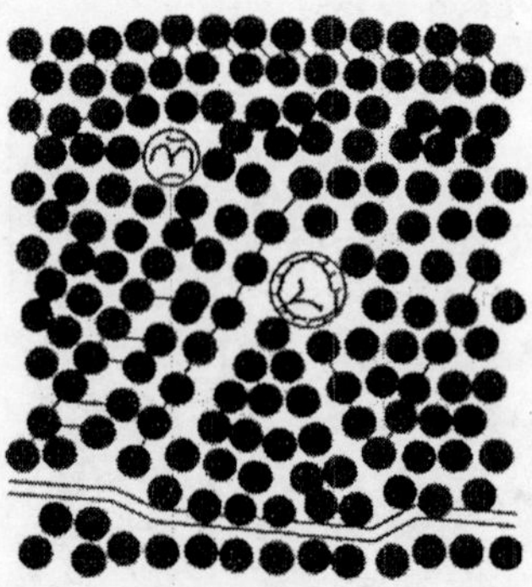

Fig. 9.12 Point effect.

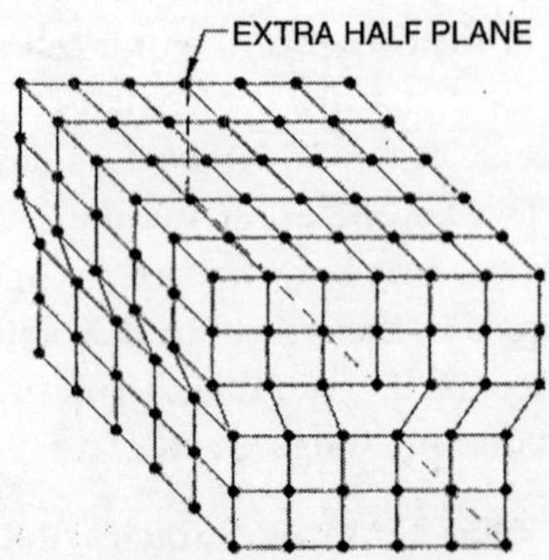

Fig. 9.13 Edge dislocation.

increases with temperature. Near the melting point there will be one vacancy for every 10000 atoms. Examples of point defects are shown in Fig. 9.12.

Line defects can be either edgeor screw dislocations. In some atomic planes, one-half of the plane may be missing and it is known as edge dislocation (Fig. 9.13). The number of dislocations and their movements are important in determining the mechanical properties of crystals in materials.

9.5 PHASE TRANSFORMATIONS

Aggregates of metal atoms exist not only in the crystalline state but also in the other two states of matter, that is the liquid and the vapour states. A phase is defined as a microscopically homogeneous body of matter.

Let us consider a simple system, a single metallic element that is, so called one component system, for example, iron. Solid iron confirms to this definition of a phase and same is true when it is in liquid or vapour form. Each of three forms of iron : solid, liquid or gas constitutes a separate and distinct phase. The phase rule put forward by Gibbs, defines the number of phases that can be in equilibrium. The equilibrium condition can exist only if $P + F = C + 2$, where P, F and C represent the number of phases, the number of degrees of freedom and the number of components in the given system respectively. The degrees of freedom are nothing but the number of such variables as temperature, pressure or concentration, that can be changed independently without changing the number of phases present. In an alloy there are terminal solid solutions, *i.e.*, phases based on the crystal structure of the components. In some binary alloys thus, crystal structure different from those of either component may form at certain intermediate phase. The concept of the solid solution is not as familiar in general as that of a liquid solutions but the meaning is exactly the same; atoms or molecules of one element may be accomodated in the structure of the other element. A solid solution may be either substitutional or interstitial : the solute atom may substitute as in Fig. 9.14(a) for a solvent atom or it may occupy an interstitial position between the solvent atoms as shown in Fig. 9.14(b). Substitutional solid solutions usually are formed between two kinds of atoms of about the same size and interstitial solid solutions usually are formed between atoms that differ greatly in size. Phase diagrams or equilibrium diagrams are important tools in the study of alloys. In brief, they are the form of a map, defining the regions of stability of the various phases that can occur in an alloy system under the conditions of constant pressure. The coordinates of these diagrams are temperature and concentration in percentage. Depending on the nature of the two metals involved, several basic reactions are described further.

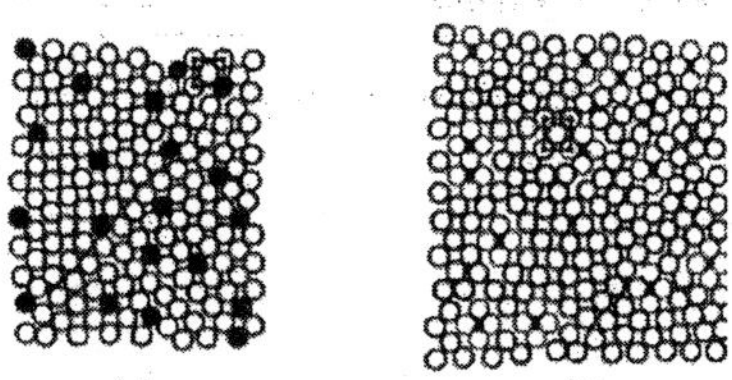

(a) (b)

Fig. 9.14

9.6 COMPLETE SOLID SOLUBILITY

A schematic solid solution diagram between two components *A* and *B* which are completely soluble in each other is shown in Fig. 9.15. The diagram is constituted with a single phase liquid zone (*L*) a single phase solid zone (*S*) and two phase liquid + solid regions (*L* + *S*). The temperature composition curves for each phase in two-phase equilibrium are the two curves that separate the single phase from the two-phase region. The liquidus is the temperature/composition curve for the liquid phase that is in equilibrium with the solid, the solidus is the temperature/composition curve for the solid phase that is in equilibrium with the liquid. The ends of a horizontal line drawn between the two curves will intercept the curves representing the compositions of the two phases at a temperature indicated by the horizontal line, *i.e.*, *X* represents the compositions of *AB* in liquid and *Y* represents the compositions of *AB* in solid.

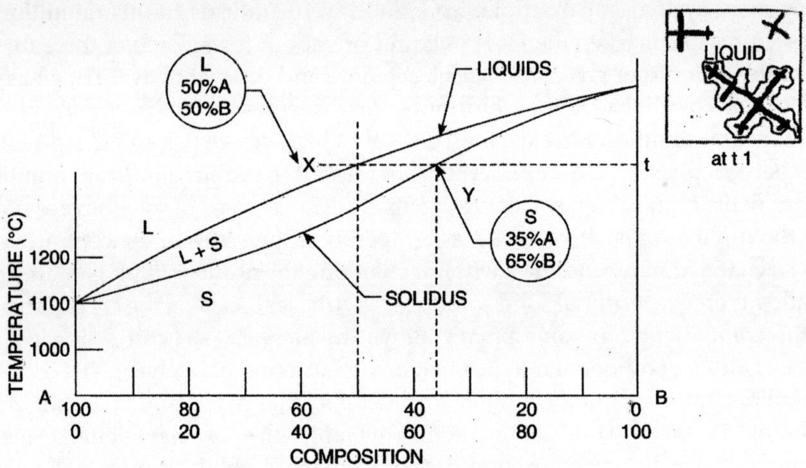

Fig. 9.15 Schematic of solid solution.

9.7 LIMITED SOLID SOLUBILITY

If metals *A* and *B* having similar melting points are only partially soluble in each other in the solid state but completely soluble in the liquid state, they form what is known as "Eutectics".

9.8 EUTECTIC SYSTEM AND SOLID SOLUTIONS

A copper-silver eutectic system in shown in Fig. 9.16. Here *AEC* is the liquidus and *ABEDC* is the solidus. In fact alloys containing less than 0.8% or more than 92% copper will solidify forming cored solid solutions in a manner exactly as has been described in the case of a simple solid solution in Fig. 9.15. The points *B* and *D* denotes the limits of solid solubility of (i) Cu in Ag and (ii) Ag in Cu respectively. All alloys between *B* and *D* solidify as 'eutectic' of Ag and Cu.

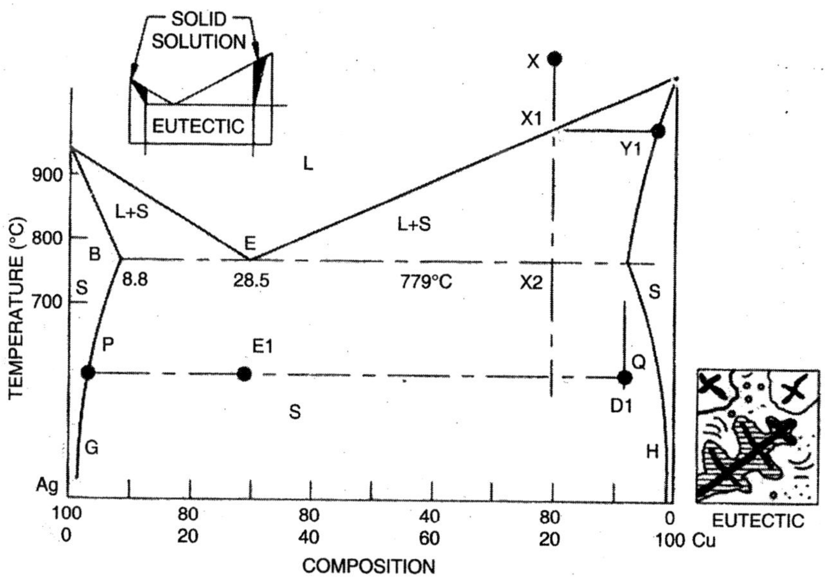

Fig. 9.16 Schematic of a copper silver eutectic system.

9.9 IRON/CARBON PHASE DIAGRAM

Since carbon steels constitute by and large for the greater tonnage of metals used, this system has been extensively studied. Allotropy is characterised by a change in atomic structure which is incited at specific transformation temperature. Four change occur in iron which give rise to certain allotropical forms known as α iron (ferrite), γ iron (austenite) and δ iron(δ ferrite).

Of these, the alpha and delta forms have the same body centered cubic structure while gamma iron has a face centered cubic structure. Unlike the phase diagrams for complete solubility (Fig. 9.15) and partially solubility (Fig. 9.16) many of the common phase diagrams of general alloys are complicated in nature. The phase diagram shown in Fig. 9.17 is for iron (Fe) and iron carbide (Fe_3C). Iron carbide is a compound of iron and carbon which is also known as cementite. The iron-cementite system shown in Fig. 9.18 is for 6.67% cementite composition. Above 6.67% of Fe_3C, the phases are not included in the diagram because of its lesser commercial significance. Figure 9.17 is not a true equilibrium diagram because cementite is not an equilibrium phase. Graphite is more stable than cementite and under proper conditions cementite decomposes to form graphite. In ordinary steels this decomposition is almost never observed because the nucleation of cementite in iron, supersaturated with carbon, occurs much more readily than the nucleation of graphite. Thus, when the carbon is precipitated from solid solution of alpha or gamma iron, the resulting precipitate is almost always cementite or some other carbide and not graphite. Once cementite is formed it is very stable and may be treated for practical purposes as an equilibrium phase. For these reasons we can use Fig. 9.17 to predict phase changes in iron/carbon alloys undergoing slowly varying temperature cycles.

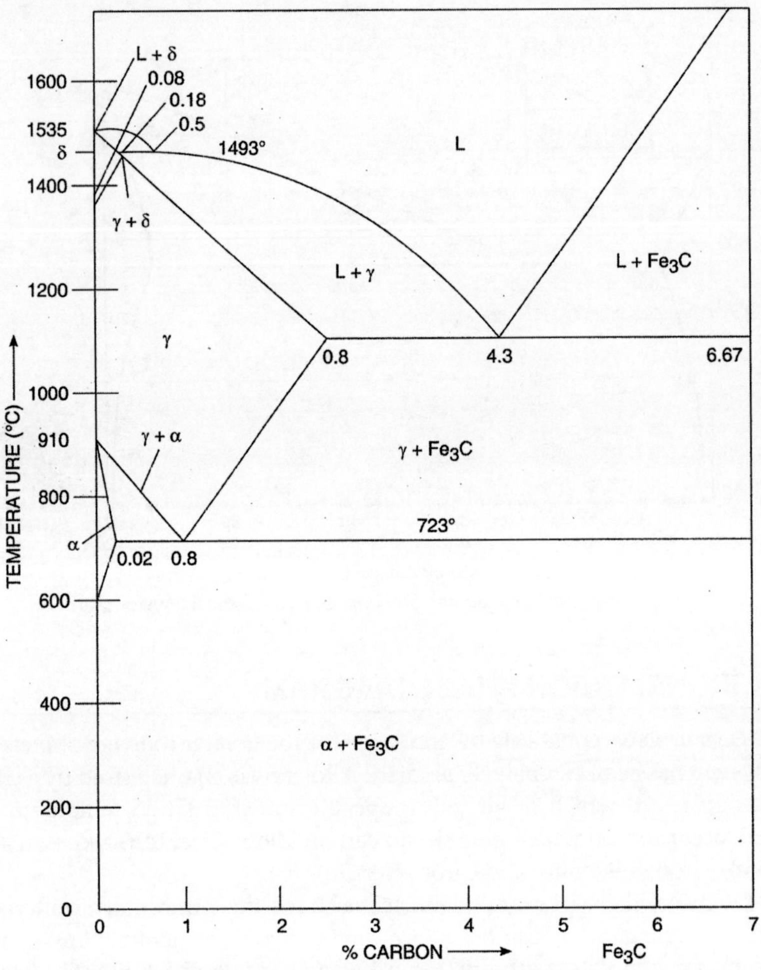

Fig. 9.17 The iron-carbon equilibrium diagram.

In Fig. 9.17 the composition is expressed on the X-axis as weight percentage of carbon. Pure iron melts at 1535°C. A 'peritectic' reaction occurs at 1493°C at 0.18% carbon as below :

$$\delta - \text{ferrite} + \text{liquid } (L) \rightarrow \text{austenite } (\gamma)$$

An eutectic reaction occurs at 1150°C with the eutectic composition at 4.3% carbon :

$$\text{Liquid} \rightarrow \gamma + Fe_3C$$

Iron carbon ($Fe - Fe_3C$) alloys containing 0–2% carbon are called steels. The approximate range of carbon content in various steels are :

(i) Mild steel – 0–0.3%
(ii) Medium carbon – 0.3 to 0.8%
(iii) High carbon – 0.8 to 1.4%

The peritectic transformation occurs at very high temperatures and in steels of very low carbon concentration. The peritectic transformation has only a lesser effect onthe structure of steels at room temperature. The face-centered-cubic solid solution or gamma phase is given the identifying name austenite. All composition containing 2% C pass through the austenite region on cooling from the liquid state to room temperature. Alloys in this group are classified as steels and composition above 2% C are classified as cast irons.

9.10 PHASE TRANSFORMATION

The microstructural products obtained by cooling a steel through the entecoid temperature (723°C at 0.8% C) are strongly dependent on the cooling rate. The structures which are mentioned in the previous paragraph are those obtained in slowly-cooled steels. The products corresponding to the change in temperature and composition are marked in the phase diagram as L, $L + \gamma$, $L + Fe_3C$ etc.

9.10.1 Austenite to Pearlite Transformation

If austenite is allowed to remain at 723°C the reaction takes place transforming the austenite to ferrite and pearlite. As described earlier, depending on the carbon percentage, the austenite will transform as pro-eutectoid ferrite and pearlite for carbon percentages below 0.8% and proeutectoid cementite and pearlite for carbon percentages above 0.8%. At 0.8% carbon, the austenite will transform completely to form an eutectoid structure which contains lamellae of ferrite and cementite. Fig. 9.18 illustrates all the three microstructures of austenite transformation products. If austenite is allowed to remain at constant temperature just below 723°C, the resultant product is the same as that given by iron-carbon diagram for a very slow, continuous cooling process. The stable phases below the eutectoid temperature are ferrite and cementite, called pearlite which consists of alternate plates of Fe_3C and ferrite with ferrite as the continuous phase.

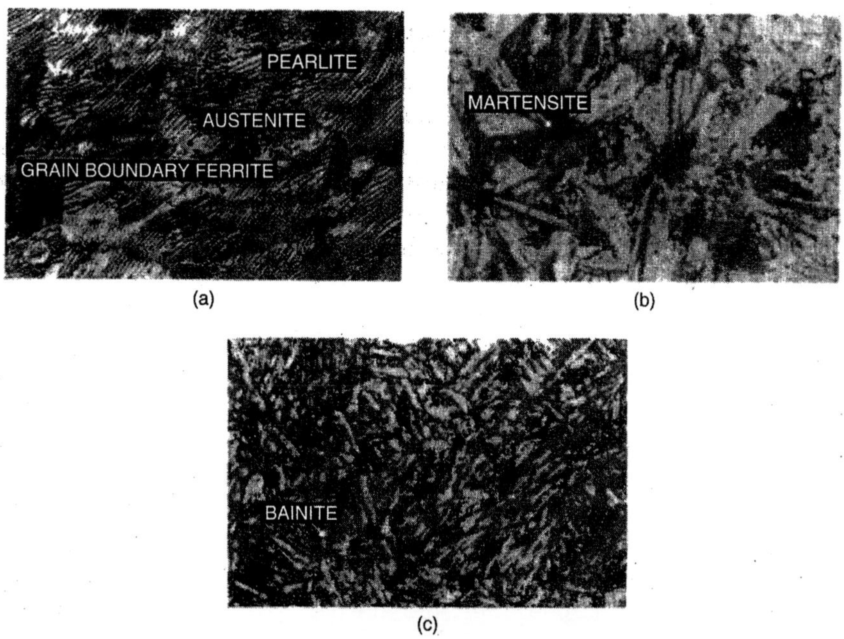

Fig. 9.18

9.10.2 Martensite

Martensite reaction is a basic time-temperature of phase transformation. Sometimes stresses may cause a martensite transformation. The FCC phase of austenite is converted into a body-centered tetragonal phase. Martensite looks like small lenses or needles or plates. Because of the manner in which martensite forms, no composition change occurs. The austenite/martensite transformation starts at a point of temperature designated as M_s and continues with the decrease of temperature down to a temperature referred to as M_f at which the reaction ceases even though some of the retained austenite may not still have got transformed. These temperatures are influenced by the composition of the steel and particularly by the carbon content.

9.10.3 Bainite

Bainite is difficult to be isolated in the structure of iron-carbon alloys because it overlaps the region of the pearlite transformation at temperatures of the order of 500°C. Steels transformed in this temperature range have structures both containing pearlite and bainite. Bainite is a mixture of ferrite and carbide (Fe_3C). The bainite reaction involves composition changes and requires diffusion of carbon. Another property of bainite reaction that differentiates it from martenstic reaction is that it is not a thermal reaction. The formulation of bainite requires time and when austenite is reacted to bainite isothermally, (at constant temperature) a typical S-shaped curve is obtained as can be seen in Fig. 9.19. The mechanism of bainite formation differs from that of pearlite and hence the resulting product does not have the alternative arrangement of parallel lamellae of ferrite and cementite found in pearlite. Bainite grows as plates in a typical martensitic characteristics.

9.11 TIME TEMPERATURE TRANSFORMATION (TTT)

Information of a very important and practical nature can be obtained from a series of isothermal reaction curves determined at a number of temperatures. Experimental counterparts of this curve are obtained by isothermally reacting a number of specimens for different lengths of time and determining the fraction of the transformation product in each of the specimen. A plotting of these data as a function of the reaction time gives the desired curve. From this curve the time required to start the transformation and the time required to complete the transformation can be obtained. Figure 9.19 shows the time relationships for the phases during isothermal transformation. The area to the left of the first S-curve corresponds to structures that are austenite. Any point to the right of S-curve represents a pearlite structure and is a mixture of two phases (cementite and ferrite). Between the two curves is the region of pearlite and austenite, with the relative ratio of these two constituents varying rom all austenite to all pearlite as one moves from left to right. Figure 9.19 shows the complete TTT diagram where the curves correspond to the transformation of austenite to bainite and martensite. Figure 9.20, because the pearlite and bainite transformations overlap in a simple eutectoid (below 0.8%) iron-carbon steel, the transition from pearlite to bainite is

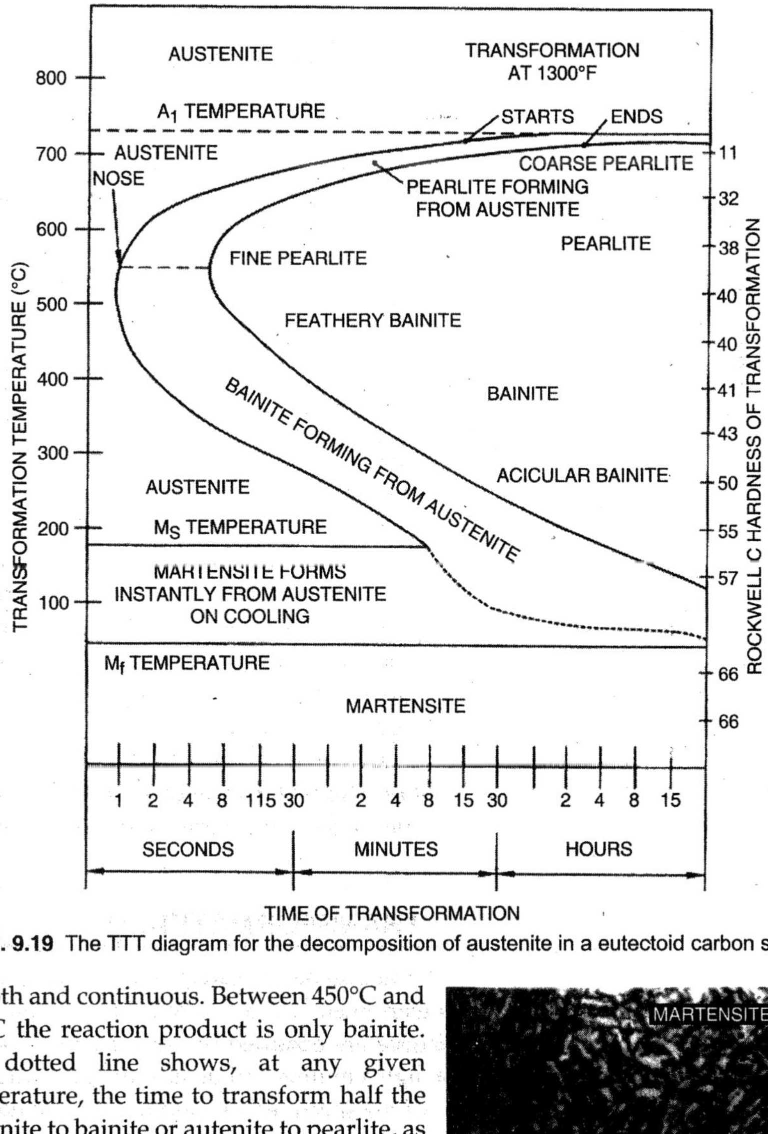

Fig. 9.19 The TTT diagram for the decomposition of austenite in a eutectoid carbon steel.

smooth and continuous. Between 450°C and 210°C the reaction product is only bainite. The dotted line shows, at any given temperature, the time to transform half the austenite to bainite or autenite to pearlite, as the case may be. The M_s temperature denotes the start of the martensite transformation and M_f denotes the completion of martensite transformation.

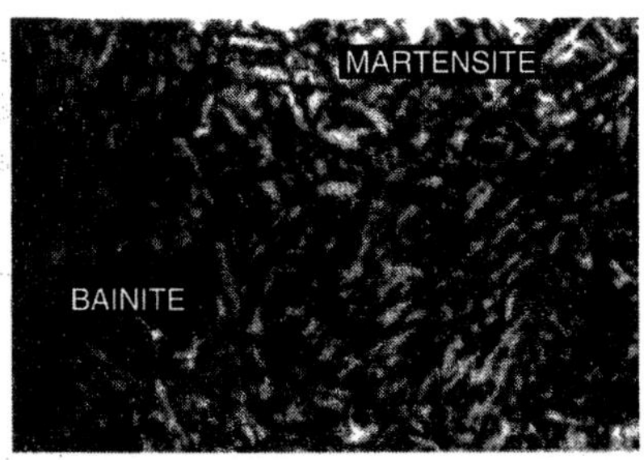

Fig. 9.20

9.12 CONTINUOUS COOLING TRANSFORMATION (CCT CURVES)

The isothermal transformation (TTT) diagram shows the time-temperature relationship for austenite transformation only as it occurs at a constant temperature but most heat-treatment and welding involve transformation on

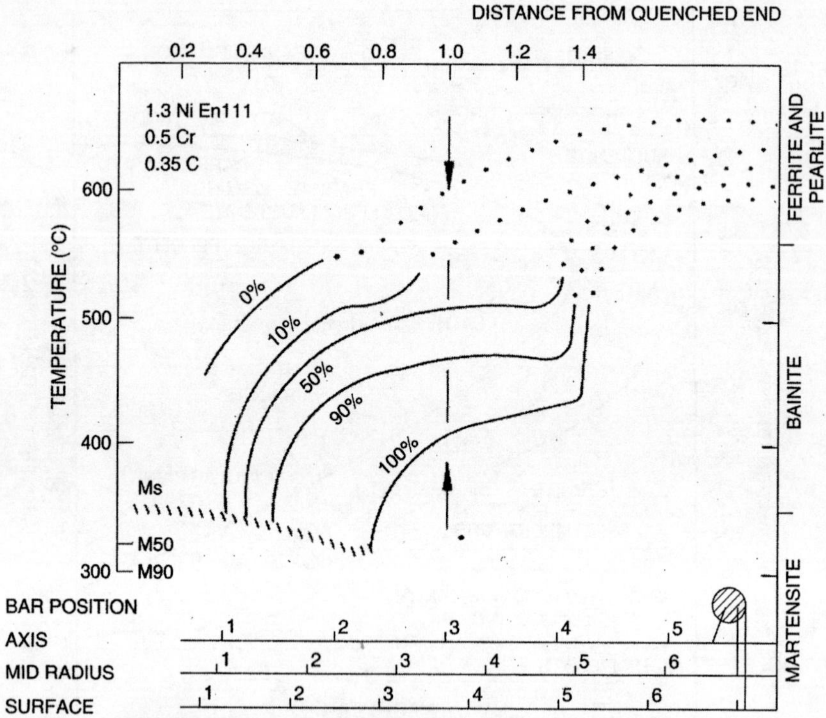

Fig. 9.21 Continuous cooling transformation (CCT) diagram.

continuous cooling. It is possible to derive from the TTT diagram another diagram which will show the transformation products under continuous cooling conditions. This is referred to as the CCT diagram or a continuous, cooling transformation diagram (Fig. 9.21). In the CCT diagram the 'nose' of the 'curves' has been moved downward to the right by continuous cooling. The critical cooling rate tangent to the nose of CCT diagram is shown as 120°C/sec. This is somewhat slower than the rate indicated by the TTT diagram. Therefore, the use of isothermal 'nose' times to determine the required cooling rates will lead to some error. However, the error will be only on the safe side indicating a slightly faster cooling rate than is actually necessary to form the martensite. Here, notice the absence of an austenite-to-bainite region in the CCT diagram. In this steel, the nose and the bainite are not formed in any appreciable quantity on ordinary continuous cooling. Depending on the cooling rate employed in practice, we can expect the different phases as per CCT diagram. Based on this the properties of the steel after heat-treatment can be derived. Such CCT diagrams have to be drawn for every type of steel.

9.13 CLASSIFICATION OF STEELS

Apart from carbon, a number of alloying elements canbe added to the steel. This together with variation in heat-treatment conditions can give rise to variety of steel grades having different mechanical, chemical and physical properties. The role of alloying elements in steel may be divided into a number of catagories. In the first catagory, alloying elements are deliberately added to steels that are not heat-treated to provide beneficial effect on strength, toughness and corrosion resistance. Typical examples are Mn, Cr, Ni, Mo and Cu. In a second group, alloying elements are purposely added to produce strengthening through heat-treatment. These include once again Mn, Cr, Ni and Mo with extra carbon. In the third catagory, alloying elements like V, Ti, Nb and Zr are added in small amounts to form carbides which give the strengthening effect. In contrast to deliberate additions, the impurity elements like S, P, As, Cu and Sb may be present which affect the strengthening and weldability of steels. Based on the chemistry, steel grades are classified into many types :

(a) Carbon steel.
(b) Carbon-manganese steel.
(c) High strength low alloy steel.
(d) Low alloy steel.
(e) Heat-resistant steel.
(f) Stainless steel.
(g) Tool steel.
(h) Electric grade steel.

The effects of major alloying elements in steel are given in Table 9.2.

Table 9.2

Element	Percentage	Effect of alloying
Aluminium	Small amount	De-oxidises, restricts grain growth, aids surface hardness in nitriding.
Boron	0.001 – 0.003%	Increases hardenability.
Carbon	≤ 1.5%	Increases strength and hardness. Decreases toughtness and ductility.
Chromium	0.5 – 2.0%	Increases hardenability and strength.
Chromium	4.0 – 18%	Increases corrosion and oxidation resistance.
Cobalt	Upto 12%	Increases hardness, improves cutting tool life.
Copper	0.1 – 0.4%	Improves corrosion resistance and strength.

❑❑❑

10

CHAPTER

WELDABILITY

INSIDE THIS CHAPTER

10.1 WELDABILITY OF CARBON AND LOW-ALLOY STEELS

Carbon and low alloy steels are classified into four groups based on their chemical composition, strength and heat-treatment condition. It is as follows :

(i) Carbon steels.

(ii) High strength low alloy steels.

(iii) Quenched and tempered steels.

(iv) Chromium-molybdenum steels.

10.1.1 Carbon Steels

This class of steel has carbon as the main alloying element carbon is present upto 1.0% maximum, Manganese (Mn) 1.65% and silicon (Si) 0.5% are other elements.

10.1.2 High Strength Low Alloy Steels

High strength low alloy steels contain a limited amount of carbon which is less than 0.2% and alloying elements like vanadium (V), niobium (Nb), titanium (Ti), copper (Cu), chromium (Cr) and zirconium (Zr) are also present. These alloying elements impart the required mechanical properties to the steel. The mechanical properties are better than carbon steels and the yield strength ranges from 36 kgf/mm^2 to 105 kfg/mm^2.

10.1.3 Quenched and Tempered Steels : (Q & T Steels)

Q & T steels contain less than 0.25% of carbon and other alloying elements like, Cr, Mo B etc. Q & T steels are strengthened by a heat treatment process, *i.e.*, quenching in oil or water to obtain martensite and then tempered to achieve the necessary mechanical properties.

10.1.4 Chromium-molybdenum Steels

Cr-Mo steels have Cr and Mo as the main alloying elements. Cr is present upto a maximum of 9% and Mo about 1% Cr and Mo give elevated temperature strength, corrosion and oxidation resistance. Due to good corrosion resistance these steels find good application in petrochemical and fertiliser plant and also for the elevated temperature applications in power plants. Welding of carbon and low alloy steels depends upon their chemical composition and their heat treatment condition. In addition, the section thickness is very important with regard to heat input, cooling rate and joint restraint. Cracking in weldment can occur when hydrogen is present and this type of cracking is called hydrogen induced cracking/cold cracking/delayed cracking/underbead cracking. The three major factors which cause cold cracking are :

(i) Presence of hydrogen.
(ii) Susceptible microstructure.
(iii) Tensile stress.

(i) Hydrogen

The sources for hydrogen are moisture in the electrode/flux/shielding gas or oil/dirt/moisture on the base metal. During welding the moisture dissociates into atomic hydrogen which dissolves in the molten metal and as the molten metal cools, the hydrogen diffuses towards the heat affected zone. This diffused hydrogen gets collected at the voids and imperfection in the metal matrix which later form hydrogen molecules. Because of this molecular hydrogen formation, enormous pressure is developed and finally results in cracking.

(ii) Susceptible Micro Structure

Welding is associated with rapid heating and cooling rates. Due to such cooling rates, hard micro-structures like martensite may be formed in the weld metal/heat affected zone. As the hardness in these regions increases, susceptibility to cracking increases. To avoid cracking, the formation of martensite is to be controlled and this can be done by controlling the cooling rates of the weldment. Preheat can control the cooling and thus the formation of martensite gets controlled. Thereby cracking is avoided. (Fig. 10.1)

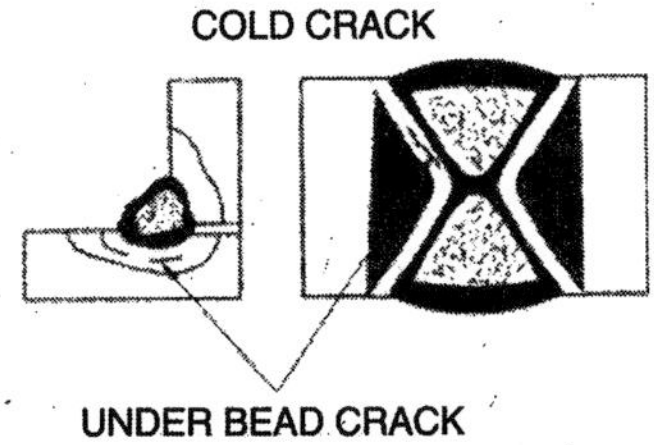

Fig. 10.1 Typical cold cracks in fillet and butt joints.

(iii) Tensile Stresses

Tensile stresses are introduced in the weldment due to the thermal contraction of weld metal/heat affected zone, the restraint due to the joint design and the fabrication sequence. The tensile stresses can be reduced by proper joint design and proper sequence of welding during fabrication. When the welding stresses in the base metal adjacent to the weld are high in a filler or a τ-joint, tearing can occur and this is called "lamellar tearing" (Fig. 10.2). This type of tearing is normally seen in steels which have higher level of inclusions. The tearing has a typical step like appearance just beneath the visible heat affected zone of the weldment. By reducing the amount of inclusions in the steel, the tendency to lamellar tearing is reduced. A proper joint design and the use of a low strength weld metal will minimise lamellar tearing.

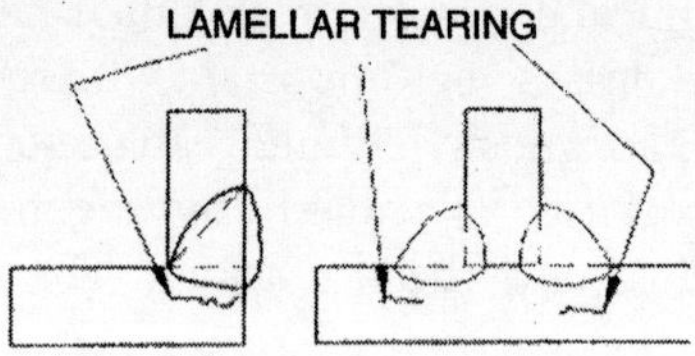

Fig. 10.2 Lamellar tearing in fillet joints.

Carbon steels are classified into four broad catagories based on their carbon content. The strength of the steel is a function of the carbon content. Tables 10.1 (a) and (b) gives the classification. During welding the steels are exposed to temperatures ranging from molten metal temperature to that of unaffeted based metal temperature, *i.e.*, room temperature or portion of the heated steel undergoes microstructural changes, resulting in changes of mechanical properties. The changes in micro-structure are dependent upon chemical composition of the steel and rate of heating and cooling. At certain cooling rates martensite may be formed. The amount of martensite formation and the increase in hardness depend upon the carbon content as well as the heating and cooling rates. When martensite formation leads to high hardness then the steel is more likely to cold crack at the heat affected zone/weld metal regions. The increase in hardness is primarily dependent upon the amount of carbon present—a higher carbon gives the higher hardness and susceptibility to cracking increases. The weldability of steel becomes poor as carbon content in steel increases.

Table 10.1(a) : Types of Carbon Steels and Their Weldability

Description	% C	Hardness	Weldability
Low carbon steel	0.15- Max.	60 HRB	Excellent
Mild steel	0.16-0.30	90 HRB	Good
Medium carbon steel	0.30-0.45	25 HRC	Fair Low hydrogen weldding practices required
High carbon steel	0.45-1.0	40 HRC	Poor Low hydrogen welding practices and preheat required.

Table 10.1(b) : Carbon Steels : Composition and Their Mechanical Properties

Steel	% C	% Mn	% Si	UST MPa	YS MPa
Structural	0.27	1.2		415-585	290
A 529	0.27				
A 36	0.29	0.8-1.2	0.15-0.40	400-550	250
Pressure vessles					
A 285 Gr A	0.17	0.90	0.15-0.40	310-450	170
A 299	0.30	0.90-1.40	0.15-0.40	520-650	275
A 515 Gr. 60	0.31	0.90		410-550	220
Pipe seamless					
A 106 Gr. B	0.30	0.29-1.06		410 Min	240 Min

Carbon has a dramatic effect on weldability of steels and even other elements to a certain extent influence the weldability of steel. The influence of other elements can be influence the weldabiilty of steel. The influence of other elements can be expressed in relation to carbon. The effect of alloying elements can be expressed in terms of carbon equivalent (CE). The international institute of welding has formulated an empirical formula to assess the steel's tendency of cold cracking. The formula is

$$CE = C + \frac{Mn}{6} + \frac{Cr + Mo + V}{5} + \frac{Ni + Cu}{15}$$

With CE of 0.40 or more, the hydrogen induced cracking tendency increases. Steels with low CE process excellent weldability. Based on the above formula the preheat temperature of a particular grade of steel can be determined. The details of preheat determination is given in IS : 9595. Steels with CE values of less than 0.35 normally do not require preheating or post heating, while CE value between 0.35 to 0.55 usually require preheating. Preheating and post heating is required when CE value is greater than 0.55.

10.1.5 Low Carbon Steels

Steels with carbon less than 0.15% and upto 0.3% can be welded easily by all fusion welding processes. Preheat is normally not required but when the section thickness is more than 25 mm. Then preheating may be required to obtain a acceptable weld joint. Mild steel contains carbon in the range of 0.25% to 0.3%. Fig. 10.3(a) shows the micro-structure of a typical mild steel. Welding of mild steel can be done by conventional welding processes. The cold cracking susceptibility is low due to the lower percentage of carbon. Preheating may be required when thickness more than 25 mm is to be welding but post heating generally is not required. The choice of electrode/filler metal is based on the strength requirement. Micro- structure of SMAW—weld metal deposit is shown in Fig. 10.3(b).

(a) Typical microstructure of mild steel

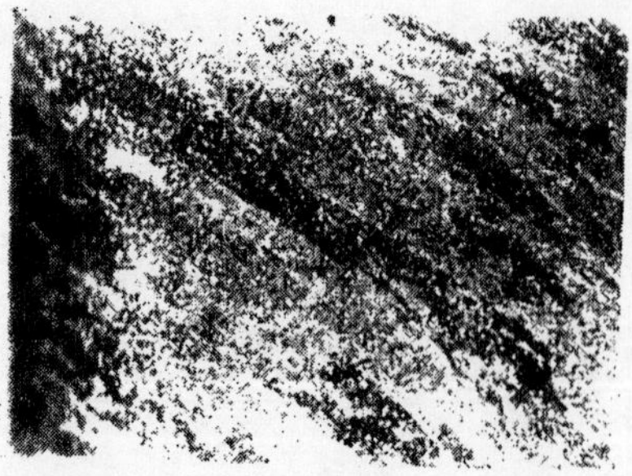

(b) Weld metal microstructure (columnar)

Fig. 10.3

10.1.6 Medium Carbon Steels

Medium carbon steels generally contain carbon between 0.3 to 0.5%. As the carbon content is more the amount of martensite formation in the weld metal/HAZ will be more and so the hardness will be more. With such condition the likelyhood of hydrogen induced cracking in the weldment is greater. Hence, low hydrogen welding practices have to be used for welding these classes of steels. Preheating and post heating may be required when the section thickness and the carbon content are more. While depositing the root run, dilution from the base metal having high carbon should be minimised otherwise cold cracking can occur. For components fabricated from this class of steels which are meant for impact dynamic loading, post-weld heat treatment is recommended as PWHT improves the toughness and reduce stresses in the weld.

10.1.7 High Carbon Steels

Welding of these carbon steels requires low hydrogen practices, as the possibility of cold cracking is high. Preheating in the range of 250°C-300°C is recommended to avoid cold cracking in the weld metal/HAZ. Post-heating may also be, required. The electrodes used should be properly baked. Welding of high carbon steels with austenetic filler metal is recommended when notch toughness is required filler metal is recommended when notch toughness is required. Welding of broken parts is to be done in the annealed condition and the weld procedures have to be developed after considering the service conditions.

10.1.8 High Strength Low Alloy Steels (HSLA)

Alloying elements lime Nb, Ti, V, Zr, Cu etc. are added to obtain a yield strength of 36 kgf/mm^2 or more, which is more than that of the conventional carbon steels. The mechanical properties (toughness) are far superior to those of carbon steels. As the carbon content is quite low, the weldability of the steel is excellent and these steels are designed to meet specific mechanical properties. Typical chemical composition and the tensile properties are given in Table 10.2(a). Micro-structure of a HSLA steel is shown in Fig. 10.4. Conventional are welding processor are used

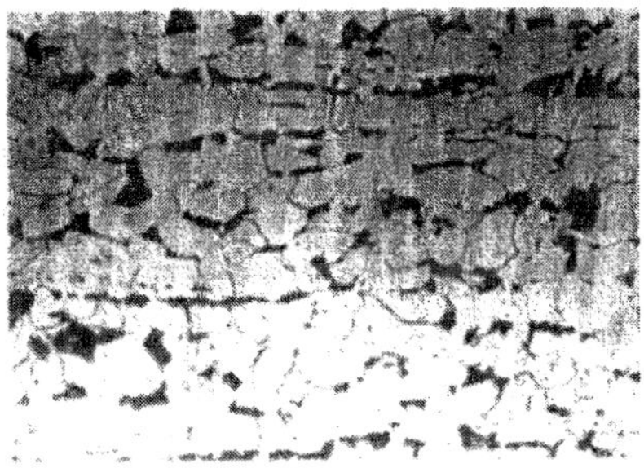

Fig. 10.4 Typical microstructure of HSLA steel showing fine grains.

for welding HSLA steels. The consumables used for welding HSLA steels are given in Table 10.2(b). Preheating may be required when section thickness is more than 25 mm, a preheat of around 100°C may be used. The filler metal selection is similar to that of carbon steels except that a low hydrogen electrode should be used.

Table 10.2(a) : High Strength low Alloy Steel—Composition and Mechanical Properties

Steel	% C	% Mn	% P	% S	% Si	% others	Mech prop. MPa
A 242 Gr 1	0.15	1.00	0.15	0.05			UST 430-480
							YS 280-345
A 588 Gr B	0.20	0.75-1.25	0.04	0.05	0.15-0.30	V 0.02-0.1	UTS 485
						Cu 0.25-0.4	YS 345
A 710	0.06	0.4-0.65	0.25	0.025	0.2-0.35	Ni 1.20-1.50	
						Cu 1-1.3	UTS 620
						Nb 0.02 Min	YS 585

Table 10.2(b) : Common Consumables for HSLA Steels

Steel	SMAW	SAW	GMAW	FCAW
A 242	E 7015	F7XX-EXXX	ER 70S-X	E7XT-1,4,5,6,7,8,11 % G
A 241	E 7016			
A 572 Gr 42	E 7018			
A 588	E 7028			
A 633 Gr A, B				
A 572 Gr 60 & 65	E 80XX-XX	F8XX-EXX	ER80S-XX	E8XTX-XX
A 633 Gr E				

Normally post weld heat treatment is not given but where dimensional stability is required or when there is a likelyhood of stress corrosion cracking during service, then post weld heat treatment is recommended.

10.1.9 Quenched and Tempered Steels (Q and T)

Quenched and tempered steels falls in one of the carbon, low alloy or any of other catagories of steels. A heat treatment which consists of austinitising, quenching and then tempering is administered. The yield, strength obtained is in the range of 36-100 kgf/min^2. The chemical composition and the mechanical properties are given in Table 10.3(a). The commonly used arc welding processes can be employed for welding of these steels. As the strength is quite high, low hydrogen practices have to be employed. Preheating should be used with caution as detrimental micro-structure may be formed under certain cooling rates which will have poor toughness. Preheating is recommended when the section thickness and joint restraint are high.

Table 10.3(a) : Chemical Composition and Mechanical Properties of Q and T Seels

Steel	% C	% Mn	% P	% S	% Si	S others	Mech prop. MPa
A 514 Gr A	0.12-0.21	0.70-1.0	0.035	0.04	0.20-0.35	Cr 0.40-0.25 Mo 0.15-0.25 V 0.03-0.08 Ti 0.01-0.03 B 0.005-0.005	UTS 895 YS 690
HY 80	0.12	0.60-0.90	0.01	0.015	0.15-0.35	Cr 0.40-0.70 Mo 0.30-0.60 Ni 4.75-5.25 V 0.05-0.10	YS 895

The heat input of the welding process is critical to obtain the optimal mechanical properties. The weld cooling rates are governed by the heat input, and this in turn decides the micro-structure of the HAZ. The heat input is very important to obtain a tough and crack free weld. The heat input is computed by the formula :

$$\text{Heat input (J/cm)} = \frac{A \times V \times 60}{S}$$

where, V = Welding voltage in volts.

A = welding current in amps.

S = welding speed in cm/sec.

The range of heat input applicable for welding is normally given by the steel manufacturer and this should be strictly follows as otherwise inferior quality weld may be obtained. The consumables used for welding for SMAW, SAW, FCAW etc. are given in Table 10.3(b). The presense of moisture in the electrode or on the base metal is highly deteremental to the weld as cold cracking may occur. Hence, the electrode and flux must be based in most cases and the base metal surface must be cleaned of oil, dirt and greese etc. If the electrode container is in the damaged condition, then the electrode must be reconditioned by backing. While welding stringer beads with minimum wearing must be used. Excessive wearing will reduce the welding speed and as a result heat treatment is not mandatory except where dimensional stability has to be maintained or stress-corrosion cracking is expected during service. PWHT has to be applied with caution. A type of cracking known as reheat/stress relief cracking may occur during PWHT at the HAZ of the weld. This type of cracking is associated with precipitation of alloy carbides at certain temperatures.

Table 21.3(b) : Consumables for Welding of Q and T Steels

Steel	SMAW	SMAW/GTAW	SAW	FCAW
A 514/A517	E1X01X-M	ER 1X0S-1	F1 XXX-EXXX-MX	E1X0TX-KX
HY 80	E1101X-M	ER110-S-1	F11XX-EXXX-MX	E11XTX-kX

10.1.10 Chromium-Molybednum Steels

Chromium-molybedenus steels contain Cr in the range of 0.5 to 9.0% and Mo in the range of 0.5–1.0%. Because of this, elevated temperature strength and corrosion resistance are obtained. Typical chemical composition of Cr–Mo steels are given in Table 10.4(a). Figures 10.5(a) and 10.5(b) shows the micro-structure of the 2.25 Cr-1 and 9 Cr-10 Mo steels. These steels find wide application in power plants, petrochemical and fertilizers industries. The common arc welding processes like SMAW, SAW and FCAW etc. can be used for welding this grade of steels. Fig. 10.5(c) shows the weld metal and HAZ microstructure of 2.25 Cr-1 Mo steel weld made with SMAW process. A preheat of 150-120°C is used to obtain crack free weld. The preheating temperature increases with increasing alloying content. As

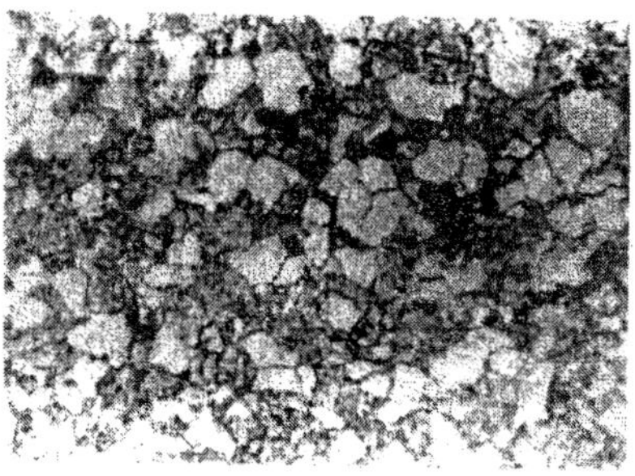

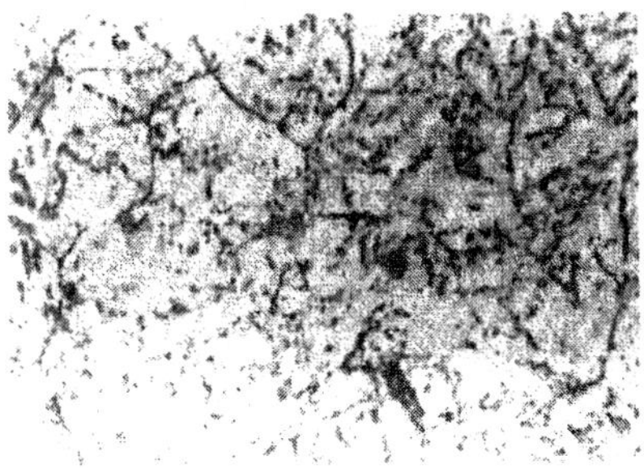

(a) Microstructure of 2.25 Cr - 1 Mo

(b) Microstructure of 9 Cr - 1 Mo

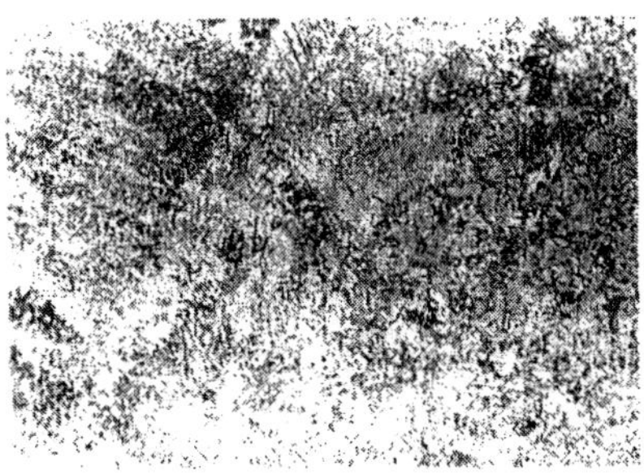

(c) Microstructure showing weld metal/HAZ regions (SMAW process)

Fig. 10.5

the steels are air hardenable low hydrogen welding practices arc to be used. The filler metal is chosen such that the chemistry of the filler metal is matching to that of the base metal. Table 10.4(b) gives the consumables used for welding of Cr-Mo steels.

Table 10.4(a) : Chemical Composition of Chromium-Molybdenum

Steel	%C	%Mn	%S	%P	%Si	%Cr	%Mo	Mech prop. MPa
A 213 T11 A335 P11	0.15	0.3-0.60	0.30	0.045	0.50-1.0	1.0-1.5	0.45-0.60	UTS 415-795 YTS 415-790
A 387 Gr 22 A 213 T22 A 335 P22	0.15	0.3-0.60	0.30	0.30	0.50	1.90-2.60	0.87-1.13	UTS 415-795 YS 210-690
A 213 T 9 A335 P9	0.15	0.3-0.60	0.03	0.03	0.25-1.0	8.00-10.00	0.90-1.10	YTS 415-795 YS 210-690

Table 10.4(b) : Common Welding Consumables for Cr-Mo Steels

Steel	SMAW	GTA/GMAW	SAW	FCAW
1.25 Cr-0.5Mo	E 80XX-B2/B2L	ER 80X-B2B2L	F8XX-EXXX-B2/B2H	E 8X Tx-B2/B2X
2.25 Cr-1Mo	E90XX-B3-B3L	ER90X-B3/B3L	F9XX-EXXX-B3/B4	E9XTX-B3.B3X
9 Cr-1Mo	E505-XX	ER505		E505T-1/2

10.2 WELDABILITY OF STAINLESS STEELS

Stainless steels are defined as iron-carbon alloys in which enough amount of chromium—a minimum of 12% is added to impart anti-corrosive property in specific environment (oxidising conditions). The stainlessness and corrosion resistance of these steels are attributed to the presence of a thin and imprevious passive oxide film on the metals surface. When exposed to atmosphere that removes the passive oxide film, e.g., reducing medium, corrosive attack takes place. The anti-corrosive mature of the stainless steel, attributed to the passiveness of the oxide film, is due to the presence of chromium uniformly dispersed in the roatrix adjoining the oxide layer. Based on room temperature and microstructure stainless steels are classified into three major groups :

(i) Austenitic stainless steel.
(ii) Martensitic stainless steel.
(iii) Ferritic stainless steel.

Since the three classes of steels differs widely in their weldability as well as physical and chemical properties, it is incorrect to discuss stainless steels as one class. Further the composition, properties and weldability aspects of three types of steels are explained.

10.2.1 Austenitic Stainless Steels

This is most widely used type of stainless steel. Because of austenitic matrix in these steels, these are used in applications involving :

(i) Low magnetic permeability (para-magnetic)
(ii) Good high temperature strength (creep and rupture strength)
(iii) Excellent toughness at sub-zero temperature (cryogenic)
(iv) High corrosion resistance (general and pitting)

Some of the standard types of wrought austenitic stainless steels (graded by American Iron and Steel Institute—AISI) with their compositions are listed in Table-10.5. There may be variations in the chemistry of steels from those given in Table-10.5 namely Extra Low Carbon (ELC) grades (C-0.03%) suffixed with the letter 'L' such as 304L and higher carbon grades (C-0.20%) suffixed with letter 'H' such as 347H. The 'L' grades have excellent corrosion resistance and cryogenic toughness. The 'H' grades are used for high temperature strength (creep and rupture resistance). Nitrogen added steels (grades suffixed with N) have higher Yield Strength (YS) and higher silicon steels (suffixed with B) have increased heat resistance. The free machining grade 303-Se with selenium and phosphorous is not easily weldable due to severe weld metal hot cracking caused by the presence of these elements in larger percentages.

Table 10.5 : Compositions of Some Wrought Austentic Stainless Steels.

Type	C	Mn	Si	Cr	Ni	Other elements
302	0.15	2.00	1.00	17.0-19.0	8.0-10.0	-
302 B	0.15	2.00	2.0-3.0	17.0-19.0	8.0-10.0	-
303 Se	0.15	2.00	1.00	17.0-19.0	8.0-10.0	0.20 P
304	0.08	2.00	1.00	18.0-20.0	8.0-10.0	0.15 Se min.
308	0.08	2.00	1.00	19.0-21.0	10.0-12.0	-
309	0.20	2.00	1.00	22.0-24.0	12.0-15.0	-
310	0.25	2.00	1.50	24.0-26.0	19.0-22.0	-
316	0.08	2.00	1.00	16.0-18.0	10.0-14.0	-
317	0.08	2.00	1.00	18.0-20.0	11.0-15.0	
321	0.08	2.00	1.00	17.0-19.0	9.0-12.0	
347	0.08	2.00	1.00	17.0-19.0	9.0-13.0	
348	0.08	2.00	1.00	17.0-19.0	9.0-13.0	

The austenitic stainless steels have generally low yield strength (25 kgf/mm^2) and medium tensile strength (around 50 kgf/mm^2) but posses high ductility (elongation even upto 70%) and taughness. They have high work hardening capability, high thermal expansion coefficient (50% greater than carbon steels) and very low thermal conductivity (one-third of carbon steel).

(a) Behaviour in Welding

These steels are considered the most weldable of the high alloy steels. Comparatively no trouble is encountered in making satisfactory welded joints if the inherent physical properties are given proper consideration to control warpage and distortions in this steels/sections. The major effort should be towards the following :

(i) Preserving the corrosion resistance of the steel at HAZ.
(ii) The prevention of hot cracking in or near the welded joints.
(iii) The control of second phase in weld (viz. delta ferrite)
(iv) Avoiding harmful embrittling effects in high temperature service.
(v) achieving tonghness in cryogenic service.
(vi) Preserving corrosion resistance in certain corrosive services.

(b) Corrosion Resistance of Welded Joint

Intergranular Corrosion (IGC)

When stainless steels are exposed to a temperature range of 450°C-750°C during heat treatment or welding, chromium based carbides are precipated from the anstentic matrix from the grain boundary regions. These carbides actually steal away the chromium from the grain boundary regions and thereby the corrosion resistance at these intergranular regions improvised in Cr gets reduced. When a condition of continuous Cr depleted path is established along the intergranular regions due to high temperature exposure, stainless steel is said to be sensitized. Sensitized stainless steel, when put in corrosive service, develops intergranular Corrosion (IGC), i.e., corrosion penetration along the grain boundary regions deeper into the material. (Fig. 10.6). Figure 10.7 shows AISI 304 steel exposed to

Fig. 10.6 Intergranular corrosion (IGC) in austenite stainless steel.

corrosive service which has undergone IGC attack, welding procedures must be tuned to avoid formation of sensitized zones in weld HAZ. To avoid 'weld decay' problem, extra low carbon varities such as 3041 or steels with strong carbide forming elements such as Ti (AISI 321) or Nb (AISI 347) can be used. These steels low carbon varieties such as 3041 or steels with strong carbide forming elements such as Ti (AISI321) or Nb (AISI347) can be used. These steels. These steels stabilise the Cr in its place so they are called "stabilized stainless steels". These steels are not be used in 'as-welded' condition in high temperature corrosive service since IGC

Fig. 10.7 Weld decay in HAZ of a AISI 304 steel.

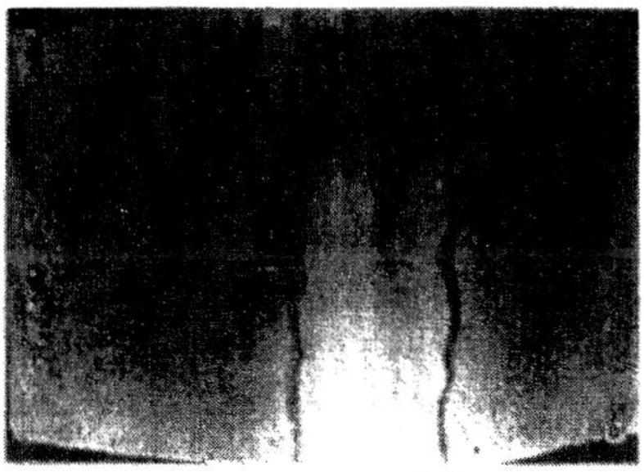

Fig. 10.8 Knife-line attack in a stabilised stainless steel.

attack takes place in the HAZ immediately near to the fusion line of weldment. On welding re-dissolution of all the carbides occur in this zone, freezing the carbon which ties up with Cr while in high temperature service, and resulting in-service fusion line. It is known as "knife-line" attack. (Fig. 10.8). There are many standardised IGC test to detect the presence of sensitization towards checking the raw materials HT condition or evaluating a welding procedure.

Stress-Corrosion Cracking (SCC)

This is another important type of corrosion attack on austenitic stainless steels which contain residual stresses due to welding fabrication. The failure appears as time multi-branched cracks usually in the HAZ of the weldment. To avoid SCC, a post weld heat treatment becomes necssary. Stress relieve at 850°C or solution annealing at high temperature in the range of 1050°C as given in Table 10.6 need to be done.

Table 10.6 : Recommended Ferrite Levels.

Application	Final values		What happens above max. value
	Min.	Max.	
Non-magnetic	0	0.1	Magnetisationpossilble
Severe corrosion (utea plants	0	0.5	Localised corrosion
Heat resisting vweldments	0	0.5	Inter-metalic types ppts.
Cryogenic application 196°C	0	0.5	Lowering of toughness and ductility
High ctemperature service up to 350°C-unstabilised weld meta	4	12	Lowering of toughness
Up to 450°C-stanbilised weld metal	6	15	Ductility and corrosion resistance
General high temperature	3	8	Sigma embrittlement
Duplex stainless steels	30	70	Toughness, ductility, corrosion resistance lowered

Hot Cracking in Welded Joints

Crater cracks, star cracks, micro-fissures and hot cracks are the various manifestations of the same basic kind of cracking namely 'hot-cracking' or when present in its early stage "micro-fissuring." Hot cracks and micro-fissures occur integranularly during the process of weld solidification. The formation of micro-fissures can be controlled by :

(i) Microstructure of the weld metal on solidification.

(ii) The composition of the weld metal especially the level of residual elements.

(iii) The amount of stress imposed on the weld.

Weld metal having a wholly austenitic weld structure is considered more sensitive to micro-fissuring than weld metals containing some ferrite (delta-ferric) in an austentic matrix. The Schaeffler diagram (Fig. 10.10) can be used for determining whether a specified weld metal composition will contains delta ferrite and if present, its approximate precentage with proper control of weld metal ferrite content, no difficulty of the weld metal micro-fissuring or hot cracking is encountered. Alloy content and residual impurity elements have a strong influence on the micro-fissuring susceptibility of the wholly austenitic weld metals. Micro fissuring can be reduced by :

(i) Small increase in carbon content.

(ii) Substantial increase in Mn content and

(iii) Incresing the nitrogen content.

(c) Delta-ferrite

Delta-ferrite cures the problem of hot cracking/microfissuring during the welding of stainless steels. Higher percentage of ferrite is not desirable in many applications. Hence, a limited range of ferrite is maintained in the welds depending on the service conditions required. The ferrite content of the weld depends upon its composition and the solidification rates. Fig. 10.9. Shows a typical austonitic weld metal microstructure, with delta-ferrite. The ferrite level can be predicted from the weld composition using empirical constitution diagrams. Those diagrams plot the ferritising power (represented by Cr or Cr equivalent) on the absciss a and the austinitising power (represented by Ni or Ni equivalent) on the ordinate.

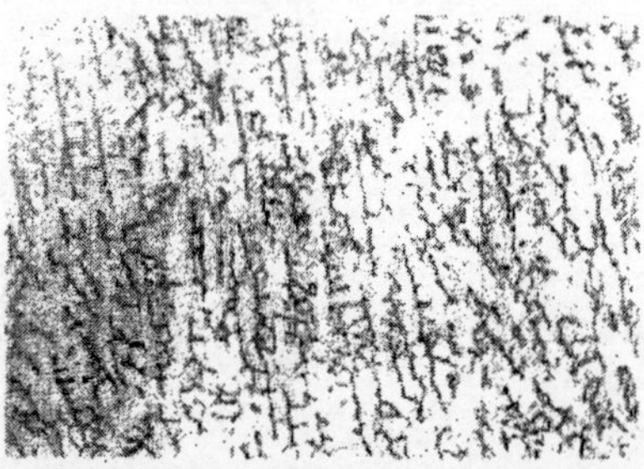

Fig. 10.9 Typical austenitic stainless steel weld metal microstructure with delta-ferrite.

The Schaeffler diagram (Fig. 10.10) plot the Cr eq. or Ni eq. from zero and indicates the ferrite percentages as measured by means of metallographic method. Delong later added to the austinitising power, the effect of nitrogen picked up during welding. Delta-ferrite being ferromagnetic, its content in an austenitic weld can be measured by means of magnetic methods. Instruments such as Magna-gauge, Ferritescope, Permascope, Severn gauge, Inspector gauge and Elcometer are available for this purpose. Though the magnetic devices are easire to use, these are only secondary measurement methods for knowing percentage ferrite. Hence, the Welding Research Council (WRC) of USA has defined an arbitrary standard-Ferrite Number (FN) to designate the ferrite content. At low

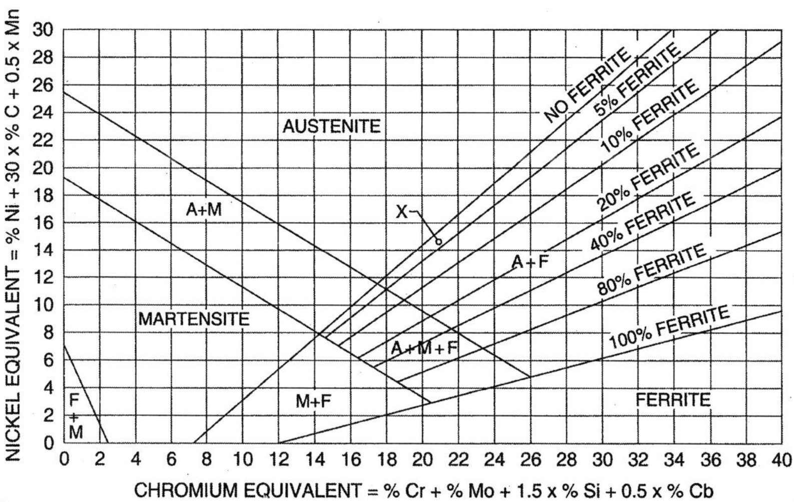

Fig. 10.10 Diagram by Schaeffler for determining approximate amount of delta ferrite in microsructure of stainless steel weld metal.

percentages (upto approximately 10%), the FN and the percentage ferrite are identical but they differ widely in higher percentages.

(d) Welding Procedures

Generally austenitic stainless steel donot require prcheating while welding, and in some cases preheating may be harmful causing increased chromium carbide precipitation (sensitisation) and warpage. However, preheating has been found to be useful in preventing stress-corrosion cracking. Interpass temperature control during welding of austenitic stainless steels is very important to avoid microfissuring. Usually, a maximum interpass temperature of 120 to 150°C is recommended. AWS classification of welding consumables used for welding various grades is shown in Table 10.7.

Tabel 10.7 : Welding Consumables for Austenitic Stainless Steels Type of Recommended AWS Filler Metals for Welding Steel.

SMAW	GMA GTA	FCAW SAW	FCAW PAW
302	E308	ER 308	E308T-X
304			
304L	E308L	ER 308L	E308LT-X
	E347	ER 347	E347T-X
309	E309	ER 309	E309T-X
309S	E309L	ER 309L	E309LT-X
	E309Cb		E309CbLT-X
310	E310	ER310	E310T-X
310S	E310Cb	ER310	E310T-X
316	E316	ER316	E316T-X
316L	E316L	ER316L	E316LT-X
321	E308L		E308LT-X
	E347	ER321	E347T-X
347	E308L		ER308LT-X
	E347	ER347	E347T-X
348	E347	ER347	E347T-X

Shielded Metal Arc Welding (SMAW)

In general, the welding correct for austenitic stainless steel covered electrode (with stainless steel core wire) is 10 to 15% lower than for the same size of carbon or low alloy steel covered electrodes. This is to avoid heat built up at the electrode tip due to poor thermal conductivity of austenitic steels. Weaving of the electrode during welding is not permitted/recommended. For optimum weld quality the bead width should not exceed two and half times the core wire diameter. The layer thickness in deep groove weld should not exceed 3 mm to avoid weld discontinuties welding is carried out with a lower arc length, i.e., lower arc voltage than for carbon/low alloy steels to avoid excessive chromium losses due to oxidation. Individual weld beads should be thoroughly cleaned with a stainless steel chipping tool and then with a stainless steel wire brush. Carbon steel tool or wire brush should not be used to avoid carbon pickup.

Moisture in the electrodes will cause porosity in stainless steel welds and hence electrodes must be stored properly to avoid absorption of moisture from the atmosphere. They should be dried in a baking over as per the recommendations of electrode manufacturer. Porosity may also be caused by improper cleaning of the base metal. Flux coating on SS electrodes and the resultent slag contain fluorides for fluxing action. The slag must be removed properly otherwise in certain environment of service, regions containing fluorides may get attacked causing service or pitting corrosion or SCC. Short arc length must be maintained during welding to avoid excessive loss of Cr by oxidation as well as to avoid too much nitrogen pickup.

Gas Metal Arc Welding (GMAW)

Argon shielding gas and DC electrode positive polarity can give good smooth welds provided the current is set in the spray tranfer region. The threshold current value for stable spray arc transfer is about 300 Amps for a 1.6 mm wire with the arc voltage in the range of 26-33 V. For welding this sheets, short-circuit transfer or pulsed spray transfer with inherent low heat input is to be employed for minimising distortion. The shielding gas used are argon or argon-oxigen (1% or 3%) mixture. Argon-oxygen can promote wetting action and arc stability by way of limited oxidation in the weld pool. While welding stainless steels, if the speed exceeds 750 mm/min, a trailing gas shield is required to protect the weld from oxidation. Using 100% CO_2 gas successful welding has been done in some parts though there may be a loss of Si and Mn and pickup of carbon in the weld metal.

Flux-cored Arc Welding (FCAW)

Direct current electrode positive polarity, FCAW is employed for stainless steel without gas shielding or with CO_2 or argan-2% oxygen mixture. The composition of electrodes is designed in such a way that there in enough latitude for the manufacturer to control the ferrite number of the diluted deposit. Long arc lengths, *i.e.,* high are voltage should not be used because they cause excessive nitrogen pick up and lowering of ferrite level in weld.

Gas Tungsten Arc Welding (GTAW)

Direct current electrode negative polarity using a droping volt-ampere characterstic or a constant current type of power source is employed for GTAW of stainless steel. Adequate supply of argon is required as a shielding gas to avoid nitrogen pickup and the resultant lowering of ferrite content is weld. Use of "consumable insert" for root welding of stainless steel can ensure uniform root penetration with high quality.

Submerged Arc Weding (SAW)

SAW process is used both for joining stainless steels and for cladding carbon/alloy steels with austenitic stainless steel. DC power sources are commonly used although AC can also be used.

Other Welding Processes

Plasma welding is very much suitable for joining this sections of austenitic steels using the 'melt-in' technique. High welding speeds can be used for continuously formed tubing. Circumferential seams of pipes can be made without backing rings or inserts. Almost all resistance welding processes can be used for welding of stainless steels viz. spot, seam, projection, flash, induction and upset welding. Friction welding offers added advantages of dissimilar welding of stainless steels with carbon, low alloy steels or non-ferrous metals. Other processes which are used for joining stainless steel are explosive welding, diffusion bonding and ultrasonic welding. Electron beam welding and laser beam welding processes give precise control during welding and laser beam welding processes give precise control during welding. High cooling rates produce narrow HAZ which is usually not sensitized for inter-granular corrosion.

(e) Post Weld Heat Treatment (PWHT)

For preserving corrosion resistance weldments of unstabilised stainless steels (viz. AISI 304, 308 etc.) are solution-treated to dissolve the grain boundary chromium based carbides and to redistribute Cr uniformly into the matrix. The solution-annealing temperatures of various grades are given below in Table 10.8. Local solution annealing is not recommended since it can produce a sensitized band slightly away from the PWHT region. Short soaking times are used to avoid grain growth. The weldment must be cooled fast (quenching for thick sections) after soaking to avoid slow cooling through temperature range of 450-750°C and resultent sensitisation. Care is to be exercised during PWHT to avoid excessive oxidation, scaling, sagging or warping.

Table 10.8 : Recommended Solution Anneal Temperatures for Stainless Steel

Grade	Temperature
301, 302, 303, 304, 304 L, 305, 308	1010-1120°C
309, 309S, 316	1035-1120°C
316 L, 317 L,	1035-1107°C
317	1065-1120°C
321 (for service temperature below about 450°C)	995-1065°C
347, 348	982-1065°C

The recommended stress-relief temperature for austenitic stainless steel is 850°C which is above the sensitization temperature range. This temperature can significantly reduce residual stress in weldments eliminating SCC problems. Lower temperature stress relief at 200-430°C has been found to improve dimensional stability by reduction of high peak stresses and reduce susceptibility to certain corrosion mechanisms.

10.2.2 Martensitic Stainless Steel

Martensitic stainles steels have moderate corrosion resistance, high strength upto 600°C, combined with good heat resistance. These steels exhibit ability to be heat treated to a very wide range of mechanical properties. These steels lack toughness and ductility in as hardened condition and generally require tempering to provide adequate toughness at the required strength level. Some of the steels classified as 'super 12% Cr steels' (12Cr-Mo-V-Ni with or without tungsten/cobalt) have excellent high temperature properties as well as high damping resistance etc. These steels are increasingly being used in power plants and refineary applications. Composition of typical weldable martensite stainless steel are given in Table 10.9.

Table 10.9 : Composition of Wrought Martensitic Stainless Steels

Type	C	Mn	Si	Cr	Ni	Other elements
403	0.15	1.00	0.50	11.5-13.0		
410	0.15	1.00	1.00	11.5-13.0		
414	0.15	1.00	1.00	11.5-13.0	1.25-2.50	
416	0.15	1.00	1.00	12.0-14.0		
420	0.15*	1.00	1.00	12.0-14.0		
422	0.20-0.25	1.00	0.75	11.0-13.0	0.50-1.00	0.75-1.25 Mo; 0.75-1.25 W, 0.15-0.3V
431	0.20	1.00	1.00	15.0-17.0	1.25-2.50	
440A	0.60-0.75	1.00	1.00	16.0-18.0		0.75 Mo

*Minimum value; all single values are maximum.

(a) Welding

While welding martensitic stainless steels, the HAZ gets hardned to a high degree which depends upon the carbon level in the steel. The steep high thermal gradients and the phase transformation to martensite can produce high internal stesses. The stresses, coupled with diffusible hydrogen from the weld metal, can easily produce the cold cracks in the untempered martensitic HAZ. Welding procedures with suitable precautions to avoid cold cracking are to be established and employed.

Two types of filler metals can be chosen for welding martensitic stainless steels—austenitic stainless steel filler wire or matching composition martensitic fillers. Using austenitic filler will produce a low Y.S. high ductile weld metal and it is advantageous from cold cracking point of view since the weld can yield plasticity during welding, minimising residual stresses. Also the weld metal does not allow diffiusion of hydrogen to the HAZ during cooling due to its lower diffusivity and higher solubility. Weldments made with these fillers are put in service in the "as welded" condition due to possible adverse effects owing to differences in physical properties.

While using martensitic steel filler wire, appropriate precautions to prevent H_2 induced cracking are essential. These included storage and handling of filler metal, cleanliness, slow cooling of HAZ during and subsequent to welding by controlling heat input, preheat and post heat, re-drying of electrodes prior to welding, non-exposure of consumable after re-baking to highly humid atmospheric condition for longer duration before welding is commenced etc. covered electrodes must be of basic or low hydrogen type to control H_2 pickup in the weld.

(b) Preheating

Normally the preheat range for martensitic stainless steel is 200-300°C. Preheating to Ms temperature range is not recommended since during cooling after welding, gross martensitic transformation can take place resulting in cold cracking. When the carbon content is above 0.2%, it is advisable to maintain the interpass temperature wherefrom the temperature is raised for the PWHT. Type 410 filler metal can be used to weld AISI 403, 410, 414 and 420 grades of martensitic stainless steels. Except for ER 410 NiMo, martensitic weld metals have low toughness in the as welded condition and PWHT needs to be done to improve the toughness.

(c) Post Weld Heat Treatment (PWHT)

Suggested PWHT requirements for martensitic stainless steels are given in Table 10.10. Two hypes of annealing treatments are applicable as PWHT for martensitic stainless steel-subcritical annealing in the range of of 595-760°C and full annealing in the range of 780-900°C. When the filler metal matches the base metal composition, a postweld Q and T-HT can be carried out to obtain homogeneous properties in the weldment.

Table 10.10 : Recommended PWHT for Martensitic Steels

Type	Subcritical anneal (°C)	Full anneal (°C)
403, 410 and 416	650-760	830-885
414	650-730	NR
420	675-760	830-885
431	620-705	NR
440A	675-760	845-900

NR—Not recommended.

(d) Welding of Typical Grade

Welding of the German grade X_{20} $CrMoV_{121}$ martensitic stainless steel is very unique in that the welding and HT cycles are combined (Fig. 10.11). Inspite of the stringent precautions required, the steel due to its very good high temperature properties is being extensively used in the world over. Free machining grades are weldable if the following precautions are followed :

(i) The process or filler does not supply hydrogen that can react with Se or S in base metal to cause porosity.

(ii) The amount of Se or S entering the weld must be made minimum by the control of dilution.

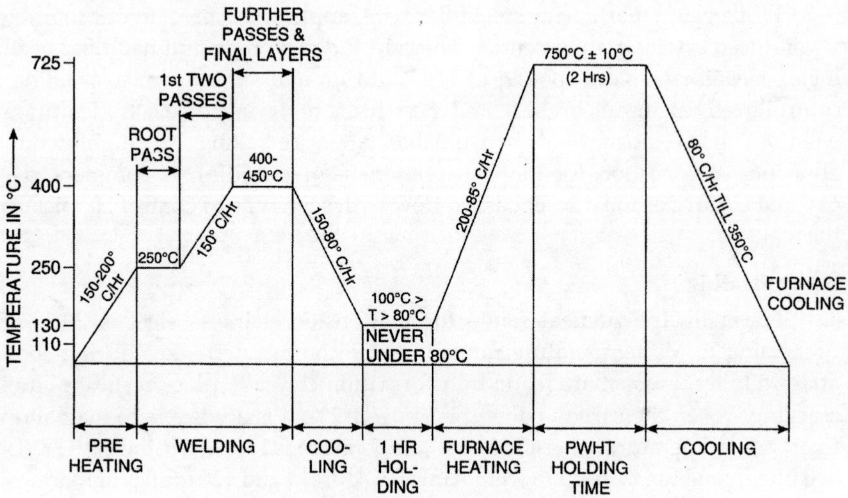

Fig. 10.11 Combined welding and heat treatment cycle for martensite X20CrMoV121 stainless steel.

(iii) The tolerance for these elements for hot cracking is to be improved by high ferrite (35-45%) filler metals. ER 312 (29 Cr-9 Ni) austenitic S.S. filler is a good choice for providing high ferrite weld metal.

(e) Welding Processes

Other than GTAW and SMAW processes, spot welding has been successfully carried out on lower carbon grades of types 403, 410, 414 and 416 grades without PWHT and high carbon types 420, 422 and 431 with a PWHT. Propensity for cold cracking in HAZ increasing with the 'as welded HAZ hardness' and hence on the carbon content of the steel.

10.2.3 Ferritic Stainless Steels

Ferritic stainless steels are non-hardenable by heat treatments. High ferritic stability is provided by the additions of the elements such as Mo, Ti, Al or Cb which inhibit the transformations to austenitie on heating. Various grades of ferritic stainless steels are given in Table-10.11. Earlier, ferritic stainless steels (AISI 430, 442, 446) were made with only Cr as a ferritiser. These had low toughness and in the as welded conditions lacked IGC resistance. Later, powerful ferritising elements such as Al and Ti were added lowering the chromium and carbon in these steels which were less costlier and had better fabrication characteristics though the toughness remained low. Recently, extra low interstitial carbon and nitrogen ferritic stainless steels (ELIFSS) have been produced. They contain powerful ferritiser such as Ti and Cb and hence IGC resistance is excellent in the 'as welded' condition. They possess good toughness and SCC as well as pitting resistance in chloride environments.

Table 10.11 : Composition of Wrought Ferritic Stainless Steels

Type	C	Mn	Cr	Si	Others
405	0.08	1.00	1.00	11.5-14.5	0.1-0.3 Al Ti min. 6xC
409	0.08	1.00	1.00	10.5-11.75	
429	0.12	1.00	1.00	14.0-16.0	
430	0.12	1.00	1.00	16.0-18.0	
434	0.12	1.00	1.00	16.0-18.0	0.75-1.25Mo
436	0.12	1.00	1.00	16.0-18.0	0.75-1.25Mo; (Cb+Ta) min 5xC%;
442	0.20	1.00	1.00	18.0-23.0	
446	0.20	1.50	1.00	23.0-27.0	0.25 N
29.4-2	0.010	0.30	0.20	28.0-30.0	2.0-2.5 Ni; 3.5-4.2 Mo; 0.02N; 0.15Cu

(a) Metallurgical Characterstics

When the interstitical content (C or N) is more or when Cr is less than 21% the ferritic stainless steels transform to austenitic structure, especially at the grain boundaries during heating in welding or heat treatment and on cooling transform to hard and brittle martensitic structure. This grain boundary envelope of martensite reduces the "as welded" ductility and tonghnes of the steel which can be regained by postweld annealing at 760-810°C. Ferritic grades with Cr > 21% develop chromium carbides at the grain boundary on heating to high temperatures leading to reduced IGC resistance. Annealing treatment at 760-815°C followed by rapid cooling can restore the corrosion resistance. Severe grain growth and the attendant loss of ductivity and toughness are the major problems during welding of ferritic stainless steels.

(b) Weldability

A preheat of 150-250°C is used for welding ferritic stainless steels due to the poor toughness of the joints which can contribute to cracking under restraint. The interpass temperature should be kept low, commensurate with the preheat level. Corrosion resistance of grades AISI 405 and 409 and low interstitial grade AISI 444 is not affected by welding and hence can be used under as welded conditions. Types AISI 430, 432, 442 and 446 are to be annealed to restore corrosion resistance after welding. Three types of filler metals are used for welding ferritic stainless steels :

(i) Austenitic (E 309 or E 312),
(ii) Matching ferritic steel, or
(iii) Nickel base alloys (E Ni Cr Fe_3)

For low interstitical grades matching composition fillers are available as bare wires. Austenitic and nickel base alloys are also used for making sound dissimilar joints with mild/low alloy/stainless steels.

(c) Welding Processes

SMAW, GTAW, GMAW, SAW and resistance seam, spot and flash welding processes are usually employed for joining these steels which can be welded also by the upset, friction, electron beam, plasma arc, laser beam welding processes.

10.3 WELDABILITY OF CAST IRON

In practical aspects it is a known fact that some metal shapes are always produced economically by casting process. Castings of various shapes and thicknesses, many a time, require weld repairs to be done on them either by foundary salvaging or for reclamation while in service.

10.3.1 Cast Iron

Like steel, cast iron is basically as alloy of iron, carbon and silicon phosphorus is also present in high percentages in cast iron compared to steel. Cast iron certain nearly 2 to 4.3% carbon whereas steel contain a maximum of 2% carbon. In steel the carbon is present as a solid solutin in iron. Cast iron is thus essentially a high carbon steel with excess carbon dispersed as seperate independent masses throughout the base metal matrix. When the molten alloy of iron and carbon freezes, carbon can both seperate out as free graphite in the form of flakes or nodules and it can also form a chemical compound called iron-carbide (Fe_3C). Free graphite separation process is known as the equilibrium process which is promoted by high percentages of carbon and silicon. Massive iron-carbide formation upon solidification is caused due to the non-equilibrium process (metal stable reaction) which is promoted by the lesser percentage carbon and silicon contents, and also due to faster cooling rates, compared to equilibrium conditions. The liquid metal which is solidified under equilibrium conditions will form various free graphite irons mainly grey cast iron and that which follows the metastable process on fast cooling will form white castiron.

When cerium or magnesium is added to grey iron metal after de-sulfurising it sufficiently, the graphite will take the shape of noduls. This iron is called "spheroidal graphite iron" (SG iron) or nodular iron or "ductile iron". The tensile strength and elongation of nodular iron is comparable to those of steels. When white cast iron is annealed for several hours at about 950°C it becomes malleable iron. White cast iron is extremely weak and brittle. Malleable iron can be bent or even formed. Table 10.12 and 10.13 respectively give the typical basic composition of cast iron types and their mechanical properties.

Table 10.12 : Chemical Composition (Typical) of Cast Irons

Iron Type	C	Si	S	P	Mn	Fe
Grey	2.0 - 4.0	1.0 min	0.3	0.5	1.2 max	Balance
Malleable	2.0 - 3.0	0.9 - 1.8	0.2 max	0.2 max	0.24 - 1.25	Balance
Nodular	3.2 - 4.2	1.8 - 2.8	0.03 max	0.1 max	0.8 max	Balance
White	2.5 - 4.0	0.4 - 1.6	0.15	0.4	0.3 - 0.8	Balance

Table 10.13 : Mechanical Properties (Ranges) of Cast Irons

Iron Type	Tensile strength	Yield strength kg/mm^2	Brinell hardness kg/mm^2	% Elongation
Grey	20 - 25	—	174 - 240	—
Malleable	36 - 40	20 - 26	140 - 280	3 - 15
Nodular	40 - 80	24 - 48	140 - 330	2 - 17

10.3.2 Uses of Cast Iron

Grey iron is used particularly for conditions of static loading such as farily stressed process equipment, domestic machinery and other applications where long life is expected duly resisting the heat atmospheric corrosion and mechanical wear. Malleable iron which is soft and ductile can withstand upto 15% elongation. It is applied where shock loads are encomtered such as universal joint yokes, automation transmission gears, crank shafts, brake drums, connecting rods, hand tools and hardware. Nodular iron displays steel like properties and can be substituted for steel components of complicated shapes.

10.3.3 Welding of Cast Iron

Welding of cast iron is apparently a common place affair. Most os the welds in almost all services, even those for, like heat resistance and corrosion resistance etc., have fairly withstood the requirements. Cast iron repair weld in moving and load bearing components have also proved their due worth. But there are also cast of repeated failures of the required joints. This may be due to lack of scientific understanding of the welding of cast iron. Actually there are no codes available for welding of cast iron other than the Do's and Dont's given by the consumable manufactures alongwith their commercial catalogues. Code AWS 5.15-82 specify the welding rods and covered electrodes for cast iron.

Many researchers have independently attempted to weld cast iron components using different welding process, consumables and techniques including specific heat-treatments. But there are no scientifically worked out and listed out procedures available yet for welding of cast iron jobs. This may be because of the reason that certain important procedural essentials like preheat cannot be prescribed for castings. The very concept of casting is thought of where the uniform thickness design does not apply. Preheat is meaningful when the thichkess is necessarily uniform. Cast iron welding requirements can be generally classified as detailed in Table 10.14. Prescription of process, procedures and consumables is based on the type of job requirements, *i.e.*, the premium attributed to a particular weld repair.

Table 10.14 : Classification of Cast Iron Welding Requirements.

Job requirement	Situation	Cast iron involved	Process and procedure
1 Repair of broken machine frames and covers etc.	Low strength Low cost Low thickiness	Grey cast iron mainly.	1 All arc welding using nickel and ferro nickel rods. 2 Braze welding (copper alloys). 3 Oxy-acetylene welding (cast iron rods) with low or no preheat. Extra reinforcement strips of steel or cast iron are welded on sometimes.
2 Foundry reclamations of fresh castings.	Medium strength High thickness	All types	1 Oxy-acetylene welding as above, with higher preheat and post heat. 2 Arc welding using nickel and ferro-nickel rods and wires, with higher preheat; peening and even stress relieving is done.
3 Repair of broken load bearing machine parts, pressure and heat containers of process plants.	High strength; cost of repair insignificant compared to production loss due to breakdown.	All types	1 All arc welding using nickel and ferro-nickel rods and wires with moderate preheat; peening, post-weld, or even interpass thermal stress relieving is done; full NDT is prescribed.

Cast iron welding for foundary reclamation is based on the ratio of the volume of defect to the volume of casting. If the defect volume is between 5 to 10% the casting is reclaimed through welding. Defect volumes in excess of 10% usually warrant rejection and remelting. The mechine frames and covers which are damaged due to transportation are required with cheaper oxy-acetylene braze welding using low heat inputs. Load bearing members and pressure parts are welded with more scientific considerations due to their critical service requirements.

10.3.4 Scientific Consideration

A high carbon "tool-steel" type base metal matrix of cast iron is welded with maximum possible prcheating and post heating to prevent hydrogen induced cold cracking at the heat effected zone of the weld. Unfortunately, when a higher heat is applied to cast irons the areas around the nodules or flakes become rich in carbon (upto 4.3%) and that causes "entectic" melting at lower temperatures arount 1130°C. Prolonged preheat and intense welding are heat promote sprawling of these carbon rich liquid zones to meet each other as shown in Fig. 10.12. This continuous liquid film will freeze as carbide (Fe_3C) or in other words as white iron. White iron is brittle and weak and crack just as the welding in going on. A steel electrode, low in carbon dilutes and brings doron the carbon ercentage of the penetration and promotes white iron, just under the bead along the fusion zone. Thus at one hand, preheating is required to soften the heat affected zone,

Fig. 10.12 Microstructure showing sprawling carbon-rich liquid zones.

martensite and at the other end, any excess heat applied to the joint will cause the fusion line (where the weld deposit just mixes with the cast iron base metal) to readily transform into a brittle carbide lack. *Weldability of castiron lies within these two opposing conditions*. Therefore,

(i) a moderate preheat is to be applied around the weld area.

(ii) the welding arc heat is to be bare minimum.

10.3.5 Preheating

When preheating is to be administred it is done in many ways. Elevating the temperature of the work-pieces immediately before a welding or cutting operation is called preheating. This treatment is employed in cutting and welding for a number of reasons and the principle ones are the following :

(i) To retard the cooling rates and avoid cold cracking in the heat affected zones of hardenable steels.

(ii) To increase the toughness of the weld joint and improve its ability to withstand adverse service conduction involving impact loading or low temperatures.

(iii) To alleviate the effects of hydrogen which enters the weld metal and the heat affected zone of the base metal.

(iv) To reduce residual stresses (internal stresses from shrinkage, phase transformation and reaction to restraint) to a desired low level.

(v) To minimize shrinkage and distortion.

(vi) To produce particular mechanical or physical properties in the steel of which the weldment is constructed.

Arc welding of cast iron is successful only if the stresses built up in the weld joint are to be relived by rigorous and timely peening. The favourable effects of hot peening (Fig. 10.13), short bead laying, *i.e.*, 50 mm and staggered welding (welding intermittently at different location within the joint) cannot be overstressed in cast iron welding. The absence of any of these can cause hot cracking while welding. A stress relieving heat treatment, after all the welding is over, is given many a time. Sometimes stress relieving heat treatments are given carefully even after depositing a few weld layers to avoid cracking.

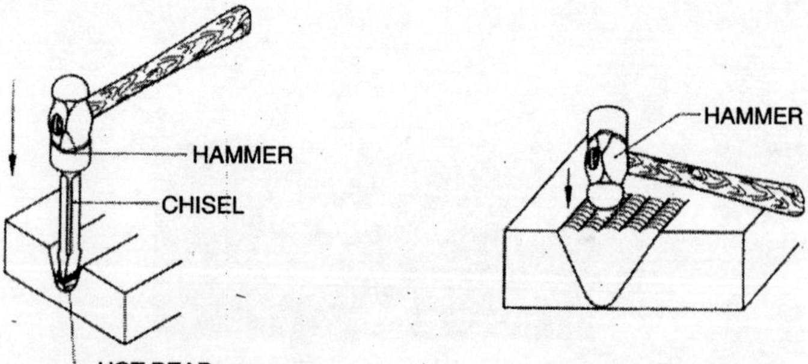

(a) NARROW GROOVED (b) OPEN BEADS

Fig. 10.13 Peening.

10.3.6 Welding Procedure

Buttering of joint faces with ferro-nickel filler and further deposition with ferro-nickel cum pure nickel by manual metal arc welding or MIG welding are the most popular cast iron welding procedures in use. (Fig. 10.14) flux-cored wire welding using Ni rod 'Fc55' is also strongly recommended by its manufactures, International Nickel Co., USA for large weld repairs. Oxyacetyline braze welding is still the cheapest welding method for cast iron, where strength, leak tighness and colour mismatch are not essential criteria. Copper-zinc alloy braze welding filler gives a cheaper deposit and is feasible for such quick and low strength applications. Oxyacetylene welding with cast filler rods is mostly resorted to in foundary repair and reclamation as the same iron melt can be poured as filler rods. Very high preheating 600°C, large scale deposition and slow cooling to avoid fusion line, white iron formation can possibly be achived on a foundary shop. In a shop repair situation, low current metal arc welding using mickel/ferro-nickel filler wires with moderate precheating and peening is the state of art technique.

LAYERS 1,3,5 PURE NICKEL (AWS E Fe_Ni_CI)

LAYERS 2,4,6 & SIDE WALLS FERRO NICKEL (AWS E Fe_Ni CI)

Fig. 10.14 Buttering with ferro nickel or pure nickel electrodes.

10.3.7 Braze Welding of Cast Irons

Braze welding is also resorted to in the welding of cast irons. The largest application of the braze welding is in the repair of grey iron castings. Conventional braze welding is done using an oxy-fuel gas welding torch or more often an oxyacetylene welding torch. Special fillers are required for use in this process of welding cast irons. The commonly used filler metals are brasses containing 60%

copper and 40% zinc. However, small quantities of tin, iron, managese and silicon are added to improve the flow characteristics, decrease volatilization of the zone etc. Table-10.15 gives the details of these fillers. For braze welding, fluxes are proprietary compounds specially developed for specific base metals and filler metals.

Table 10.15 : Filler Metals for Braze Welding Cast Iron using OAW Process

Descri-ption	AWS		Chemical composition of filler metals							
	Specifi-cation	Classifi-cation	C	Si	Mn	Ni	Mo	P	S	Ce
Gray iron	A5.15	RCI	3.25 - 3.50	2.75 - 3.00	0.60 - 0.75	Trace	Trace	0.50 - 0.75	0.10 max	—
Alloy gray iron	A5.15	RCI-A	3.25 - 3.50	2.00 - 2.50	0.50 - 0.70	1.20 - 1.60	0.25 - 0.45	0.20 - 0.40	0.10 max	—
Ductile iron	A5.15	RCI-B	3.25 - 4.00	3.25 - 3.75	0.10 - 0.40	0.50 max	—	0.05 max	0.03 max	0.20 max

(a) General Process Applications

The greatest use of braze welding is in the repair of defective or broken cast iron and steel parts. As large components can be repaired in-citu, significant cost saving could be realised through this method.

(b) Metallurgical Considerations

Braze welding reduces two of the most serious problems encountered during the repair of cast irons viz. The formation of martensitic cast iron and the contraction of casting during welding. The first is lessened to a degree because the broze filler melts at a lower temperature which inhibits the formation of martensite. The contraction problem is partially overcome by lower temperatures and partly by the yielding to bronze as cooling takes place. The bond between the filler and base materials is much the same as obtained in conventional brozing. The base metal is heated to a temperature which is sufficient to allow the wetting of the surface by the molten filler thus producing a metallurgical bond. Cleanliness of the surface is very important to obtain wetting. Wetting is the prime action that initiates bonding. Atomic diffusion takes place between the filler and the base material in a narrow zone at the interface.

(c) Braze Welding Procedure

Generally, fixturing will be required to weld parts together in the alignment for braze welding. However, fixturing may not be necessary in repairing cracks and other defects in cast iron parts unless the part is broken and separated.

(d) Joint prepration

Joint designs are similar to those used in oxy-acetylene welding. Single or double 'V' grooves with included angle between 90-120° are recommended for thickness above 2.4 mm. Square grooves may be used for lower thickness. The surfaces to be wedded should be cleaned thoroughly of dirt, dust, grease, paint, oil etc. Graphite smears left after machining should also be removed. This is accomplished by heating the casting to a dull red colour and wire brushing the parts when they attain a black colour. Castings soaked in oil should be "furnaced" or flame heated to burn the oil. Further wire brushing may be done to remove any residue. The most effective method to remove graphite from the broken or the machined surfaces is by resorting to cleaning by immersion in an electrolytic molten salt bath.

(e) Preheating

Preheating may be required to prevent cracking in large cast iron parts. It is dependent on the size of the parts to be welded. Preheating may be applied either locally or throughout the part. Temperatures in the range of 425°C-480°C are recommended for cast irons. Slow cooling after welding will help minimize the developement of thermal stresses in the casting.

10.3.8 Special Techniques for Welding Cast Iron

Wide grooved joint designs are typical in cast iron (Fig. 10.15) joining. Welding of each cast iron job calls for a unique approach. In metal arc welding procedures, the application of preheat requires utmost care. Large castungs will induce cracks at welds if sufficient preheat is not applied. Castings of low thicknesses may crack under excess preheat. Therefore, preheat is not recommended as a rule for welding

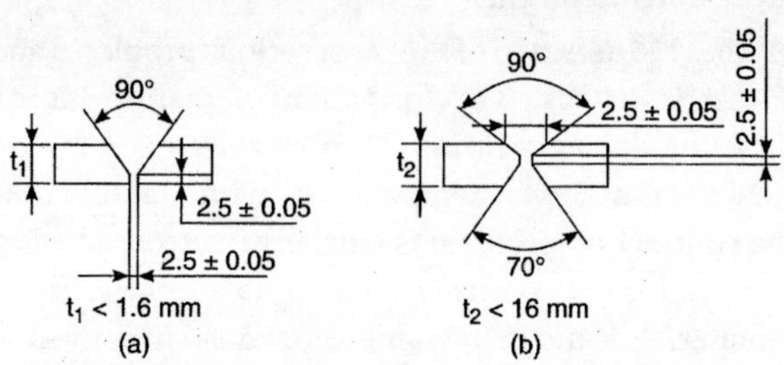

Fig. 10.15 Joint designs for full penetration butt welds.

low thicknesses castings. Extra reinforcement members could be welded on extremally at convenient locations to aid the joint for carrying loads after repair has been carried out. These can be included as part of the buttering layer as shown in bolt-reinforced butterings. (Fig. 10.16). Heating of adjacent sections can be helpful

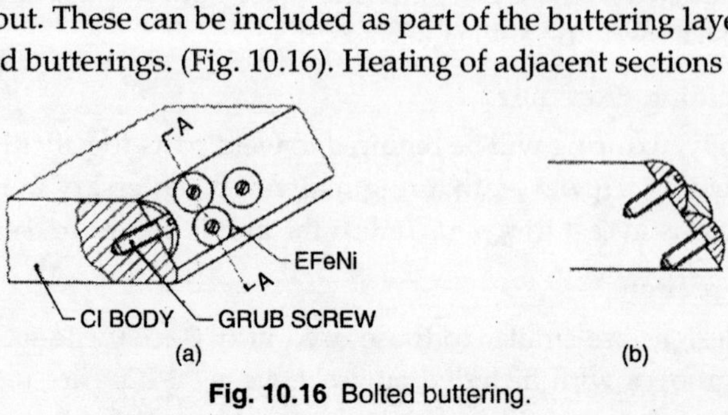

Fig. 10.16 Bolted buttering.

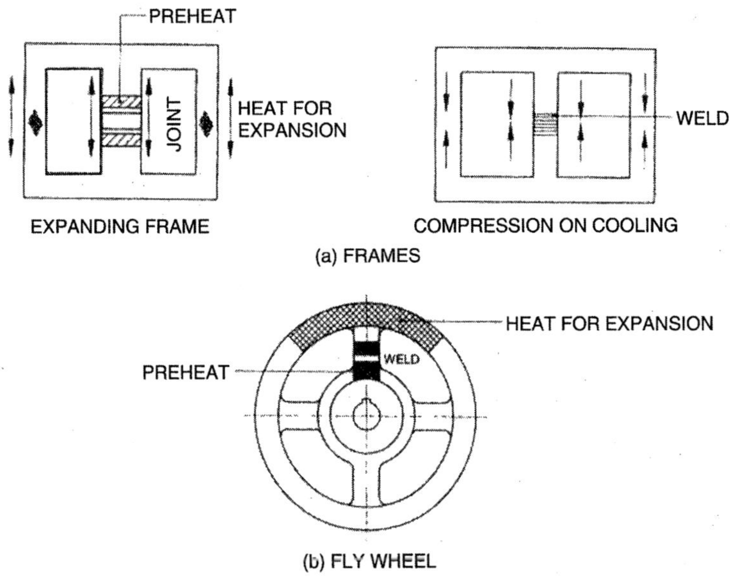

Fig. 10.17 Preheating techniques.

in expanding the weld joint gap before welding and later inducing a compressive force on to the gaps as shown in Fig. 10.17.

SUMMARY

Ferro-nickel manual metal arc welding, sometimes with alternate passes of pure nickel, is presently the most common cast iron welding procedure at the shop floor. A preheat of 150-250°C seems to be an optimum value for thicknesses ranging from 8 to 40 mm. Below 8 mm preheating may promote cracks. Above 40 mm preheating may be increased to 300°C. A heat input of around 1.0 kilojoule per mm of weld is the optimum arc heat. The welding parameters can be summarised for 8 to 40 mm thick cast sections as shown in Table 10.16.

Table 10.16 : Typical Welding Parameters for 8 to 40 mm Thick Cast Irons

Electrode diameter (mm)	Current (Amps)	Voltage (Volts)	Welding speed range (mm/sec)
2.5	70	20	1.4 to 1.75
	90	22	2.0 to 2.5
3.15	100	24	2.5 to 3.0
	120	26	3.2 to 4.0

Typical preheat and post heat are as follows :

(i) Preheat : 150-250°C, Interpass temperature : 350°C (maximum).

(ii) Post heat : 200°C for several hours.

(iii) Post weld stress relieving heat treatment is done, if possible.

10.4 WELDABILITY OF ALUMINIUM

Aluminium is one of the most widely used metals in the world. It possesses and exciting range of properties. Moreover, aluminium combines with alloying elements like copper, manganese, silicon, magnesium and zinc and form a very useful series of alloys.

(a) Important Properties

(i) Aluminium is a light weight metal, its density is about 27 gm/m^3. It is about 1/3rd as light as steel.

(ii) While pure Aluminium has a low strength of 7 kgf/mm^2, its alloys are moderately strong. Some alloys have strength as high as 45 kgf/mm^2 in the heat-treated condition.

(iii) The above two properties together provide it with high strength to weight ratio, which makes it suitable for aerospace applications.

(iv) Some of the alloys have excellent tonghness at low temperature making them suitable for cryogenic (below 0°C) applications.

(v) Some alloys have excellent corrosion resistance.

(vi) Aluminium and its alloys have high thermal conductivity.

(vii) Aluminium and its alloys have high electrical conductivity.

(b) Applications

(i) Household furniture and utensils.

(ii) Containers, tanks and vessels.

(iii) Antomobile structures, busbodies, road and railways tankers and wagons.

(iv) Buildings and other architectural structures.

(v) Portable bridges.

(vi) Aircraft, missiles and other aerospace components.

(vii) Radiators and other heat exchangers.

(viii) Electrical conductor cables and busbars.

10.4.1 Aluminium Alloy System

Aluminium alloys are classified on the basis of the principal alloying element present in a particular alloy. These alloys are designated by a 4-digit numbering system by Aluminium Association (AA-USA) or by a 5-digit numbering system by Bureau of Indian Standards (BIS) as shown in Table-10.17.

Table 10.17 : Aluminium Alloy Designation System

Alloys	As per AA Standards		As per Indian Standards	
	Designation	Example	Designation	Example
Unalloyed Aluminium	1 xxx	1060	1 xxx	19800
Aluminium-Copper	2 xxx	2219	2 xxx	24345

Aluminium-Magnaese	3 xxx	3003	3 xxx	31000
Aluminium-Silicon	4 xxx	4043	4 xxx	43000
Aluminium-Magnesium	5 xxx	5083	5 xxx	54300
Aluminium-Magnesium-Silicon	6 xxx	6061	6 xxx	65032
Aluminium-Zinc	7 xxx	7005	7 xxx	74530

(a) Alloy Classification

The above mentioned alloys can be classified into two groups :

(i) Non-heat treatable.

(ii) Heat treatable.

Non-heat treatable alloys include unalloyed aluminium (IXXX), Al-Mn (3XXXX), Al-Si (4XXX), Al-Mg (5XXX) series of alloys. These alloys cannot be strengthened by heat treatment. Al-Mg alloys are the strongest among non-heat treatable aluminium alloys. Heat treatable alloys include Al-Cu (2XXX), AG-Mg-Si (6XXX) and Al-Zn (7XXX) series alloys. These alloys are strengthened by heat treatment process called age hardening.

(b) Temper Designation

The condition of an alloy is indicated by its temper designation which is written after the alloy designation. For example, 6061-0 means 6061 alloy in annealed condition, 3003-H1 means 3003 alloy in cold worked condition. Similarly 7005-74 or 7005-76 means 7005 alloy is in age hardened condition (both T4 & T6). As per Indian Standard age-hardened condition is indicated by temper designation WP, for example 65032 WP.

10.4.2 Welding

Aluminium and its alloys are welded extensively for various applications. Welding of aluminium is not really difficult but an understanding fo the characteristic of aluminium and its alloys will definitely help to make a good weld. Some of the welding characteristics of aluminium are as follows :

(i) Aluminium oxidizes very quickly. Even at room temperature a thin film of oxide can form on a clean and fresh surface. At high temperatures, as we find in welding, aluminium oxide film forms very quickly. The melting point of aluminium oxide is much higher than aluminium itself. it does not dissolve in the molten metal, as a result, aluminium oxide film lead to difficulties in fusing and making a good weld. Therefore, it is necessary that the welding process should be able to clean and remove oxides during the course of welding.

(ii) Aluminium has a high thermal conductivity. Heat given to a weld groove easily spreads to surrounding areas. Therefore, it requires a higher heat input than steel for welding the same thickness.

(iii) Aluminium does not change its colour when heated, so the welder would require practice to judge the formation of the proper weld pool.

(a) Weldability of Aluminium Alloys

All aluminium alloys are not equally weldable, some are easily weldable, some can be welded with care and caution and some are not suitable for welded application. In general, the non-heat treatable alloys of 1XXX, 3XXX and 5XXX series are easily welded.The risk of cracking is less in these alloys and the "as welded" strength will be same as in the annealed condition. However, the conditions are different in the case of heat treatable alloys. Some heat treatable alloys are easily welded but there are many heat treatable alloys which are either difficult to weld or cannot be welded. The chances of cracking are more in these alloys. When the heat treatable alloys are welded in the age-hardened condition, there will be a loss of strength in the HAZ region. The strength can be brought back by a post weld heat treatment but this is normally not done because post weld heat treatment may cause distortion and warpage. Instead, the "as welded" strength is accepted. The "as welded" strength will normally be more than the strength in the annealed condition. Some of the easily weldable alloys are :

(i) Non-heat treatable alloys : 100, 3003, 5052, 5083.

(ii) Heat treatable alloys : 2219, 6061, 6063, 7005, 7039.

However, there are other alloys also which are weldable.

(b) Filler Metals

Aluminium filler alloys are also classified in the same manner as the base material. These filler alloys are mainly available from 1XXX, 3XXX, 4XXX and 5XXX alloys. One alloy type base metal is sometimes welded with another alloy type filler metal. For example, alloy 6061 is welded with either 4043 or 5356 filler alloys. Similarly 7XXX series base metals are welded with 5XXX series filler alloys. The filler alloy is selected based on the property which is required to a greater extent in the weld strengthen ductility or anodic colour matching or salt water-corrosion resistance or less chance of cracking. The recommended filler metals for some aluminum alloys are given in Table 10.18.

Table 10.18 : Recommended Filler Metals for Some Auminium Alloys

Base metal	Filler metal	Remarks
1100	4043, 1100	The first named filler metal is for higher strength while the second named filler metal is for more ductility.
2219	2319, 2319	-do-
3003	4043, 1100	-do-
5052	5356, 5654	-do-
5083	5183, 5356	-do-
7005	5039, 5356	-do-
7039	5039, 5356	-do-
6061	5356, 4043	First named filler metal for higher strength and ductility. Second named filler metal for salt water corrosion resistance and last cracking tendency.
6063	5356, 4043	-do-

(c) Welding Processes

A summary of welding processes and their applications in aluminium welding is given in Table 10.19.

Table 10.19 : Summary of Welding Processes and their Applications

Sl. No.	Welding/joining process	Brief description and application in welding aluminium
1	Oxyacetylene welding	Slow process. It required preheating and also flux to remove oxides. This flux is corrosive. It must be removed throughly after welding. Suitable for non-critical application.
2	Shielded metal arc	It requires preheating. Electrode coating contains the flux which removes oxides. Electrode connected to positive polarity (DC, EP). Suitable for non-critical and repair applications.
3	TIG welding	Maximum-used process for welding aluminium. To be discussed in detail.
4	MIG welding	Used widely for welding aluminium. To be discussed in detail.
5	Plamsa arc welding	Good quality welding possible. Suitable for thick plate welding.
6	Electron beam welding	Good quality welding possible. Used for critical components for aerospace application. Equipment cost is very high.
7	Electric resistance spot and seam welding	Suitable for welding sheets up to about 2 mm thick. Restricted to Lap type joints.
8	Solid state welding. (Friction, Flash-Butt, Explosion, Diffusion)	No fusion takes place during joining. Suitable processes to weld dissimilar combinations like aluminium to steel, aluminium to stainless steel, aluminium to copper etc.
9	Brazing and soldering	Suitable for some alloys. Brazing flux is very corrosive. It should be removed thoroughly after brazing. Brazing is done without flux in vaccum for some critical applications.

10.4.3 Preparation of Welding

(a) Joint designs

Jiont designs are made based on the design of product, section thickness, property requirements and welding process to be used. Some typical joint designs for are welding of aluminium are given in Fig. 10.18.

JOINT DESIGN	APPLICATION
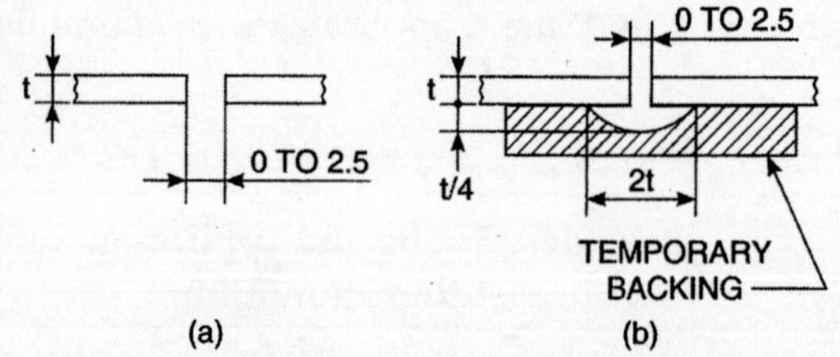	TO WELD THICKNESS OF' 1.6 TO 3.2 mm FOR MIG WELDING USE OF TEMPORARY BACKING IS PREFERRED
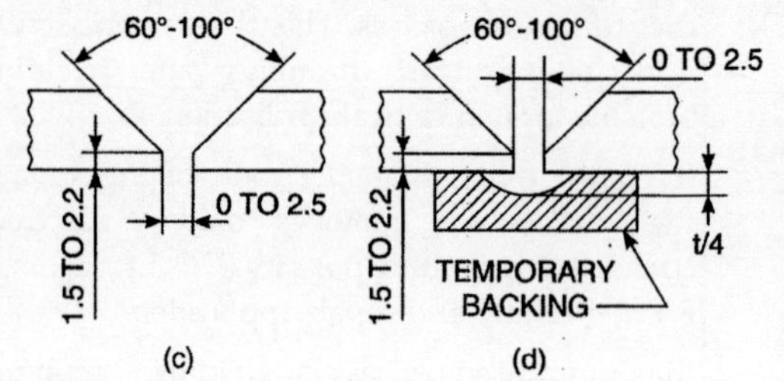	TO WELD THICKNESS OF' 4.8 TO 9.5 mm FOR MIG WELDING USE OF TEMPORARY BACKING IS PREFERRED JOINT DESIGN (D) MAY BE USED FOR HIGHER THICKNESS ALSO
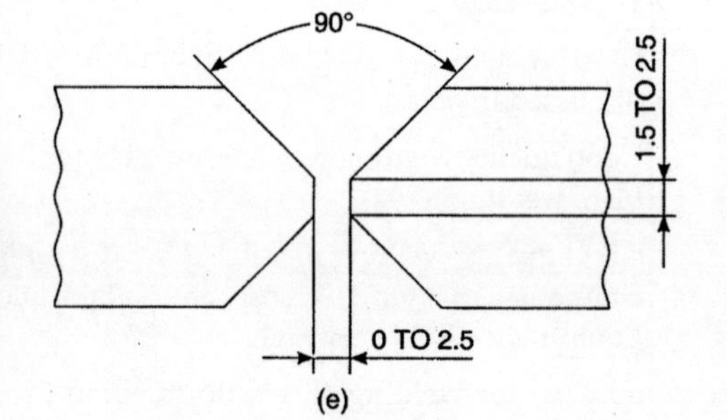	TO WELD THICKNESS OF 10 TO 20 mm

Fig. 10.18 Typical joint designs for arc welding of aluminium.

(b) Cleaning of Joint Area

The surfaces to be joined must be cleaned to obtain good wetting and bonding between the filler metal and the base metal. The joint groves and the adjacent areas must be free of dust, oil. grease, paint and of course, thick oxide film. The basic cleaning procedures are :

(i) Cleaning of dust, oil, grease, paints etc. These are cleaned by wiping and acetone.

(ii) Cleaning of the oxide film.

(i) Mechanical Method

The oxide layers on the surface can be removed by filing, wire brushing or grinding. Rotary wire brush (wire brush wheels) are quite effective in removing light oxide film. The wire brushes themselves should be kept clean and they should have preferably stainless steel bristles.

(ii) Chemical Method

Thick oxide layers can be removed by dipping the surfaces in chemical baths. First, the surfaces are dipped in a 5% solutions of sodium hydroxide for 10 to 60 seconds, then they are thoroughly washed with water. Next, they are dipped in a concentrated nitric acid for 30 seconds and then thoroughly washed with water. The cleaned surfaces are to be protected till the welding is over.

10.4.4 Tungsten Inert Gas Welding (TIG)

Tungsten inert welding is the most widely used process for aluminium welding. The welding current used the alternating current (A.C.) and hence it is generally termed as AC TIG welding process :

(a) Power Source

The power source for AC TIG welding has three main parts : the transformer, the high frequency unit and the D.C. suppressor. The transformer supplies the main welding power. The HF remains continuously ON during AC TIG welding. The D.C. suppressor produces current balancing by cutting off the D.C. component of the welding current.

(b) TIG Welding Torch

Gas cooled TIG welding torches can be used upto 200 amperes. Above 200 amperes, water cooled TIG welding torches should be used.

(c) Tungsten Electrode

Either pure tungsten or zirconiated tungsten electrodes are used in AC TIG welding. The thoriated tungsten electrodes are not preferring AC TIG welding. The tip of tungsten electrode should be rounded. The electrode may be ground sharp in the usual manner. Then the arc is struck on a scrap plate and welding is continued for a little time using the required welding current. The tip fo the electrode will become rounded which helps in maintaining a smooth are during welding.

(d) Shielding Gas

Argon is the inert gas normally used in aluminium welding. Argon is slightly heavier than air. The shielding quality and the arc stability are good with argon. Sometimes an argon helium mixture is used to weld plates of high thickness. By adding helium we get higher arc energy and higher penetration. Mixing ratio is normally 75% argon and 25% helium.

(e) Welding Technique

(i) After striking the arc, give a slightly rotating motion to the TIG torch to make a weld pool of proper size. Feed the filler rod to the leading edge of the weld pool. After feeding, do not take out the filler do too much away. Keep the top tip of the rod close enough to get the argon protection from the torch. Move the weld pool forward.

(ii) Angles of holding the torch and the filler rod are important. See Fig. 10.19.

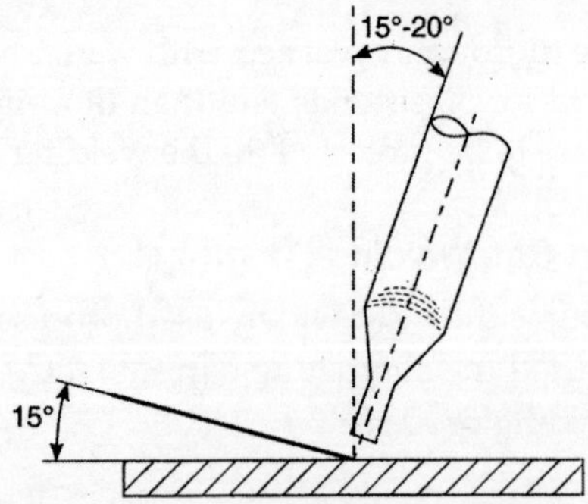

Fig. 10.19 Proper angles of feeding the filler wire and holding the torch forehand welding for aluminium.

(iii) Use always the forehand technique for all positions of welding.

(iv) Weld with stringer bead. In filling passes slight weaving may be done but wide weaving is not to be done.

(v) Temporary backing plates can be used for better control of penetration. Root purging can also be done through the backing plates to protect the root from oxidation. Temporary backing can be made from stainless steel or copper.

(vi) As the welding progresses, the plate picks up heat and it often becomes necessary to reduce the welding current to avoid excess penetration and formation of wider bead.

(vii) At the end of welding do not switch off the arc abruptly because it will form a depression called crater in the weld. Craters may contain craks and pores. To avoid craters use run off plate and switch off the arc on run off plate. Another method of avoiding craters is to reduce the current at a low amperage at the end of the weld and then switch off. Keep feeding the filler rod to the end. Any extra metal deposited is ground off.

(f) Welding Conditions

Welding conditions vary from work-to-work. In each case several parameters are to be considered to make a satisfactory welded joint. Welding current and speed of welding are two important operating parameters as these two parameters fix the heat input. If the speed of welding is increased the effective heat input is reduced. This enhances the mechanical properties so for heat treatable alloys it is better to use higher speed of welding maintaining full penetration of the joint.

Aluminium has a high thermal conductivity and the welding current is alternating type in AC TIG welding. For these reasons, a higher current is required to weld an aluminium plate with AC TIG welding than a steel plate of same thickness with DC ENTIG welding. The diameter of tungsten electrode, the internal diameter of the shielding nozzle and the argon flow rate will also be higher. Other important conditions are the cleanliness of the base metal, filler metal and also the impurity level of argon gas. If the base metal and filler metal surfaces are not free from oxides, moisture and oily matter, they will surely lead to porosity in the weld metal. Similarly the argon supply should be free from moisture, oxygen and hydrocarbon to make a radiographically clean weld. A general guideline for AC TIG welding of aluminium is given in Table 10.20.

Table 10.20 : Guideline for Manual AC TIG Welding of Aluminium

Plate thickness (mm)	Welding position	Joint type	Current (Amp)	Electrode Dia. (mm)	Nozzle size (10 mm) (mm)	Argon flow ate LPM	Filler rod Dia. mm	Number of passes
1.6	F	Square butt	70-100	2.4	8.0	10	2.4	1
	H, V	Square butt	70-100	2.4	8.0	10	2.4	1
	O	Square butt	60-90	2.4	8.0	13	2.4	1
3.2	F	Square butt	120-150	3.2	9.5	10	3.2	1
	H, V	Square butt	110-140	3.2	9.5	10	3.2	1
	O	Square butt	110-140	3.2	9.5	13	3.2	1
4.8	F	60° Single Vee	180-220	4.0	11	12	4.0	2
	H, V	60° Single Vee	160-200	4.0	11	12	4.0	2
	O	60° Single Vee	170-200	4.0	11	12	4.0	2
6.35	F	60° Single Vee	220-240	4.8	12.7	15	4.0	2
	H, V	60° Single Vee	220-240	4.8	12.7	15	4.0	2
	O	60° Single Vee	210-250	4.8	12.7	18	4.0	2

10.4.5 MIG Welding of Aluminium

Besides TIG welding, MIG welding process is also widely used to weld aluminium. In MIG welding, direct current (DC) is used and the electrode is connected to positive polarity (DCEP). Here the aluminium filler wire itself acts as the electrode. It is fed continuously throuygh a welding gun by a wire feeder. The arc forms between the filler wire and the work-piece. The filler wire tip melts and the molten droplets transfer across the arc to the weld pool. The arc energy is used much more efficiently in MIG welding, hence thicker plates can be welded by this process without preheat as compared to TIG welding process. Argon is the shielding gas used in MIG welding of aluminium. The oxide film automatically breaks up from the surface with DCEP power in argon shielding. Helium gives higher arc energy so helium is sometimes mixed with argon to get higher arc energy to weld plates of higher thicknesses.

Welding Techniques

(i) Always use forehand technique for all positions of welding.

(ii) In vertical position welding, it should always be vertical-up mode welding.

(iii) Adjuste the current and voltage such that only 'spray' type of metal transfer takes place. In MIG welding of aluminium, the globular and short-circuiting modes do not produce good quality welds.

10.4.6 Defect in Aluminium Welds

(a) Cracks

(i) Solidification Cracks

These cracks are caused by shrinkage stresses developed when the weld metal cools and contracts. To avoid this type of cracking, heat input should be reduced so that the weld pool size is also reduced.

(ii) Hot Short Cracking

This type of cracking depends on the composition of the weld metal which may have the hot short tendency. In addition the risk of this type of cracking also depends on the shrinkages stresses. To avoid this type of cracks the following steps are necessary :

(a) Select a filler alloy which will take the weld metal composition out of the hot short range.

(b) Reduce the dilution of the weld metal by the base metal.

(c) Use a higher welding speed of that the heat input and weld pool size gets reduced.

(b) Porosity

Porosity is the most noticed defect in aluminium welds. It is caused by hydrogen which dissolves in the liquid weld metal at the high temperature of a welding arc. Hydrogen comes to the weld pool from the arc heat. So, the material coming in the weld zone must be very clean and free of hydrated oxides, moisture, oil, grease etc. Not only should the base metal surfaces be clean but also the filler rod surfaces must be clean. This is much important in MIG welding because in MIG welding large lengths of filler wire are fed to the weld cool and hence chances of adding these impurities are more.The filler wires should be kept clean and stored in moisture proof packing. The moisture content of the shielding gas is also another contributing factor, so the quality of this gas should be checked and high purity gas should be used to obtain radiographically clean welds.

Besides the quality of the shielding gas, the shielding gas nozzle size and flow rate of the gas are important factor. Enough inert gas coverage should be provided to the hot weld pool.

10.5 WELDABILITY OF COPPER AND ITS ALLOYS

Copper and copper alloys are important industrial materials. They have some unique properties because of which these are used widely for several industrial applications.

10.5.1 Properties of Copper

(i) Copper has its own typical colour by which it is easily recognised. on account of this colour, copper is used for many architectural and decorative purposes.

(ii) Formability and malleability of copper are very good. It can be easily formed into different shapes.

(iii) Electrical conductivity is very high in copper. It is widely used in electrical conductor cables, wires and electrical machines.

(iv) Thermal conductivity is also very high for copper. Many types of heat-exchangers, radiators etc. are made of copper.

(v) Copper has good corrosion resistance properties against fresh and salt water, alkaline solutions other than ammonia and many organic chemicals. Copper tubes are widely used in refrigerator systems. Copper nickel alloys are used for marine applications. (Ship-hull etc.)

10.5.2 Copper Alloys

Many well known alloys are made by alloying copper with zinc, tin, aluminium, nickel and silicon. Small amounts of phosphorus, arsenic, berillium, cadmium, chromium, magnanese, boron, lead and sulphur are also added for specific purposes. Some of the important grades of copper and copper alloys are given in Table-10.21.

Table 10.21 : Important Grades of Copper and Copper Alloys

Commonly known	Alloy type	Nominal composition	Thermal conductivity in comparative scale
Oxygen free copper	High purity copper free of oxygen	99.95 Copper	100
Tough pitch copper	High purity copper with little oxygen.	99.90 Copper 0.04 Oxygen	100
Deoxidised copper	High purity copper deoxidesed with phosphorous.	99.90 Copper, 0.02 Phosphorous	87
Commerical bronze	Low zinc brass	90 Copper, 10 Zinc	48
Red brass	Low zinc brass	85 Copper, 15 Zinc	40
Cartridge bras	High zinc brass	70 Copper, 30 Zinc	31
Munz metal	High zinc brass	60 Copper, 40 Zinc	31

Admiralty brass	Tin brass	71 Copper, 28 Zinc, 1 Tin	28
Naval brass	Tin brass	60 Copper, 39 Zinc, 1 Tin	30
Nickel silver	Copper-zinc-nickel	55-65 Copper, 27-17 Zinc, 18-10 Nickel	8 to 12
Phosphor bronze	Copper-zinc-phosphorous	95 Copper, 5 Tin, 0.2 Phosphorous.	18
		90 Copper, 10 Tin 0.2 Phosphours	13
Aluminium bronze	Copper-aluminium	91 Copper, 6 to 8 Aluminium, 1.5 to 3.5 Iron, 1 Manganese.	17
		82 Copper, 10 Aluminium, 5 Nickel, 3 Iron.	10
Silicon bronze	Copper-nickel	97 Copper, 3 Silicon	9
Copper nickels	Copper-nickel	90 Copper, 10 Nickel	
		70 Copper, 30 Nickel	

10.5.3 Welding Behaviour of Copper and Copper Alloys

The most important factor to remember while welding copper is that the thermal conductivity of copper is very high, about six to seven times of that of carbon steel so the heat moves away rapidly from the joint to the adjoining areas and its becomes difficult to get proper fusion and penetration. Therefore, a lot of heat is required to get proper fusion and penetration in copper. Thermal conductivity of copper alloys is much lower than that of copper but have greater thermal conductivity than steel. A comparative sale of thermal conductivity is used to indicate relative thermal conductivities of different grades of copper and copper alloys. This is also given in Table-10.21. It is intersecting to find that the thermal conductivity of steel as per this scale is only 13 whereas for oxygen free copper it is as high as 100.

10.5.4 Requirement of Preheat

As mentioned above a considerable amount of heat is required to weld copper and copper alloy—now the question is this can we increase the welding current by any amount to get the extra heat required; the answer is—No. It would be difficult to control the weld pool if the current is very high and the weld will be defective. The safer way of providing the extra heat is by preheating the joint and its adjoining areas by gas burners or electrical resistance heating pads. The level of preheat depends on the type of alloy, its thickness, the welding process and the shielding gas to be used. The welding current has a direct effect on the degree of preheat required. Attempts to use high currents in order to reduce the level of preheat may produce bad quality welds with gouging effect and undercuts.

10.5.5 Filler Metals

Filler metals for welding of copper and copper alloys are available in the form of arc welding covered electrodes and bare wired for almost all the grades of copper and its alloys. Some filler alloys containing zinc are available for oxy-fuel gas welding and brazing. Various grades of filler metals for welding of copper and copper alloys are given in Table-10.22.

Table 10.22 : Filler Metals for Welding of Copper and Copper Alloys

Covered Electrode	Bare wire	Common name	To weld base metal
ECu	ERCu	Copper	Coppers
ECuSI	ERCuSi-A	Silicon Bronze	Silicon bronzes, brasses
E CuSn-A	ERCuSn-A	Phosphor bronze	Phosphor bronzes, brasses
ECuSn-C			
ECuNi	ERCuNi	Copper-Nickel	Copper-Nickel Alloys
ECuAl-A2	ERCuAl-A2	Aluminium bronze	Aluminium bronzes, brasses, silicon bronzes, manganese bronzes
ECu-Al-B	ERCuNi-A3	Aluminium bronze	Aluminium bronze
ECuNiAl	ERCuNiAl	–	Nickel-aluminium bronzes
ECuMnNiAl	ERCuMnNiAl	–	Magnanese-nickel-aluminiu mbronze
	RBCuZn-A	Naval brass	Brass,Copper
	RCuZn-B	Low fuming brass	Brasses,manganese bronzes
	RCuZn-C	Low fuming brass	Brasses-manganese bronze

10.5.6 Welding of High Purity Coppers

There are three basic grades of high purity copper :

(i) Oxygen-free copper
(ii) Tough pitch copper
(iii) De-oxygenated copper

Oxygen free copper and tough pitch copper are primarily used in electrical industry for current carrying purposes. These grades are difficult to weld. Welded joints usually have lower strength and ductility due to the formation of pores and micro-fissures, particularly in reducing atmospheres as in the case of oxy-acetylene welding. This is due to the oxygen present in the tough pitch coppers and oxygen picked up by oxygen-free coppers during welding. De-oxygenated copper is the weldable grade of copper. Copper easily forms oxides and also dissolve oxygen at high temperatures. This reduces the strength and ductility in the base metal heat affected zone and forms porosity in the weld metal. Because of this reasons all the grades should not be kept in the preheated condition for a long time and welding should be completed quickly. Filler rods must contain sufficient deoxidisers like phosphorus, silicion and manganese. preheating is required for thicknesses as low as 4 mm.

10.5.7 Welding Processes and Techniques

Copper is best welded by TIG and MiG welding processes. TIG (DCEN) is preferred for lower thickness and all position welding. MIG is more suitable for higher thicknesses. Proper inert gas shielding is important to avoid oxygen pick up by the weld metal. The nozzle size should be adequately large. Adequate fixturing and tack welding are required to prevent distortion. It is useful to deposit a relatively thicker root bead to reduce the chances fo cracking. Both forehand and backhand techniques can be used for TIG and MiG processes. Forehand technique is generally preferred. Stringer or narrow-weave beads should be deposited. Wide weaving should not be done. For vertical welding use vertical up-mode. For MIG welding parameters are set for spray transfer. Argon is the preferred shielding gas. Argon-helium mixture of helium by itself can be used for higher thicknesses. Preheat level may very form 100 to 600°C depending on the thickness being welded, refer Fig. 10.20. Copper filler metal can be used as given in Table 10.22. For higher strength phosphor bronze or silicon bronze filler wire can be used in which case electrical conductivity will be considerably lower.

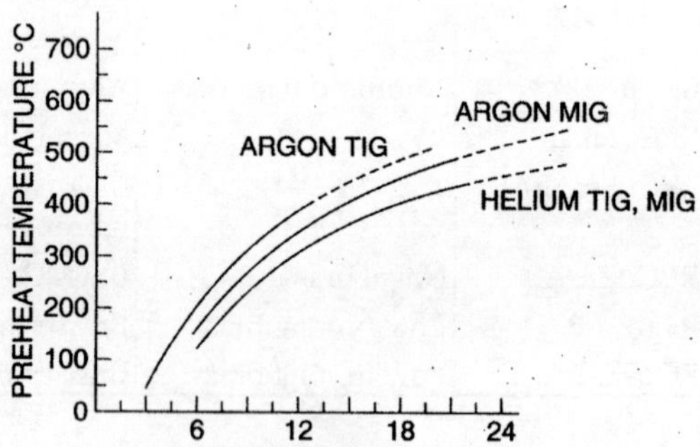

Fig. 10.20 Preheating for copper welds.

SWAW (DCEP) process can also be used but the quality of welds will be inferior to TIG or MIG welded joints. Copper, silicon bronze, phoshor bronze electrodes can be used. Copper can be welded with oxyacetylene welding process although it is a slow process. High preheat is required. Backhand welding with flux and neutral flame is preferred. Copper and silicon bronze filler rods can be used.

10.5.8 Welding of Brass (Copper-Zinc Alloys)

The main problem in welding brass is that zinc tends to evaporate from the molten weld metal as fumes. This zinc fuming makes the weld porous. Low zinc brasses containing less than 20% zinc, like commercial bronze, red brass etc. are easier to weld. High zinc brasses containing more than 20% zinc, like cartridge brass or Munz metal are more difficult to weld due to service zinc fuming problem.

Welding Processes and Technique

For the welding of brass TIG, MIG, SMAW and oxyacetylene welding processes can be used. To minimise zinc fuming, the arc shold not be directed to the base metal. In the TIG welding the arcis directed onto the filler rod. In MIG and SMAW, the arc is directed onto the weld pool. Preheat of 100-300°C can be employed. The preheat temperature will be somewhat lower for high zinc brasses. Preheat may notbe required for MIG welding of high zinc brasses. Due to zinc fuming problem,the filler metals do not contain zinc in the arc welding of brass. In other words, we do not use the brass rod as a filler in TIG or MIG welding fo brass or a coated brass electrode for SMAW. Instead phosphor bronze or silicon bronze or aluminium electrode for SMAW. Instead phosphor bronze or silicon bronze or aluminium bronze filler rods or covered electrodes are used. For oxyacetylene welding brass rods can be used as filler.

10.5.9 Welding of Phosphor Bronze

Phosphor-bronzes are basically copper-tin alloys with about 0.2 percent phosphorus. They are also known tin bronze. These alloys have a tendency to hot crack. Hot peening of each layer in multipass welding reduced welding stress and risk of cracking. Phosphor bronzes can be welded by the TIG, MIG and SMAW welding processes. Oxyacetylene welding is normally not used as the risk of cracking is more due to the wider heat affected zone and slower cooling rates. Preheat and interpass temperature should not exceed 200°C otherwise the risk of cracking would be more. Stringer bead and fast welding speed are required to keep the weld pool size small.

The TiG welding (DCEN) process is suitable for welding the thin sheets, repair of castings and rebuilding the worn out surfaces. MIG welding is preferred to join large phosphor bronze components and thick sheets. It is beneficially to keep some rot gap in the joint fit up. Phosphor bronze filler metals are used both for TIG and MIG welding processes (Table 10.22). In SMAW (DCEP) stringer beads are deposited to get better mechanical properties. Preheat and interpass temperature is maintained as mentioned above. Covered arc welding electrodes of phosphor bronze can be used.

10.5.10 Welding of Silicon Bronze

Silicon bronzes are copper-silicon alloys containing 1.5 to 3.5 percent silicon.These alloys have good weldability due to the low thermal conductivity, good de-oxidation of the weld metal by the silicon and protection given by the resulting slag. Preheating is normally not needed and the interpass temperature should also be kept below 100°C. Rapid cooling should be done just below the solidification temperature to avoid the risk of cracking. TIG welding (DCEN or AC) can be used for all position welding. MIG welding would be advantageous for thicker sections. Each pass should be cleaned by using a wire brush. Silicon bronze filler metals can be used. SMAW can be used for welding in flat as well as in vertical or overhead positions. Covered electrodes of silicon bronze can be used. Aluminium bronze electrodes can also be considered for some applications.

10.5.11 Welding of Aluminium Bronze

Aluminium bronzes are widely used in heat exchanges, piping and vessels.There are two types of aluminium bronzes available, namely alpha (single phase) and alpha-beta (two phase) alloys. The single base alpha alloy are prone to hot short cracking in the HAZ if the alloy contains less than about 8 percent aluminium. TIG, MIG and SMAW processes can be used to weld aluminium bronzes, MIG welding being the preferred one. Oxyacetylene welding process is not normally used to weld aluminium bronzes.

Due to the presence of aluminium in the alloys, an aluminium oxide coating forms on the weld surface which needs cleaning. For this reason AC TIG welding process is sometimes preferred using argon shielding. DC TIG (EN) can also be used with argon, helium and argon-helium mixture. Preheat and interpass temperature should be below 150°C for aluminium content of less than 10 percent. MIG welding is more suitable for relatively thick sections. Proper inert gas shielding is important. Forehand technique is required. Argon or 75 argon-25 helium gas mixtures are used. This process can be used for all positions welding. For overhead and vertical positions, mostly globular trnasfer mode is used while for flat position it is spray trnasfer. Aluminium bronze filler metals can be used.

The SMAW process can be used to weld aluminium bronze both in wrought and cast forms. Electrode coatings take care of the aluminium oxides. Stringer or narrow weave beads can be deposited using a short arc length. Each bead must be cleaned thoroughly before depositing the next bead. For thick sections preheating upto 200°C can be applied. Aluminium bronze covered electrodes can be used. Aluminium bronzes with higher aluminium content and high thickness may require preheat upto 625°C and fan cooling may be necessary to avoid cracking. Oxyacetylene welding is not recommended for the welding of aluminium bronze.

10.5.12 Welding of Copper-Nickel Alloys

These alloys are widely used in marine applications like ship hulls pipings.The thermal conductivity of copper-nickel alloys is similar to that of carbon steel. These alloys are easily welded by TIG, MIG and SMAW processes. Preheating is not required. The interpass temperature should also be kept low. Cupro-nickels are prone to oxidation which results in porosity. Surface areas in the weld zone should be thoroughly cleaned any contamination with lead, sulphur and phoshorus will lead to inter-granular cracking in the weld and HAZ. In order to get porosity free welds, keep short arc-length and avoid contamination with air. A filler metal of 70 percent copper and 30 percent nickel composition is normally used for all grades of copper-nickel alloy base metals. The filler metal contains 0.20-0.50 percent titanium as a de-oxygenerator. This must be ensured to avoid porosity in the weld metal. Argon shielding is preferred in TIG and MiG welding. Argon-helium mixtures can be used thick sections. Backing plates made of carbon, graphite or steel should not be used. Use copper or cupro-nickel as the backing material.

TIG welding (DCEN) is suitable for all position welding. Proper control of arc length and inert gas shielding are necessary to obtain defect free welds. Alternating current (AC TIG) can also be used for mechanised welding ensuring proper arc length control. Welding without a filler should be avoided. In MIG welding of copper-nicked alloys, spray or short-circuiting type metal transfer may be used. Spray transfer is preferred for thick sections. For thin section either short-circuiting type metal transfer or spray transfer in pulsed mode is preferred. Covered cupro-nickel electrodes are used for manual metal are welding of all grades of copper-nickel alloys. The elecrode should be manipulated carefully to maintain a short arc as well as to produce a good bead shape. Weaving upto three times the core wire diameter is allowed. For root run, usually a stringer bead with minimum weaving is preferred. The slag must be completely removed from each bead before the next bead is deposited.

The oxyacetylene process is restricted to minor repair welding only. Cupronickel welding rods can be used with a slightly reducing flame. Proper fluxing is necessary. Preheating is not required.

10.6 WELDABILITY OF NICKEL AND ALLOYS

Weldable grade nickel and nickel alloys can be generally classified as wrought, precipitation-hardenable and heat resistant alloys. Nickel and nickel alloys are commonly used when corrosion resistance is prime consideration. There is no standard specification and designation system for nickel alloys. They are known by trade names and suffix numbers. A few of the common trade mark names are :

(i) Model : Ni-Cu alloy.

(ii) Inconel : High Ni-Cr-Fe alloy.

(iii) Incoloy : Ni-Fe-Cr alloy with other elements.

(iv) Hastealloy : Ni-Mo-Fe alloy.

Welding of Ni based alloys is similar to that of stainless steels but with some exceptions. Ni-based alloys forms on their surface an oxide coating and hence, it should be removed before welding. Ni alloys are susceptible to embrittlement at low or high welding temperatures due to the presence of lead, sulphur and phosphorus.

10.6.1 Welding Techniques

(a) Joint Design

The joints used for Ni based alloys should provide for proper accessibility. The joint opening must be sufficient to permit the torch, electrode or filler metal to extend to the bottom of the joint. The characterstics of nickel-alloy weld metal necessitate joint designs different from those used for other common materials.

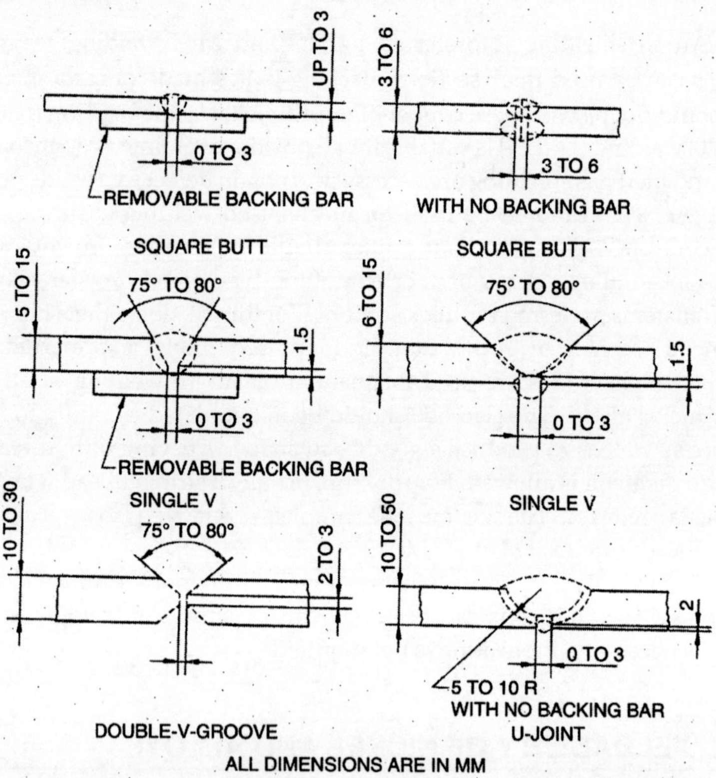

Fig. 10.21 Typical joints designs for arc welding of Ni, Ni-base alloys.

Since the molten weld metal is sluggish in nature, the joint design should be open enough to provide space for manipulation of the torch or filler metal. Nickel alloys have an inherent lower weld penetration which necessitates the use of smaller land in the root of the joint. Sopme of the joint designs used are shown in Fig. 10.21.

(b) Welding Process

Shielded metla arc welding, gastungsten arc welding and gas metal arc welding are the generally used welding processes for Ni and Ni-based alloys.

(c) Cleaning

For the all above mentioned welding processes the surface to be joined has to be cleaned throughly before the start of the welding. At high temperatures Ni and its alloys are susceptible to embrittlement. At high temperatures Ni and its alloys are susceptible to embrittlement by sulphur, phosphorus, lead and other low melting substances. These substances may be produced by greease, oil, paint, cutting fluid, marking crayans etc. which are used during processing and fabrication of the alloy. Hence, it be comes mandatory that the joint be thoroughly cleaned before welding. The cleaning method to be used depends on the composition of the substances to be removed. If hard tenacious oxide is present it should be removed by grinding, abrasive blasting or machining.

10.6.2 Shielded Metal Arc Welding

Shielded metal are welding can be used for welding nickel and nickel alloys. The electrode compositionis collected so as to be similar to the base metal. Table-10.23 gives the base material filler material combination for nickel and nickel alloys for using SMAW.

Table 10.23 : Base material filler material combination for nickel-based alloys using SMAW process

Filler material-AWS Classification	To weld base metal
ENi Cu-7	Nickel-Copper alloys to themselves, surfacing steel with Nickel-Copper alloys Welding Nickel - Copper alloys to steel
ENi Cr Mo-3	Nickel-Chromium-Iron-Molybdenum alloys of composition range upto inconel 625
ENiCr Fe 2	Dissimilar joints involving carbon steel, stainless steel, Nickel, Ni band alloys
ENi Cr Fe 3	Nickel, Ni band alloys
ENi CR Mo 4	Nickel-Chromium-Molybdenum alloys to themselves or to steel
ENi-Mo	Nickel-Molybdenum alloys to themselves or to other Nickel Cabalt, iron based metal

Welding Technique

SMAW is done using DCEP, in flat position, where ever possible. The electrode is held at an inclination of 75-88 to the weld puddle and ahead of it. A short arc should be maintained. Welding can also be done in vertical or over-head position. While welding the bead can be deposited using the weaving technique.Before each bead is deposited remove all the flux and slag for multi-pass welds.

10.6.3 Gas Tungsten Arc Welding

Solid strengthened or precipitation hardenable grades of cast and wrought nickel alloys can be welded using this process. Filler metal is added. DCSP is used for both manual and machine welding. The shielding gas used is either argon or helium or a mixture of the two. For the welding of pure nickel argon with small quantity of hydrogen (around 5%) may be used to avoid porosity. Pure tungsten electrodes or tungsten alloyed with thorium or circonium is used. The filler metal used have similar composition as of base metal.

Welding Technique

When filler metal is used, the wire diameter should berelated to the base metal thickness. During welding the hot end of the filler metal must be kept under the shielding gas. The shortest are length should be maintained for complete penetration welds. Grooved-backing bars have to be used to permit local gas shielding.

10.6.4 Gas Metal Arc Welding

High nickel and nickel copper alloys can be joined by gas metal arc welding and special welding procedures. Precipitation-hardenable alloys such as monel-K 500 can also be welded. Spray, globular and short-circuiting metal transfer can be used. For gas metal arc welding of nickel alloys by spray transfer, argon can be used as a shielding gas. 15 to 20% helium is added to argonfor welding nickel alloys. Table 10.24 gives the filler metal selection chart for welding nickel alloys.

Table 10.24 : Base material filler material combinations for nickel, nickel-based alloys for GTAW, GMAW process

Filler material-AWS Classification	To weld base metal
ER Ni-1	Wrought and cast forms of commercially pure Nickel
ER Ni-1	Nickel-Copper alloys
ER NiI Fe Cr-1	Nickel-Iron-Chromium-Molybdenum-Copper type alloys
ER Ni-Cr 3	Nickel-Chromium-Iron alloys
ER Ni Mo-1	Nickel-Chromium-Molybdenum base metals to steel or other alloys
ER Ni Cr Mo-3	Nickel-Chromium-Molybdenum base metal, for cladding steel with Nickel-Chromium-Molybdenum weld metal

Welding Techniques

For good welding electrode holder must be positioned vertically along the centre line of the joint. Some shight inclination is permitted to allow a better view of the work. Optimum arc length should be maintained. In short-circuiting transfer, if manipulation is faulty, then lack of fusion can occur.

10.6.5 Welding of Precipitation Hardenable Alloys

Precipatation hardenable alloys like Monel K-500 and Monel 502 require special care due to their susceptibility to cracking. Cracks can occur in the HAZ on ageing or in service at temperatures above the ageing temperature. Any cold-worked component should be annealed before welding. At-Ti hardened alloy should be solution—annealed after welding and before precipitation hardening. The stresses created by repair or alteration during welding must be relieved by rapid heating to the solution annealed temperature range prior to re-ageing. Precipitation hardenable alloys are generally welded by gas tungsten arc process. However, shielded metal arc and gas metal arc welding are also used. The heat input during welding should be moderately low.

10.6.6 Welding of Heat-resistant Nickel Base Alloys

Nickel-based heat-resistant alloys are solid solution strengthened and others are strengthened by the precipitation of a second phase from a super saturated solid solution. Most of the solid solutions nickel based alloys like Inconel 600, Haste alloy B and Haste alloy C are used for applications ranging from cryogenic to 430°C or higher. The heat-resistant nickel based alloys are commonly welded by gas tungsten arc or shielded metal arc processes. Gas metal arc welding is used for joining sections of more than 10 mm in thickness. Generally, a root pass is made by the gas tungsten arc and the filler pass by gas metal arc welding. The Ni-Cr and Ni-solid solution alloys are readily welded in the annealed condition. No heat-treatment is needed after welding in order to improve corrosion resistance.

10.6.7 Quality

Weld defects such as cracks, porosity, inclusions and incomplete fusion are usually not acceptable in weldments of nickel based heat-resistant alloys. In SMAN, stringer beads help to avoid cracking by minimising heat input and weld stress. Non-destructive testing methods arc adopted on all weldments. Destructive tests are done on test samples only. For weldments subjected to pressure in service, leak detection tests are carried out.

10.7 WELDABILITY OF TITANIUM AND ALLOYS

Titanium is not just another structural metal, it possesses some special properties. It is sometimes called an exotic material. It is a reactive metal. Titanium and its alloys are used in chemical, aerospace, marine and defence applications. Titanium and most of its alloys are weldable. These materials are quite expensive.

10.7.1 Properties of Titanium and Its Alloys

(i) Pure titanium is a silver cxoloured metal.

(ii) Pure titanium has a tensile strength of about 25 kgf/mm^2 only. Titanium alloys are very strong. Many alloys have strength in the range of 100-150 kgf/mm^2 in the heat-treated condition.

(iii) Titanium has high resistance to corrosion against salt and oxidising acid solution. It is fairly resistant to mineral acids also.

(iv) Titanium is lighter than steel but slightly heavier than aluminium. Its density is 4.5 gm/cc.

10.7.2 Titanium and Its Alloys System

(a) Titanium

Pure titanium exists in a metallurgical phase called alpha upto 885°C and above this temperature it changes to beta phase. Commercially pure titanium has a moderate tensile strength of 25 to 55 kgf/mm^2 depending upon the presence of impurities like oxygen, nitrogen, hydrogen, carbon and iron. At high temperatures titanium can dissolve a fair amount of oxygen, nitrogen, hydrogen, carbon and iron. The presence of these elements in pure titanium increases the strength of the metal but reduces ductility. Hence, these elements are generally considered as impurities. Oxygen, nitrogen, hydrogen and carbon form interstitical solid solution with titanium and these are often referred to as interstitial elements. Since, they lower ductility in titanium, the presence of these interstitial elements in many titanium and titanium alloy grades are restricted to very low levels. These grades are referred to as extra low interstitial grades (ELI). The iron content in titanium should be restricted to below 0.05 percent, otherwise the resistance to corrosion decreases. Commercially pure titanium is weldable but extra care should be taken during welding so that contamination of the weld and heat affected zone by oxygen, nitrogen, hydrogen, carbon and iron does not take place.

(b) Titanium Alloys

There are three types of titanium alloys available based on the micro-structure :

(i) Alpha titanium alloys.
(ii) Alpha-beta titanium alloys.
(iii) Beta titanium alloys.

Some of the important alloys are listed in Table 10.25.

Table 10.25 : Some of the alloys of titanium

Alpha alloys	Alpha-beta alloys	Beta alloys
Ti - 0.2 pd	Ti - 3 Al - 2.5 V	Ti - 13 V - 11 Cr - 3 Al
Ti - 5 Al - 2.5 Sn	Ti - 6 Al - 4 V	Ti - 11.5 Mo - 6 Zr - 4.5 Sn
Ti - 5 Al - 2.5 Sn (ELI)	Ti - 6 Al - 4 V (ELI)	
Ti - 7 Al - 4 Mo		

(i) Alpha and near Alpha Titanium Alloys

These alloys have good corrosion resistance, creep strength and toughness. Some of the alloys have very good toughness even at low temperatures. The extra low interstitial (ELI) version of alloy Ti-5Al-2.5 Sn retains high toughness even at –250°C. These alloys have good weldability and they are normally welded in the annealed condition only.

(ii) Alpha-Beta Alloys

Alpha-beta alloys are used in annealed as well as heat-treated condition. The heat treated alloys have high strength to weight ratio. The strength and ductility of alpha-beta alloys mayt change significantly after welding. Most of the alpha-beta alloys suffer a loss of ductility in the weld and HAZ. These alloys are often welded with a pure titanium or alpha-titanium alloy filler metal. This gives more ductility to the weld metal but the strength becomes low. Among the alpha-beta alloys. Ti-6Al-4 V alloy, particularly its ELI grade (Ti-6 Al-4 V-ELI) has the best weldability. Ti-6 Al-4 V can be welded in the annealed condition or in the solutionised and partially aged condition.The ageing is completed during post-weld stress-relieving heat teeatment. The alpha-beta alloys having more of alpha phase are easier to weld. It is difficult to weld alpha-beta alloys which have more beta phase due to the increased cracking tendency in these alloys.

(iii) Beta Alloys

Beta alloys can be welded in the annealed or solution heat-treated condition. In the "as welded" condition, the welds will be ductile though the strength will be low. For many applications, these alloys are used in the "as welded" condition. To obtain higher strength in the weldment, sometimes, alloys are welded in the annealed condition then they are given cold working by peening or planishing, and after the cold working, solutionising and aging heat treatments are given. This process gives higher strength with adequate ductility in the weld joint.

10.7.3 Welding

Titanium has great affinity to react with oxygen and quickly forms a layer of titanium oxide on its surface. At high temperatures as encountered in welding, titanium can dissolve good amounts of oxygen, nitrogen, hydrogen and even its own oxides layer from the surface. Since the presence of these elements embrittle titanium, care should be taken in the preparation stages and during welding such that weld and HAZ areas in hot condition do not come in contact with atmosphere containing these elements. As iron decreases the corrosion resistance in titanium, care should be taken to preventiron pickup during welding. Hence, preweld surface preparation and cleaning as well as protection of the weld and the adjoining areas during the welding thermal aycle are important basic steps to make good welds in titanium.

10.7.4 Cleaning

Before welding, the groove and the adjacent areas should be thoroughly cleaned of oily matter, grease, dust and dirt. Cleaning can be done by using acetone. Carbon tehachloride, trichloro ethylene or other chlorine containing solvents should not be used to clean titanium. Chlorides can cause stress corrosion cracking. Removal of oxide layers can be made by mechanical methods like grit blasting, draw filing, wire brushing or gridnding. This is then followed by pickling in a solution of 2 to 4 percent hydrofluroric acid (HF) and 30 to 40 percent nitric acid (HNO_3) in water. After pickling the parts should be thoroughly washed and cleaned in water preferably in distilled water free of chlorides.

10.7.5 Filler Metals

In general, titanium base metals are welded with filler metals of the same mominal compositions. The filler metals are avzailable in the form of bare wire and rods. Pure titanium filler wires are available in different grades where the level of impurity like oxygen, nitrogen, hydrogen and iron vary from grade to grade. To emphasize on the need to control impurities, the approximate range of each impurities in pure titanium filler metal grades are given here in weight percentage.

Carbon–0.1 to 0.25, Hydrogen–0.005 to 0.008, Nitrogen–0.012 to 0.020, Iron–0.10 to 0.30. Apart from filler metals of pure titanium grades, matching filler metrals for titanium alloy grtades are also available.

10.7.6 Welding Processes

Inert gas shielded are process like TIG, MIG and plasma arc welding, and high energy-density processes like electron beam and laser beam welding processes can be used for welding titanium and its its alloys. Out of these processes TIG welding is the most widely used method of welding titanium. Electron beam welding is also used in the fabrication of air craft and space components.

(a) TIG Welding

TIG Welding of titanium is carried out in direct current electrode negative mode. Electrode negative mode is also termed as straight polarity mode. 1% to 2% thoriated tungsten electrode is used. The argon flow rate through the mozzle is controlled to obtain a smooth, lamellar flow. Care should be taken to avoid turbulent flow which will entrap air and cause contamination to the weld. Tungsten electrode extension is kept at the minimum length required for clear visibility of the weld pool and easy feeding of the filler wire. Extra projection of the electrode or excessive arc length may lead to improper shielding of the weld pool. The tip of the filler wire should be kept close to the nozzle so that it gets inert gas protection. The welding machine should have prefllow and postflow facilities. When the arc is stopped, at the end of the weld the TIG torch shgould be held at that position far a few seconds with the post flow of gas as this will protect the last spot of the weld. It is a good practice to check the quality of the argon gas by spot check before starting the weld run. The arc is struck on a clean scrap plate of titanium and once a molten purddle is farmed, the arc is switched off but the torch is held in position with the post flow until the spot cools down. If the gasis free from contamination, the weld spot will have a shining, silver coloured appearance.

(b) Protection during Welding

As metioned above titanium can be easily embrittled by the pick of oxygen, nitrogen and hydrogen at high temperatures during welding, if the weld and its adjacent areas are not protected from atmospheric contamination. The inert gas shielding provided by the TIG torch protects the molten weld pool but as the welding progresses and the TIG torch moves forward, the just solidified areas become exposed to atmosphere. The inert gas flow from the TIG torch cannot provide protection over a very large area outside the molten weld pool. Therefore, it becomes necessary to make arrangements to protect the just solidified weld and its adjoining areas with inert gas coverage till these areas cool down to below 350°C. There are two methods normally followed to provide the required inert atmosphere protection during the welding of titanium :

(i) Welding in inert atmosphere chambers.

(ii) Welding in the open air using auxillary shielding.

(i) Welding in Chambers

In this method welding is carried out inside a closed chamber filled with argan or helium. The work pieces, the filler wire, The TIG torch, the auxillary hand tools are all kept inside the chamber. Then the air in the chamber is purged out and the chamber is fillled back with the inert gas. The chamber are either of a flow-purged type or a vacuum-purged type. In flow-purged type chambers, the chamber air is purged out by continuously blowing in argon or helium. It requires a large volume of gas, about five to ten times volume of the chamber to purge effectively.

In vacuum purged chambers, the chamber air is taken out by using a vacuum pump. The chamber is then filled back with the inert gas. In both types of chambers, the inertness of the chamber atmosphere checked by making a weld bead on a scrap plate of titanium. The colour of the weld bead would be like shining silver when the chamber atmosphere is fully inert. The inertness of the chamber atmosphere can be checked with analysing equipments. Both the flow-purged and vacuum purged chambers contain glove ports on the chamber wall. The glove ports are fitted with sleeves and gloves. In manual welding, the welder inserts his hands through the glove ports to carry out welding of the work-piece already kept inside the chamber. Mechanised welding can also be carried out inside the chamber. The limiting factors are the size of the job that can be accommodated inside the chamber and the cost of purging with inert gas. Everytime the chamber lid is opened for loading and unloading of the work-piece, the chamber has to be purged.

(ii) Welding in Open Air

In TIG welding of titanium in the open air method, additional shielding arrangements are made to protect the hot weld bead, its adjacent areas and also the under side (root side) of the weld. The protection on top side is provided with what is known as trailing shield. The trailing shield travels closely behind the TIG torch providing a diffused flow of argon to protect the immediately solidified weld and its adjoining areas. The length of the trailing shield depends on the heat input and the welding speed used. Fig. 10.22 (a) and (b) shows a systematic arrangement of the application of the trailing shield. To protect the root side from atmospheric contamination, usually a grooved backing bar with purging facility is used. The groove in the backing bar should be continuously purged with argon so that the hot root cannot come in contact with the atmospheric oxygen and nitrogen. The colour of weld bead indicates the effectiveness of the shielding. A properly shielded weld will have a shining silver colour. A light bronze or golden or straw colour indicates a slight cotamination. Normally this surface can be brushed off with a stainless steel brush. More and more contamination through improper or ineffective shielding will produce a light blue colour followed by a dark blue colour and finally severe contamination will make the weld bead surface white and white flakews will be on the surface. This type of weld will be highly brittle and unacceptable.

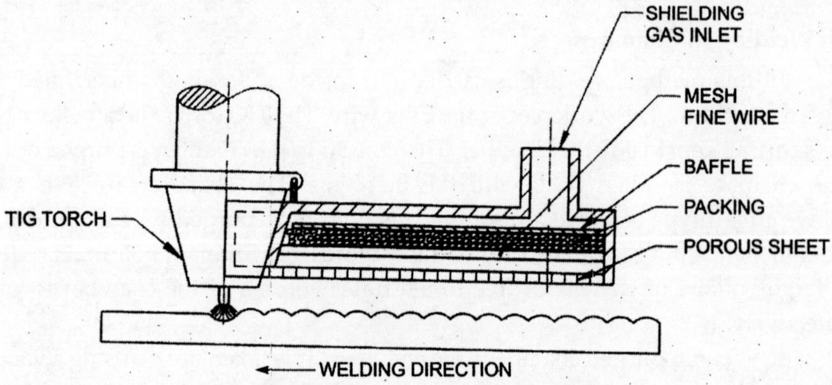

(a) TIG welding of titanium with trailing shield

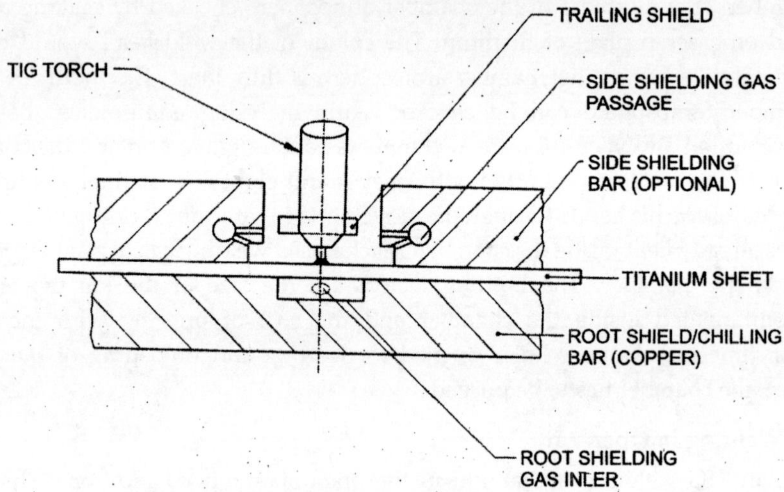

(b) Setup for open air TIG welding of titanium

Fig. 10.22

10.7.7 Porosity in Titanium Welds

A porosity problem quite often arises in the welding of titanium. Ion proper cleaning of the base metal and filler is a major source of proosity in titanium welds. Any trace of hydrocarbons like oily matter or even finger prints can cause porosity in the weld. The presence of moisture in the arcing zone can lead to formation of pores. Normally a more fluid weld pool and a slower cooling rate will provide more time for the disolved gas to escape and thereby reduce porosity.

10.8 WELDABILITY OF COBALT ALLOYS

Cobalt based alloys are generally used for applications involving corrosion and erosion, especially, high temperature seizure service. Hard facing applications such as wear cum impact resistance also requires cobalt alloy for surfacing.

10.8.1 Types of Alloys

The commercially available cobalt based alloys, produced or deposited by welding, contain a combination of two or more of the elements : Ni, Cr, Mo which form a second phase with the cobalt base to form a hard matrix. Cobalt based alloys can be classified into two types :

(A) Heat resistant alloys.

(B) Hard facing alloys.

Table 10.26 gives the composition of weldable grade cobalt based alloys :

Table 10.26 : Chemical composition in Wt. %

S. No.	Alloys designation/ trade name	C	Ni	Cr	Mo	Fe	Co	Cb	N	Others
1	HS-21	0.25	3.0	27	5	1	62	–	–	–
2	HS-25 (L-605)	0.10	10	20	15	–	55	–	–	–
3	HS-31 (x-40)	0.50	10	25	–	–	57	–	7.5	–
4	HS-188	0.08	22	22	–	1.5	53	–	14	0.08La
5	S-816	0.40	20	20	4	–	48	4	4	–
6	UnCo 50	0.06	–	28	2.2	20	49	–	–	–
7	W I 52	0.40	1	21	–	2	63	2	11	–
8	Stellite 1	25	–	30	–	–	55.5	–	12	–
9	Stellite 6	1.1	–	28	–	–	66.9	–	4	–
10	Stellite 12	1.3	–	29	–	–	61.7	–	8	–

1 to 7 : Heat-resistant alloy; 8 to 10 : Hard facing alloy.

(A) Heat-resistant Alloys

These alloys are used for applications involving high temperatures service and corrosion resistance. Both wrought and cast forms are available. Wrought alloys can be fusion-welded using GTAW, GMAW and SMAW processes. Cast alloys are difficult to weld and generally, only repair welding is carried out.

(i) Welding Processes and Techniques

(a) **Joint Design :** Joint design and weld grooves for these alloys are similar to those required for Ni-based heat resistant alloys. Cobalt alloy weld metal does not flow and wet the base metal, like steel, and hence the joint must have an open configuration so as to permit electrode or filler metal manipulation and placement of the weld metal. All the joints should be designed for full penetration. The grooves should be machined well to ensure accurate fit up.

(b) **Gas Tungsten arc Welding :** Straight polarity (DCEN) gas tungsten arc welding is generally used for cobalt alloys. Usually a small diameter thoriated tungsten electrode which can carry adequate current is recommended for use. When very small diameter electrodes are used with very large current densities, excessive electrode erosion will take place and the tungsten inclusion will result in the weld metal 12.5 mm and 19 mm thick plates have been welded successfully using single V-groove (70° included angle) and 3.2 mm root opening. Argon was used as shielding gas and good weldability was obtained.

(c) **Plasma Arc Welding Process :** Cobalt based alloys can be joined using the plasma welding process. The constricted arc permits deeper penetration than that by GTAW process. The keyhole technique with a single pass can be used for square groove welds upto 8 mm thickness. This sections can be welded using the melt-in technique.

(d) **GMAW Welding :** Wrought alloys can be welded using the automatic gas metal arc welding process, using reverse polarity but the controlled heat input. The shielding gas used is argon with 1% O_2. GMA Welding of Haynes Alloy-25 can be performed using both short-circuit and spray transfer modes and with an argon-helium-CO_2 gas mixture.

(e) **SMAW Welding :** SMAW, using coated electrodes and reverse polarity (DCEP) can be done in flat position with rapid travel using bck hand technique with minimum weaving. Generally cobalt alloys are welded in solution-treated condition using electrodes of composition similar to the base material.

(f) **Resistance Welding :** Similar and dissimilar combination of cobalt based alloys can be welded using resistance spot, seam and projection welding processes.

(g) **Brazing :** Furnace brazing using high purity reducing or inert atmosphere can be done for cobalt alloys.

(h) **Quality :** In the case of welding cobalt based heat resistant alloys the following precautions are to be observed :

(i) Residual elements like S, Pb, P, Bi have to kept at very low levels for the alloys being welded.

(ii) The weld joint and adjacent areas have to be thoroughly cleaned. No sand Blasting should be done. The oxide layer on the plates should be removed by grinding using aluminium-oxide or silicon carbide wheels.

(iii) The copper backing plate when used for welding should be preferably nickel or chrone plated to avoid copper contamination.

(iv) Other quality requirements such as precaution for weld defects, shape of bead for welding of cobalt based alloys are similar to those of Ni-based alloys.

(v) The yield and tensile strength of the weld metal are mostly equivalent to those of the annealed base metal.

(B) Hard Facing Alloys

These alloys are used generally where erosion and impact wear are jointly encountered. Cobalt-based, hardfacing alloys welding consumables are commercially available in the form of rods, coated electrodes, tubular wires, solid wires and powders. The applications, deposition processews and available product forms of these hard facing alloys are given in Table 10.27. Stellite 6 is the toughest and the least hard. Stellite 12 is harder than stellite 6, and stellite 1 is the hardest and most abrasion resistant of the three.

Table 10.27 : Comparison of data for commonly used stellite alloy welding

S. No.	Alloy/trade name	Available product form	Deposition process	Application areas	Service conditions
1	Stellite 1	CR, BR, P	Oxyacetylene GTAW, PAW	Screw components, pump sleeves.	Abrasion and solid erosion valves.
2	Stellite 6	CR, BR, TW, SW	Oxyacetylene GTAW, SAW	Valves, pumps screw components, knives, erosion shields.	Antigalling resistance cavitation erosion, moderate abrasion.
3	Stellite 12	CR, BR, TW, SW, P	GTAW, SMAW GMAW, PAW	Turbine blades, valves, pumps, knives, IC engine valves.	Good resistance to abrasion.

CR - Coated rod; BR - Bare cast rod; P - Powder; TW - Tubular wire; SW - Solid wire.

(i) Welding Techniques

Common welding techniques like oxyacetylons, gas tungsten arc, shielded metal arc welding, gas metal arc welding are mostly used for depositing stellite. Each process has its own advantages and charactersitics.

(a) Oxyacetylene Welding : The surface to be deposited is thoroughly cleaned before welding by grinding or machining. Figure 10.23 shows the recommended design of the base metal for stellite deposition. A standard type of oxyacetylene torch is used for depositing stellite and a proper tip size has to be selected. Before stelliting the base metal has to be preheated. Small jobs are preheated locally with the torch and big jobs are preheated using gas burners or charcoal in a brick enclosure. While depositing following important parameters have to be taken care of :

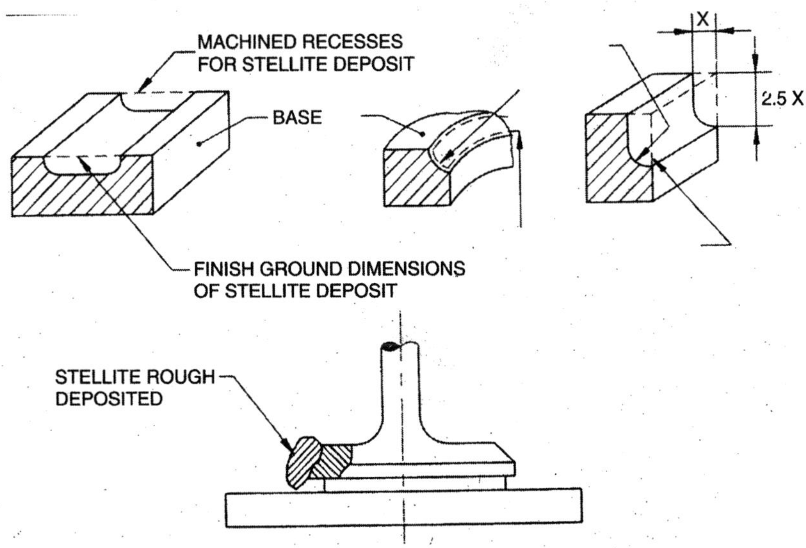

Fig. 10.23 Recommended base metal designs.

(i) Flame adjustment
(ii) Sweating
(iii) Depositing.

(i) Flame Adjustment : Normally for depositing stellite a 3x feather flame is used which is carburising in nature and removes the oxides. For a 3x flame. The length of the excess acetylene (bluish coloured) feather is three times. The length of the bright inner cone. This flame adds carbon to the base metal as temperatures rises so that the melting point of base metal surface is lowered.

(ii) Sweating : The ewxcess acetylene flame is used to produce the phenomenon of sweating i.e., an extremely thin layer of the steel base material is melted to give a watery glazed appearance. A weating surface is necessary for stellite rod deposition, to allow the depositied metal to spread freely. For obtaining the seating the torch is directed at an angle of 30° to 60° to the job surface. The sweating distance should be 3 mm ahead of the deposit.

(iii) Depositing : The flame is withdrawn such that the end of the stellite rod is brought between the inner core of the flame and the sweated steel base metal. The tip of the inner cone touches the rod and the rod is slighly touch as the sweating area. Thus the end of the stellite rod will melt and form a puddle on the sweat surface and then will flow and spread on the sweat surface. After sweating the stellite rod is removed from the flame and the flame is directed into the puddle. This allows the molten stellite

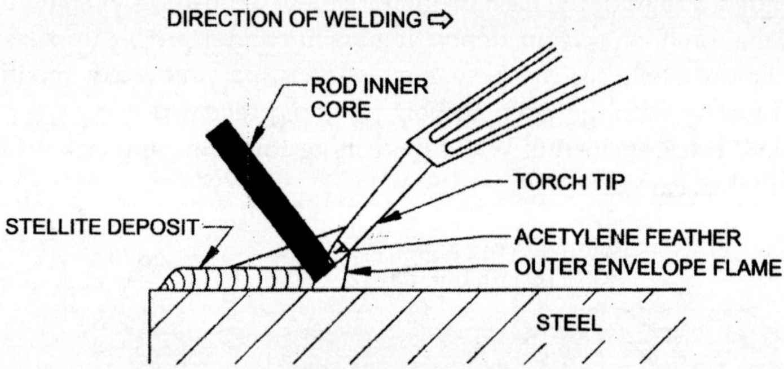

(a) Back hand method

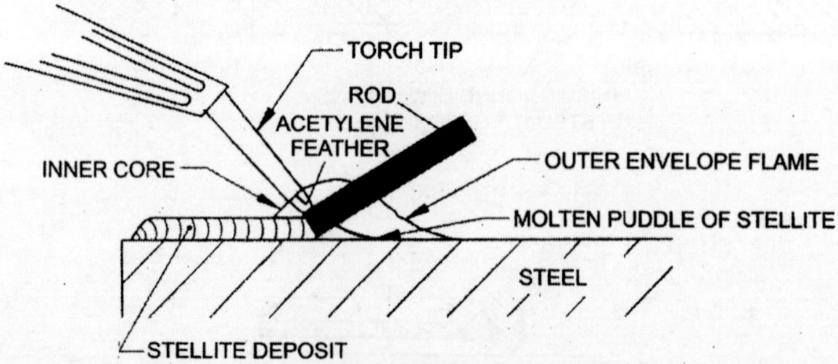

(b) Regular method

Fig. 10.24 Techniques for stellite deposition.

to spread uniformly. Now the rod is returned and more stellite is melted into. The process is repeated in sequance until the desired thickness of the first layer has been deposited. If a deposit thickness of more than 3 mm is required, the flame is repeatedly moved back over the job to melt a thin layer of the deposited stellite to smoothen out the high spots behind the molten puddle. The front edge of the puddle should be allowed to solidify. If a second pass is to be made the stellite layer should not be melted to the base metal, only the surface layer of the stellite should be melted. If it is melted deeper, dilution of stellite by iron from the base metal will take place and reduce the wear resistance of the deposit. Stellite can be deposited by two different techniques. Refer Fig. 10.24.

1. The normal method, where the deposition progress towards the hand holding the stellite rod.
2. The back hand method where the deposition progresses towards the hand holding the torch (i.e. the torch moves backward).

(iv) Cooling : As soon as the deposition is completed, the entrire job is to be placed in a chamber cntaining powdered mica or slaked lime or powdered asbestos or any similar insulating material so that the jobs will cool slowly. Slow cooling is essential to rpoduce a crack-free deposit. Never cool a job that has been heated in a furnace or by a torch by immediately dipping it in the water. This will rise to quench cracks and high interval stresses.

(b) Arc Welding Processes : Stellite deposits, free from dewfects, can be deposited by the electric arc process. In the metal arec process the stellite deposited has move dil;ution compared to with oxyacetylene process. The first layer has more dilution which reduces the wear resistance but improves the toughness. The second layer will be harder and the third layer, if deposited, will be almost undiluted.

(c) Shielded Metal Arc Welding : Both DC and AC power can be used for stellite deposition. DCEP is mostly used for stellite deposition. A flux-coated stellite rod is used to get a smoother and flatter bead. The deposition parameters depends on the size of the rod used. When AC power is used, an OCV of a minimum 100 V is preferred. The current required depends on the size of the rod used. While depositing the following must be observed carefully :

(i) Electrode should be held nerly vertical.
(ii) Maintain an arc length of around 4 to 6 mm for a 6 mm diameter rod.
(iii) Do not weave. Use only stringer bead technique. Weaving may cause slag entrapment.
(iv) When the width of deposit required is greater than that which is possible, the second bead should overlap the first bead by about 1/3 of the width of the bead. (Fig. 10.25).

Fig. 10.25 Single layer.

(v) Before laying the second bead clean the first bead thoroughly.

(vi) For depositing a second layer, each bead of the second layer should cover the line of fusion between the previous two beads below. (Fig. 10.26)

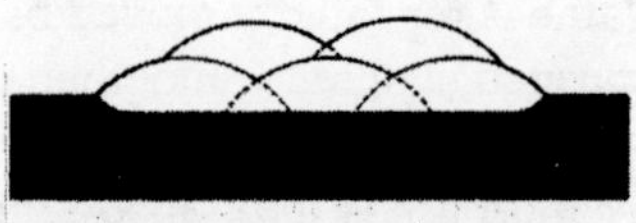

Fig. 10.26 Multilayer build up.

(vii) For stellite deposited by SMAW, cooling does not need close attention.

(d) Gas Tungsten Arc Welding : In this process the source of heat is more localised than in oxyacetylene process. The dilution of stellite is less than that of SMAW process. Austenitic stainless steel stabilized with titanium can be hard faced with stellite using this process. The job to be deposited should be free from sharp corners. The job has to be preheated to reduce the melting of the base metal and to prevent contraction cracking. DCEN is preferable. The deposition is done using the backward technique. The torch is moved from left to right with the arc directed on the rod and the molten pool of stellite instead of the base metal. A foot operated HF unit is used to control the arc to decrease the heat slowly so that the surface of the molten pool solidifies faster. It is preferable to cool the component slowly in dry insulating powder and carry out subsequent heat treatment.

(e) Quality : Quality is an important aspect during deposition of stellite by the gas or arc welding processes. The following are some important aspects :

(i) In oxyacetylene hard facing, the impurities present on the base metal have to be thoroughly removed. During hardfacing, if oxide is produced, it has be dislodged by prodding or rubbing with the electrode tip to prevent formation of gas and porosity due to the reaction of oxide and carbon in the hard facing alloy.

(ii) The properties and quality of deposits, deposited by the arc welding depends on the type of hard facing alloy, the base metal composition, the welding process and the method employed and also on number of layers deposited and other welding parameters. The selection of welding process for a particular application depends to a greater extent on the ability to limit dilution while achieving high rate of deposition.

❑❑❑

11

CHAPTER

WELD DESIGN – SYMBOLS AND ESTIMATION

In a welded structure loads are transferred from one member to other member through welds placed in a joint. The joints are broadly classified into the following six types :

(i) **Butt Joint :** Joint between two members ailgned approximately in the same plane.

(ii) **Tee Joint :** Joint between two members located approximately at right angles to each other in the form of 'T'.

(iii) **Lap Joint :** Joint between two overlapping members in parallel planes.

(iv) **Edge Joint :** Joint between the edges of two or more parallel or nearly parallel members.

(v) **Corner Joint :** Joint between the members located at right angles to each other.

(vi) **Cruciform Joint :** A cruciform joint is defined as the joint in the shape of a cruciform with the fillet.

These joints are illustrated in Fig. 11.1. Different type of weld joints can be made by employing the following types of welds viz. Groove welds, fillet welds and plug welds. Sometimes combinations of the above types of welds are plug welds. Sometimes combinations of the above types of welds are used to make a joint depending on the requirements of the strength and load conditions.

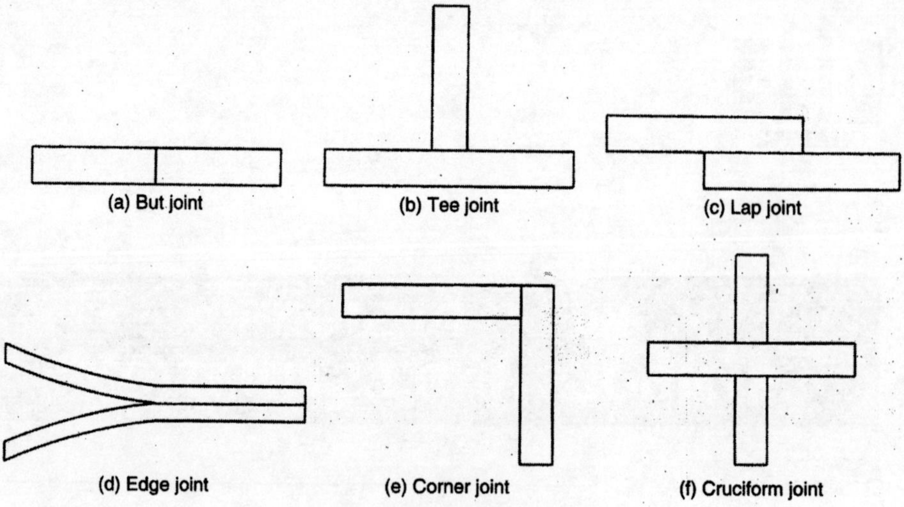

Fig. 11.1 Types of joints.

11.1 GROOVE WELDS

A weld made in a groove between the work-pieces is known as groove weld. If the weld is made on one side of the work-piece, then it is known as a single groove weld. If the weld is made on both sides of the work-piece, then it is known groove weld. If the weld is made on both sides of the work-piece, then it is known as double groove weld. Groove welds of different types are used in many combinations. Figure 11.2 and 11.3 shows various combinations of single groove welds and double groove welds respectively. The square groove weld is economical, provided one is able to get suifficient penetration and strength. But its use is limited to the thickness of the work-piece. When the thickness of the joint increases, the edges of the work-piece are prepared to a particular geometry to provide accessibility for welding and ensure sufficiant soundness. From the economic considerations the groove angle and the root gap are selected in such a way that a minimum amount of metal is consumed, at the same time giving sufficient accessibility for depositing sound weld metal. A typical V-groove weld with its nomenclature is shown in Fig. 11.4.

11.2 FILLET WELDS

A fillet weld is defined as a weld of approximately triangular cross-section joining two surfaces of work-piece approximately at right angle to each other in a lap joint, in a T-joint or in a corner joint. Fillet-welded joints are comparatively easy to make than groove-welded joints from the point of view of fit-up and edge preparations. Sometimes intermittent fillet welds are also used to avoid over-welding and distortion problems. A typical fillet weld with its nomenclature is shown in Fig. 11.5. A fillet weld is specified by the leg size, which is generally the side of the

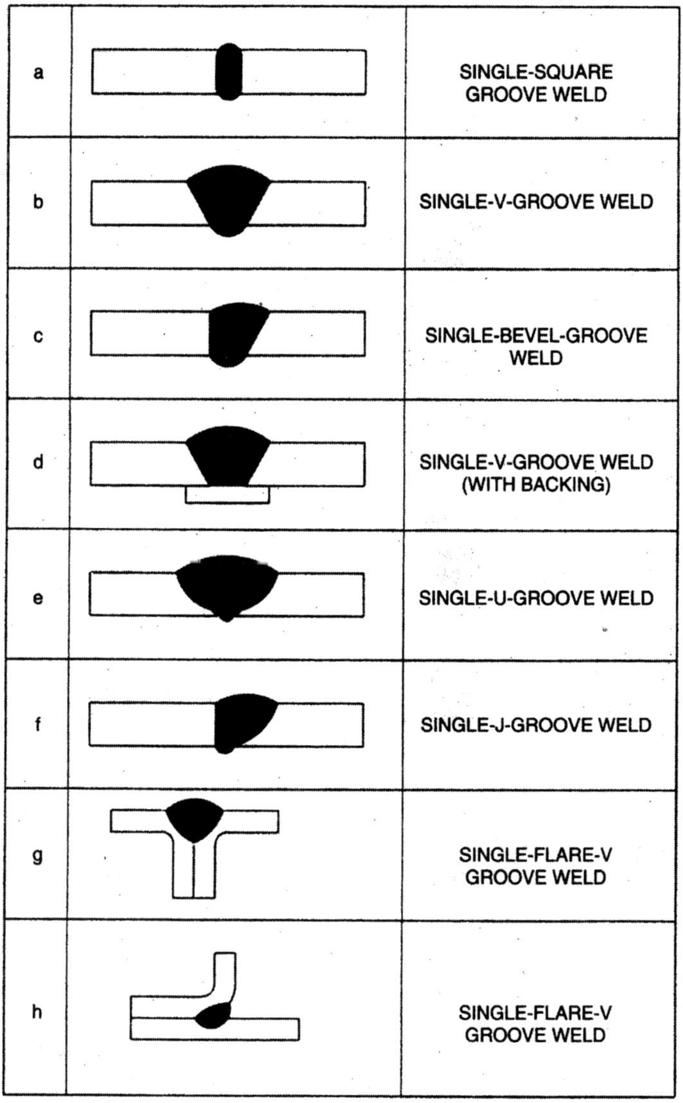

Fig. 11.2 Single groove welds.

largest right angled triangle inscribed in it or sometimes a fillet is specified by its throat size. The throat of the fillet weld is defined as the shortest distance between the root of the weld and the face of the weld. The strength of the fillet weld is dependent on the effective throat and the length of the weld. Generally fillet welds are used for transfer of shear forces parallel to the axis of the weld and transfer of forces perpendicular to the weld axis. Fillet welds are generally used in Tee, corner and lap joints due to its simplicity and economics.

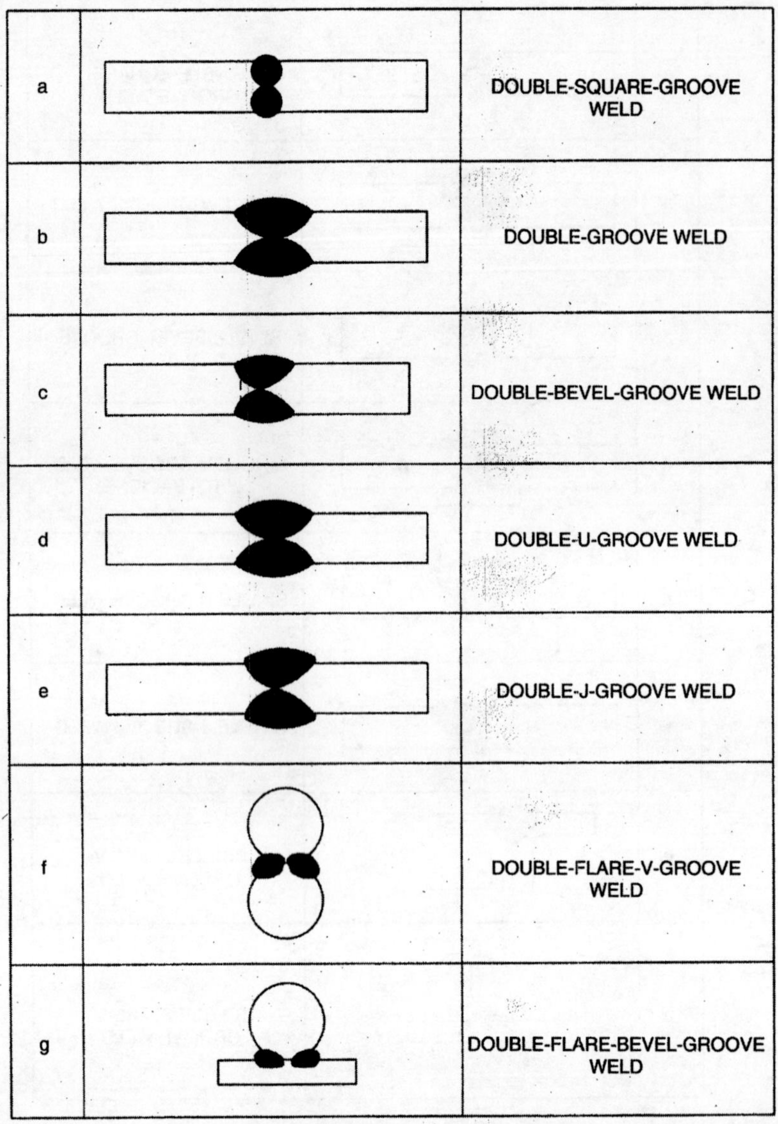

Fig. 11.3 Double groove welds.

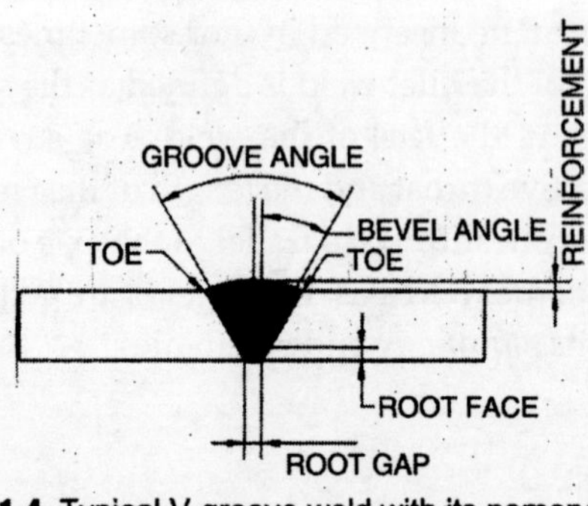

Fig. 11.4 Typical V-groove weld with its nomenclature.

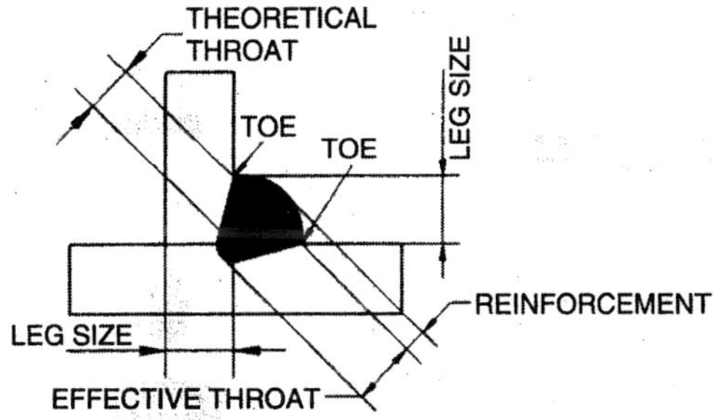

Fig. 11.5 Typical fillet weld with its nomenclature.

11.3 PLUG WELD

A weld deposited in a hole made in one of the members of a joint, fusing that member to another member is called plug welding (Fig. 11.6).

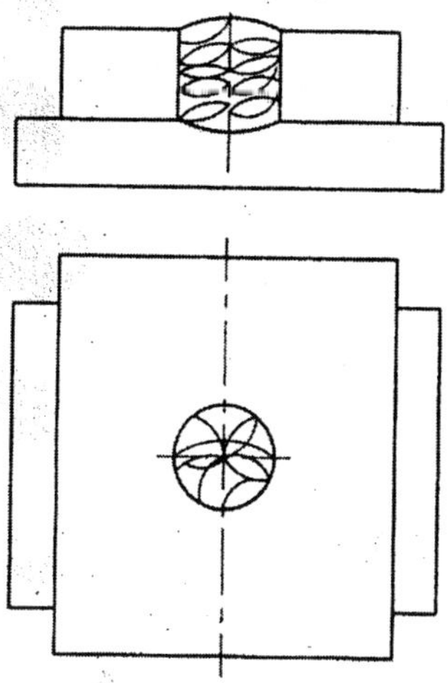

Fig. 11.6 Plug weld.

11.4 WELDING SYMBOLS

Standard Welding Symbols are used to indicate welding information in the fabrication drawings. These symbols convey the design requirements envisaged by the designers to the shopfloor personnel in a concise manner. Generally a welding symbol indicates the type of weld, groove design, weld size, length of weld, welding process, contour of the faced and root and other information. Further welding symbols are made complete by denoting the shape of the external surface. See Figs. 11.7 (a) and 11.7 (b).

Welding symbol including the weld size, length of weld etc. is represented as shown in Fig. 11.8. It comprises the following :

(i) an arrow line for each joint.

(ii) a dual reference line, consisting of two parallel lines, are continuous line and the other dashed or discontinuous.

SL.NO.	TYPE	DETAILS	SYMBOL
01	BUTT WELD BETWEEN PLATES WITH RAISED EDGES		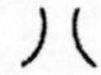
02	SQUARE BUTT WELD		
03	SINGLE-V-BUTT WELD	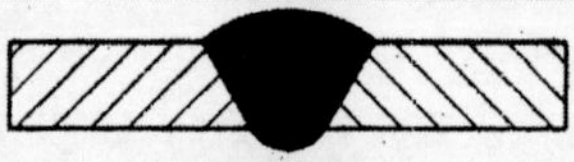	
04	SINGLE-V-BUTT WELD WITH BROAD ROOTFACE	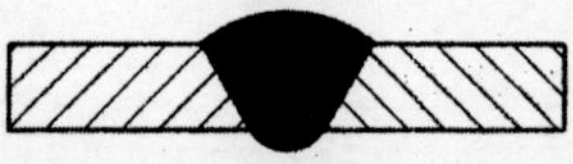	
05	SINGLE -BEVEL BUTT WELD		
06	SINGLE -BEVEL- BUTT WELD WITH BROAD ROOT FACE	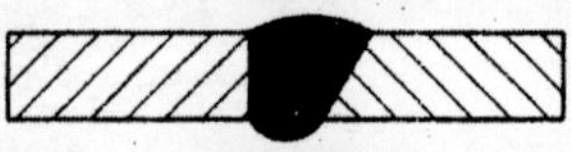	
07	SINGLE -U BUTT WELD		
08	SINGLE -J BUTT WELD	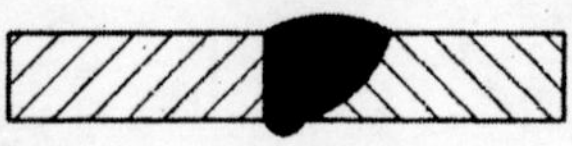	
09	BACKING RUN OR BACKING WELD OF SEALING RUN		
10	FILLET WELD	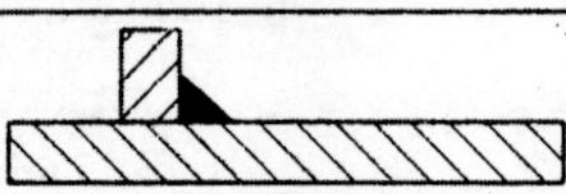	
11	PLUG WELD OR SLOT WELD	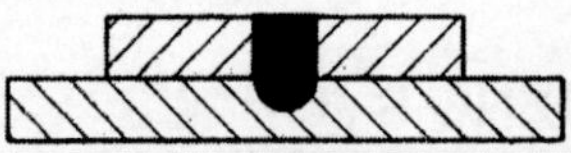	

Fig. 11.7(a) Basic welds symbols.

S. No.	Shape of weld surface	Symbol
I	Convex	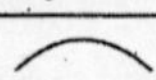
II	Concave	
III	Flat (normally finished flush with the adjacent surface)	

Fig. 11.7(b) Weld surface symbols.

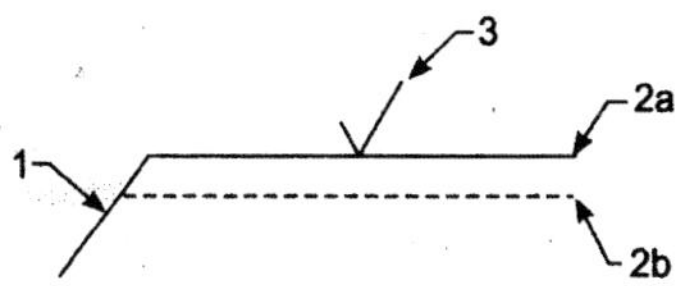

1. Arrow line
2a. Reference line (Continuous line)
2b. Identification line (Dashed line)
3. Weld symbol

Fig. 11.8 Method of representing a welding symbol.

11.5 REPRESENTATION BY SYMBOLS

A certain number of dimensions and conventional signs are used to represent a weld.

(i) Arrow line
(ii) (a) Reference line (continuous)
     (b) Identification line
(iii) Weld symbol (dashed line).

11.5.1 Position of Symbols

Figure 11.9 explains the meaning and use of the 'arrow side' of the joint and the 'other side' of the joint. The symbol may be placed either above or below the reference line as per the following :

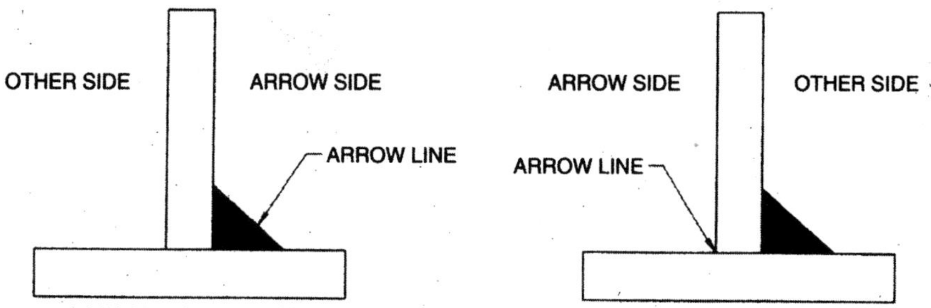

(a) Weld on the arrow side

(b) Weld on the other side

Fig. 11.9 T-weld joint.

(i) The symbol is placed on the continuous line of the reference line if the weld is an arrow side of the joint.

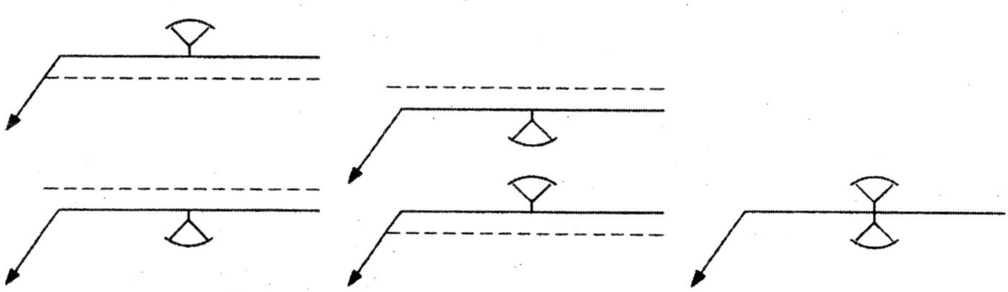

(i) To be welded on the "arrow side"

(ii) To be welded on the "other side"

(iii) To be welded on "both sides" (Symmetrical welds)

Fig. 11.10 Positioning of symbols.

(ii) The symbol is placed on the dashed line of reference line if the weld is on the other side of the joint.

(iii) For symmetrical welds, i.e., welded on both sides, the dashed line of the reference line need not be shown.

Figure 11.10 explains all above.

11.5.2 Weld Dimensions

Weld size and weld length (both in mm) are placed in the welding symbol as shown in Fig. 11.11. For groove welds the weld throat size is the distance from the surface of the part to the bottom of the penetration. For fillet welds there are two methods to indicate the weld size. Usually the leg length of the fillet is used as the fillet size. When the leg length is used for the fillet weld size, the letter 'z' should be placed in front of the weld size. In another method the throat thickness is used for fillet weld size. The letter 'a' should be placed in front of the weld size to denote the 'throat'. Typical examples of showing fillet sizes are indicated in Fig. 11.12.

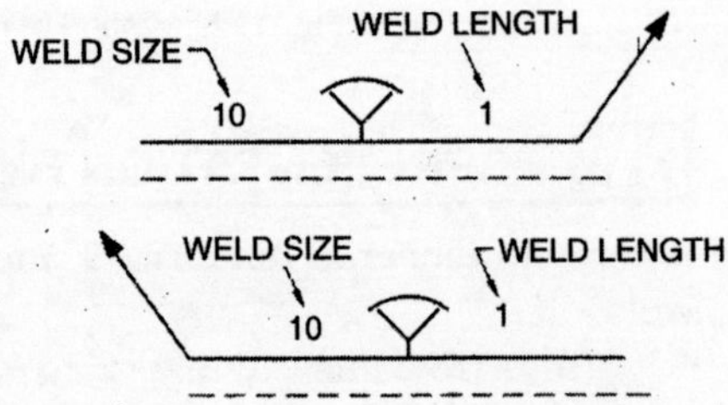

Fig. 11.11 Location of size, symbol and length.

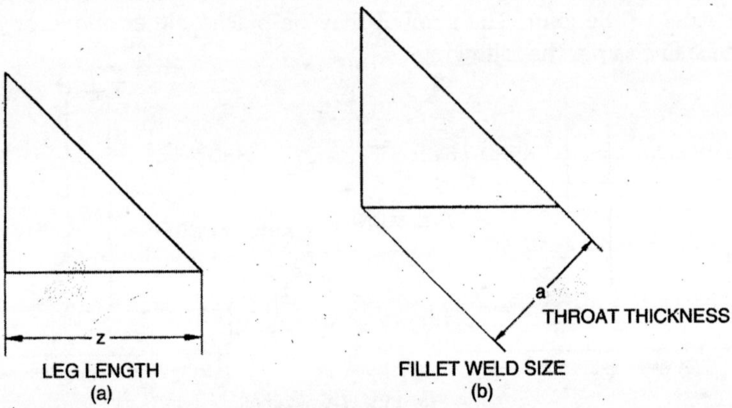

Fig. 11.12 Fillet weld size.

11.5.3 Peripheral Welds

When a weld is to be done all around the mating component, the symbol 'circle' is to be shown as illustrated in Fig. 11.13.

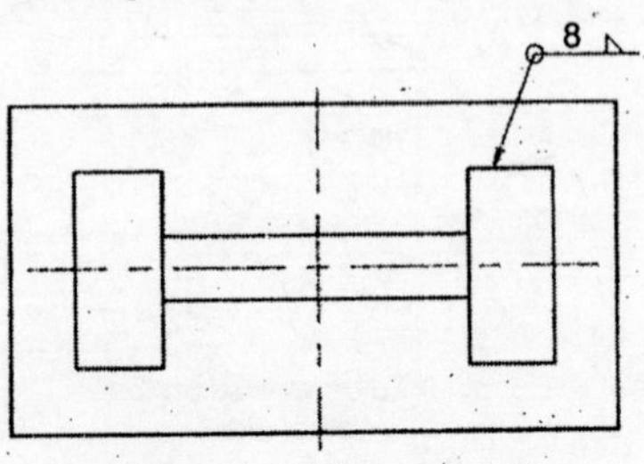

Fig. 11.13 Peripheral weld.

11.5.4 Site Weld

Sometimes the same drawings are used for both shop fabrication and for site fabrication. In such cases site welds have to be separately identified by a flag symbol as shown in Fig. 11.14.

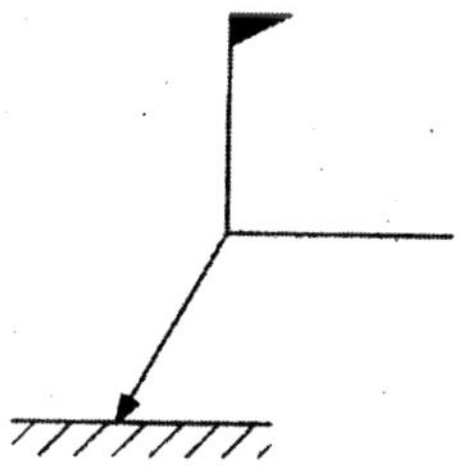

Fig. 11.14 Indication of site weld.

11.6 STATIC STRENGTH OF THE WELDED JOINTS

Usually weld metals are selected such that they are stronger than the base metal and one can calculate the strength of a butt-welded joint using the strength of the parent material. For this, throat thickness and nature of loading should be known. The load carrying capacity of a butt weld is decided by the following equation :

$$P = a \times l \times \sigma_{all},$$

where,

P = applied load in kn

a = weld throat size in mms.

l = length of weld in mms.

σ_{all} = allowable stress of the weld metal in N/mm^2.

The throat of a full penetration weld is the same as the thickness of the part being joined whether under axial load, shear or bending. If the design is checked for stress, then there is no need to do special calculations for the butt weld.

11.7 FILLET SIZE CALCULATIONS

The throat of a fillet weld is assumed to be in shear for all types of load. The shear stress in the throat is the load divided by the throat area as shown in Fig. 11.15.

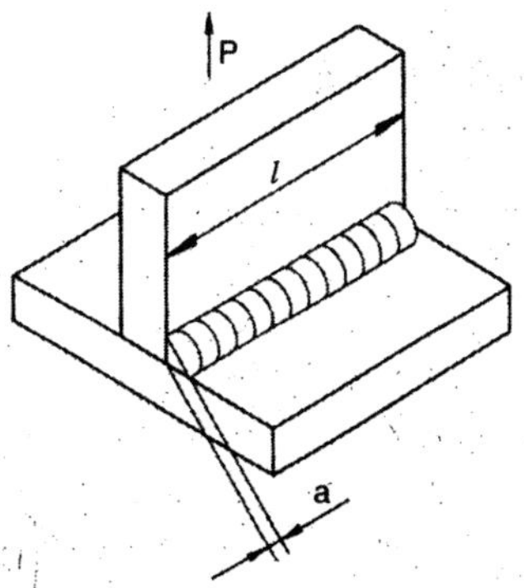

Fig. 11.15 Shear stress in the throat.

$$P = a \times l \times \tau_{all}$$

where

P = applied load in N

a = throat size in mms.

l = length of weld in mms.

τ_{all} = allowable shear stress of weld metal in N/mm^2

Example : Calculate the fillet size required for the plate shown in Fig. 11.16.

Weld throat area $= 100 \times 2 \times a$

Allowable shear stress = 115 N/mm^2 steel for mild steal electrodes.

Throat size $= a$

Length of weld $= 2 \times 100$ mm.

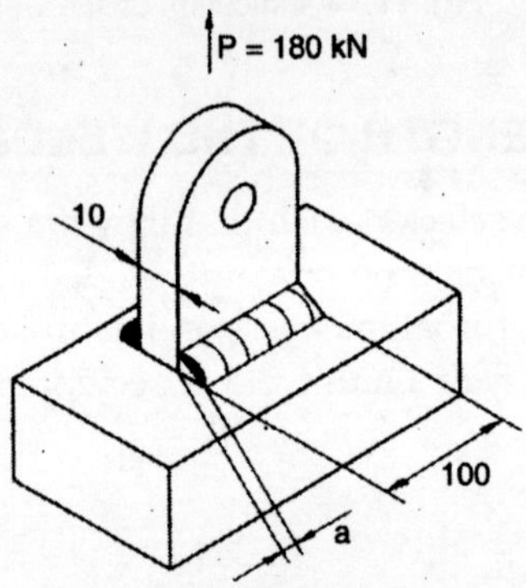

Fig. 11.16 Fillet size.

Weld size 'a' can be calculated using the equation

$$P = a \times l \times \tau_{all}$$

$$180000 = a \times 2 \times 100 \times 115$$

So,
$$a = \frac{180000}{2 \times 100 \times 115}$$

$$= 7.83 \text{ mm}$$

$$\text{Leg size} = 1.4 \times 7.83 = 10.95 \text{ mm}$$

So, recommended weld size = 12 mm. leg size.

11.8 EDGE-PREPARATIONS

The recommended edge preparations shown in Table 11.1 are intended primarily for shield are welding in the flat positions. Since overhead and vertical welding require manipulation of the electrodes, easy access to the vertical welding require manipulation of the electrodes, easy access to the root of weld is necessary for welding in these positions. This is achieved by increasing the groove angle or increasing the root gap. In the case of horizontal position butt welds the edge preparation may have to be symmetrical with respect to the horizontal plane with a narrow bevel angle for the lower section to be welded. Normally single and double

'V' preparations are machined or flame-cut. Single and double 'U' and 'J' preparations have to be machined.

Table 11.1 : Typical edge preparations

Sl. No.		Thickness t, mm	Root gap G, mm	Root face R, mm	Angle α
1	Open square butt weld (Welded from both sides) 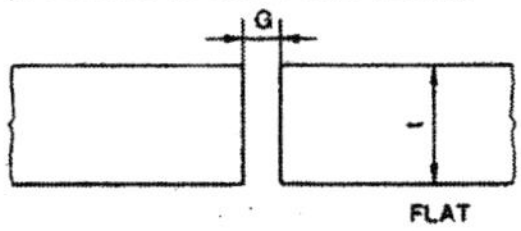	3 - 6	3	-	-
2	Open square butt weld with backing 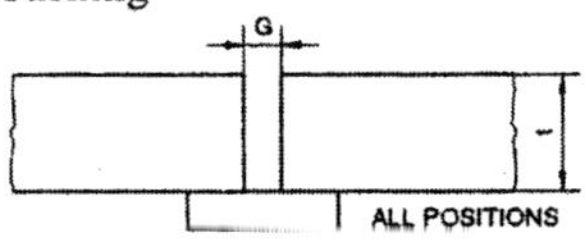	3 - 5 5 - 8 8 - 16	6 8 10	-	-
3	Single V Butt weld 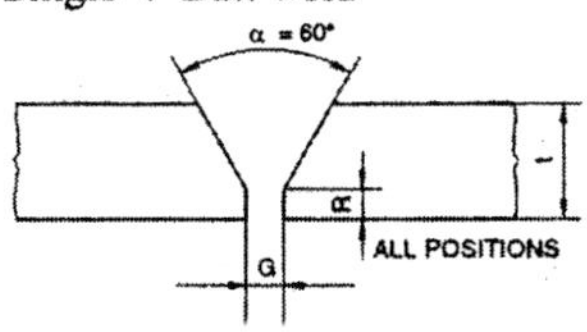	5 - 12 Over 12	2 2	1 2	60° 60°
4	Single V Butt weld 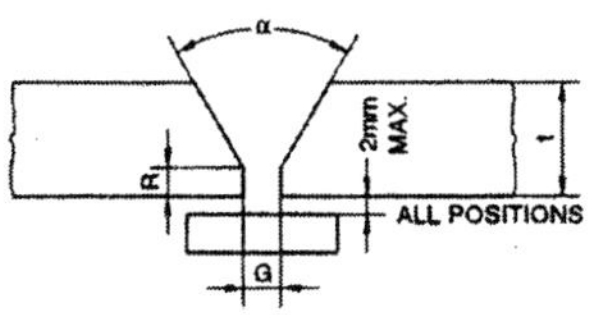	Over 10	6 10	0 0	45° 20°
5	Double V Butt weld 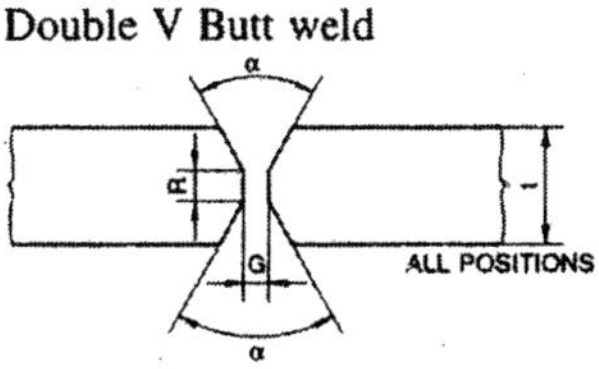	Over 12	3	2	60°
6	Asymmetric double V Butt weld 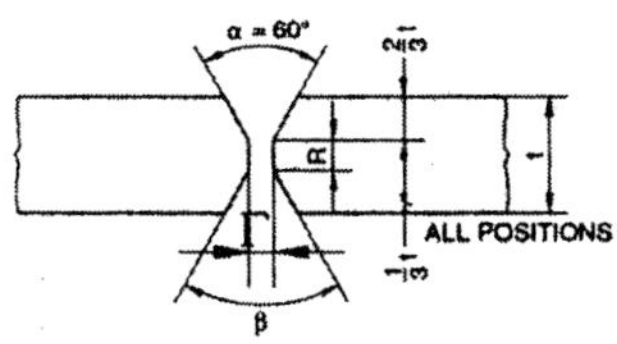	Over 12	3	2	60° Angle β = 60° This may be increased to 90° to facilitate gouging

<table>
<tr><td>7</td><td>Single U Butt weld
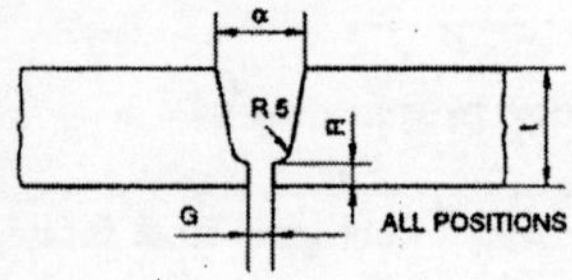
</td><td>Over 20</td><td>2</td><td>3</td><td>20°</td></tr>
<tr><td>8</td><td>Double U - Butt weld
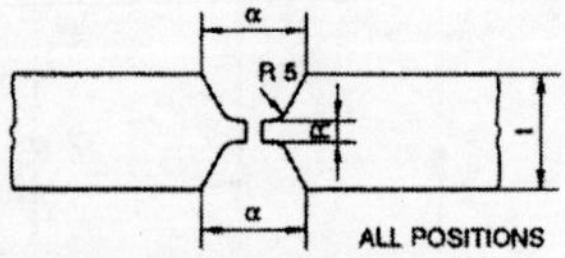
</td><td>Over 40</td><td>2</td><td>3</td><td>20°</td></tr>
<tr><td>9</td><td>Single bevel butt weld
(Welded both sides)
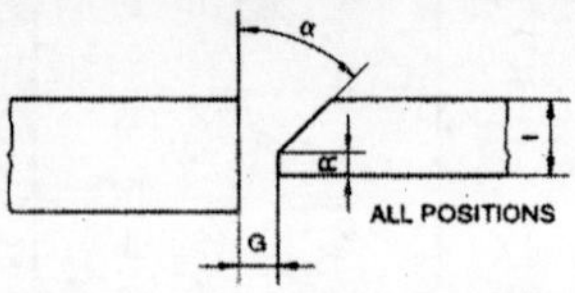
</td><td>5 - 12
Over 12</td><td>3
3</td><td>1
2</td><td>45°
45°</td></tr>
<tr><td>10</td><td>Double bevel butt weld
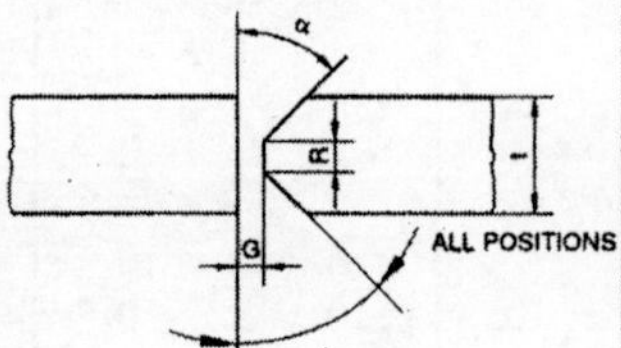
</td><td>Over 12</td><td>3</td><td>2</td><td>45°</td></tr>
<tr><td>11</td><td>Single J - Butt weld
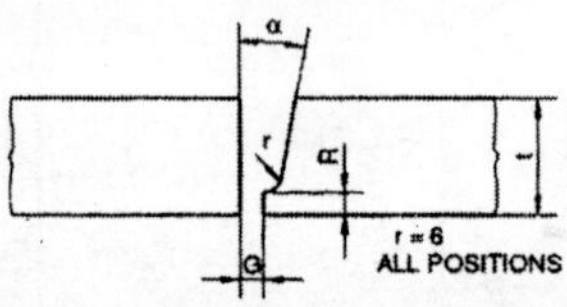
</td><td>Over 20</td><td>3</td><td>3</td><td>20°</td></tr>
<tr><td>12</td><td>Double J - Butt weld
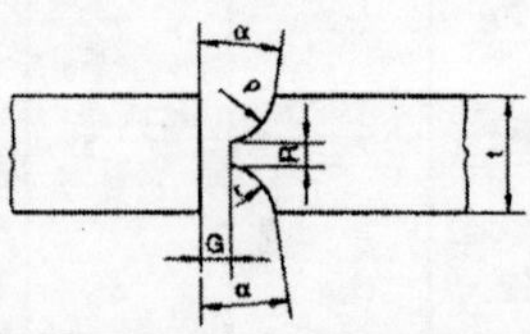
</td><td>Over 40</td><td>3</td><td>3</td><td>20°</td></tr>
</table>

□□□

12

CHAPTER

WELDING DISTORTION AND CONTROL

❑ **INSIDE THIS CHAPTER**

12.1 DISTORTION

The change in dimensional shape that occurs in a fabricated structure after welding is called distortion. This change is often observed soon after the removal of jigs, fixtures, clamps, if any, that were used while welding the structure. This permanent change that occurs in the shape due to inelastic strains, induced while welding is called distortion.

12.2 CAUSES

During welding and subsequent cooling, thermally induced shrinkage strains build up in the weld metal and the base metal regions. The stresses resulting from these shrinkage strains combine and react to produce bending, buckling and rotation. The fundamental dimonsional changes that occur during welding are :

(i) transverse shrinkage perpendicular to the weld line.
(ii) longitudinal shrinkage parallel to the weld line.
(iii) angular distortion-rotation around the weld line.

These dimensional changes are shown in Fig. 12.1.

12.2.1 Transverse Shrinkage

The transverse shrinkage is due to the rotational movement of the unwelded portion of the assembly. It is a function of heat input and location of tack welds. Transverse distortion occurs perpendicular to the weld line. It will be predominant

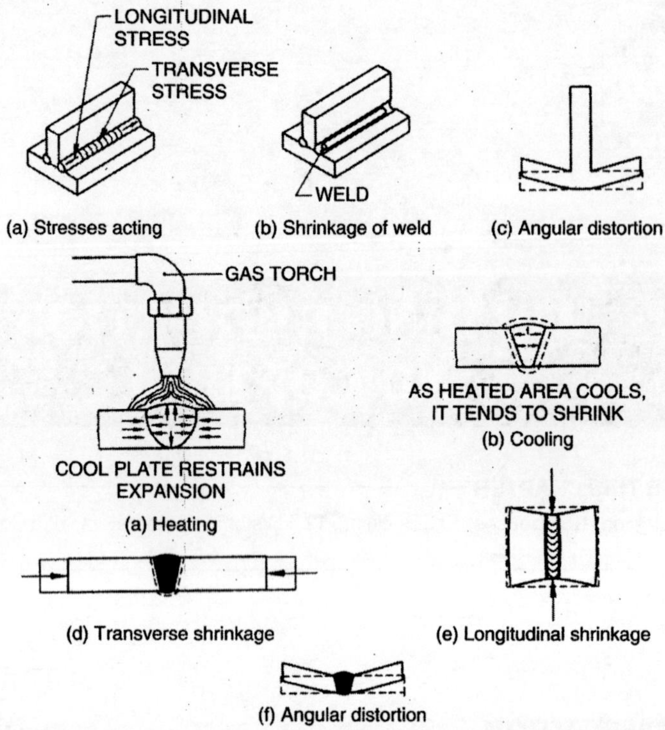

Fig. 12.1 Transverse and longitudinal shrinkage in fillet and butt welds.

when length to width ratio is very much less. The shrinkage formula for transverse distortion of butt welds is :

$$S = \frac{5.16\, A_w + 1.27\, d}{t},$$

where,

A_w = area of cross-section of the weld in mm^2

t = thickness in mm.

d = free distance or root opening in mm.

S = transverse shrinkage in mm.

12.2.2 Longitudinal Shrinkage

Longitudinal shrinkage occurs parallel to the weld line. When the weld line does not concide with the neutral axis of the weld structure, the longitudinal shrinkage of the weld induces bending movements resulting in longitudinal distortion of the structure. Longitudinal distortion will be predominant if the length of welded assembly is more than the 10 times of the width of the assembly. This longitudinal shrinkage is influenced by the area of cross-section of the weld, length of weld, thickness and stiffness of the section. The relation for longitudinal shrinkage is given by,

$$S = \frac{0.005\, L^2\, A_w\, d}{I},$$

where,

S = longitudinal Shrinkage in mm.

d = distance between the neutral axis and weld centre line in mm.

I = moment of inertia about the neutral axis.

A_w = area of cross-section of weld in mm^2

L = length of continuous weld in mm.

Refer Fig. 12.2 for longitudinal shrinkage.

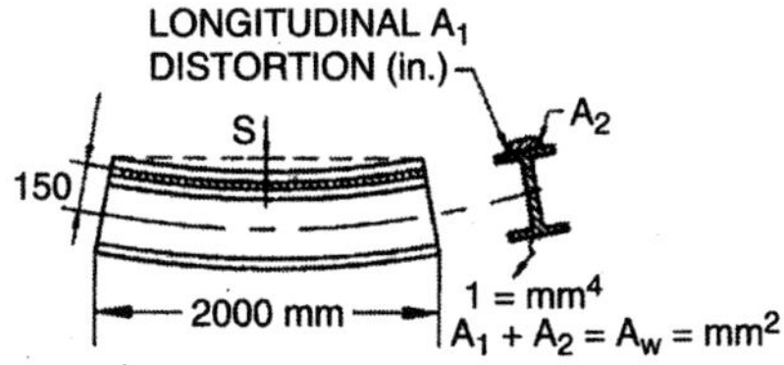

Fig. 12.2 Welded 'I' beam.

12.2.3 Angular Distortion

Angular distortion occurs due to the rotation around the weld centre line. Mostly the angular distortion depends on the area of the weld cross-section, weld metal deposited, the thickness of the plate and the sequence of welding. The relation of angular distortion is given by,

$$S = \frac{0.02\, W \,.\, a^{1.3}}{t^2}$$

where,

S = angular distortion in mm.

W = flange width in mm.

a = weld size in mm.

t = plate thickness in mm.

Refer Fig. 12.3 to illustrate angular distortion.

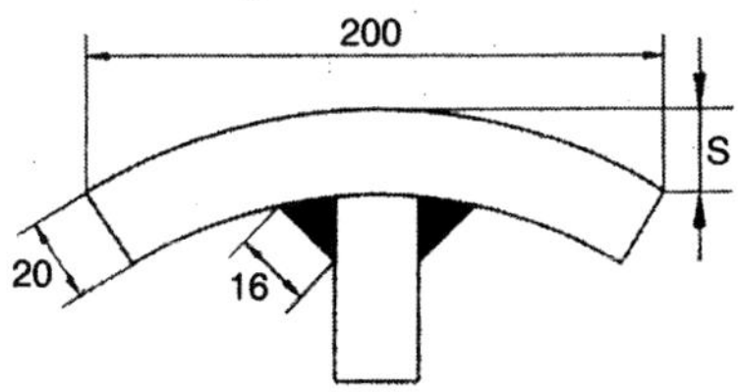

Fig. 12.3 T-beam weld.

12.3 BUCKLING

Buckling is the total deformation due to the localised non-uniform heating and cooling particularly in plates of less thickness. When this plates are welded, residual compress stress occur in areas away from the weld which cause buckling. Buckling due to welding increases directly as the thickness decreases.

12.4 DISTORTION PARAMETERS

12.4.1 Structural Parameters

The important structural parameters are :

(i) geometry of the structure
(ii) plate thickness
(iii) joint type

The geometry of the weld joint and shape of the structure largely contribute to the resistance offered to the total shrinkge and hence distortion. The stiffness of the section offers resistance to longitudinal shrinkage and distortion. The longitudinal distortion also depends on the distance '*d*' between the neutral axis of the cross-section and the weld axis. The larger this distance, the larger is distortion. The transverse shrinkage and distortion are also resisted as the shape or mass of the metal is larger. As the thickness of the material decides the cross-sectional area for resisting the distortion, it influences all the three types of distortion. When the thickness is larger, the distortion will be lesser. In the case of thin plates a phenomenon called buckling is also observed. The basic configuration of the weld joint types such as butt, fillet and lap also contribute to weld distortion. The distortion is larger as the included angle between the groove faces increases. In general, an unsymmetrical double 'V' joint is associated with lesser distortion comparaed to a symmetrical double 'V' joint, without back gouging. The ratio between larger to smaller area of the joint is normally kept as 70:30 for a thickness less than 25 mm and 60:40 for thickness above 25 mm. In such cases the smaller area of the joint is completed prior to the larger portion of the joint.

12.4.2 Material Parameters

The physical properties of the base and filler metals influence the weld distortion. Some of the properties affecting the distortion are :

(i) thermal expansion coefficient.
(ii) thermal conductivity.

The thermal expansion coefficient increases with the increase in temperature. For instance austenitic stainless steels with higher thermal expansion coefficient and lower thermal conductivity distort more than ferritic steels as these two physical properties promote distortion.

12.4.3 Fabrication Parameters

The fabrication parameters are :

(i) Welding processes.
(ii) Welding procedures.
(iii) Assembly sequences.

In general, the heat input rate, the heat affected zone width, depth of penetration and speed of welding vary with different fusion welding processes. Use of a faster welding process helps in minimising the distortion due to associated narrow heat affected zones. Since most of the high speed welding processes are associated with deeper penetration, it helps in minimising distortion. The use of a larger diameter of electrode in the first pass helps greatly in minimising the distortion. Iron powder and coated electrodes should be preferred compared to the other types since more metal can be deposited at a lesser time minimising the distortion. The effect of the welding process and associated heat inputs on distortion in a fabricated structure using the three common fusion welding process viz SMAW, CO_2 and SAW is shown in Table 12.1.

Table 12.1 : Effect of welding processes

S. No.	Process	Number of welds	Total number of passes	Total heat J/mm	Maximum distortion	Central (mm)
					X-axis	Y-axis
1	SMAW	10	8 + 12 + 8 = 28	62157.42	– 2.375	+ 7.875
2	CO_2 (GMAW)	10	12 + 12 + 6 = 30	39990.26	– 2.375	+ 6.375
3	SAW	10	8 + 8 + 6 = 22	35424.10	+ 0.625	+ 4.125

It can be seen that the welding procedure adopted for the fabrication of the structure very much influences the distortion, and hence, control and manipulation of the various procedural variables can minimise distortion to a large extent. The procedure technique includes :

(i) Material preparation procedures.
(ii) Preheat condition.
(iii) Deposition techniques—heat input.
(iv) Fit up of the joint.
(v) Degree of constraints.

The method employed to prepare the materials for welds are gas cutting, gouging and soon. Since the residual stresses are present in gas cut plates, the plates may require correction by localised heating before usage. Similarly oxygen or air arc gouging increases the shrinkage during repair welding. The remedy is proper use of low heat input gouging electrodes with proper techniques. Preheat is usually applied to reduce the cracking tendency of the weld. It may also be applied for reducing the angular change in a fillet weld. This reduction is due to a lesser thermal gradiant between the weld and the base metal zones. The effect of heat input on the distortion is given in Table 12.1. Using identical welds, locations and weld sequences, in the three processes with the heat input only variable, it was found that the SAW process produced lesser distortion than CO_2 and SMAW process. The SMAW process produced maximum distortion. The joint fit up or the root opening has considerable effect on distortion. The effects of root opening on transverse shrinkage and distortion is well known. As the root opening increases the shrinkage, the overall distortion increases.

The number of runs and the sequence employed to deposit each layer of weld metal contributes considerably to the distortion. As the welding progresses, the heat input into the weld metal and the restraint change, imparting different amounts resistance to shrinkage and distortion. A minimum number of passes should be used for depositing the weld metal from a distortion prevention angle, though microstructure wise the other way is beneficial. Tack welding the joint edges keeps the under restraint condition, offering resistance to distortion. The placement of tack welds also plays an important role in minimising the distortion.

Lesser the plate thickness, closer should be the tack welds to avoid angular distortion and buckling of the thin plates.

12.4.4 Assembly Parameters

By controlling the assembly methods employed to build the structure, distortion can be minimised. Usually two methods of assembly procedures are adopted :

(i) By estimating the amount of distortion likely to occur during welding and then assemblying the job with the members preset to compensate for distortion is called presetting.

(ii) By assemblying the job so that it is nominally correct before welding and by employing some form of restraint to minimise the welding distortion after welding called fixturing.

Figure 12.4 (a) and 12.4 (b) show simple examples of the presetting and prebending methods applied to fillet and butt welds. The amount of preset

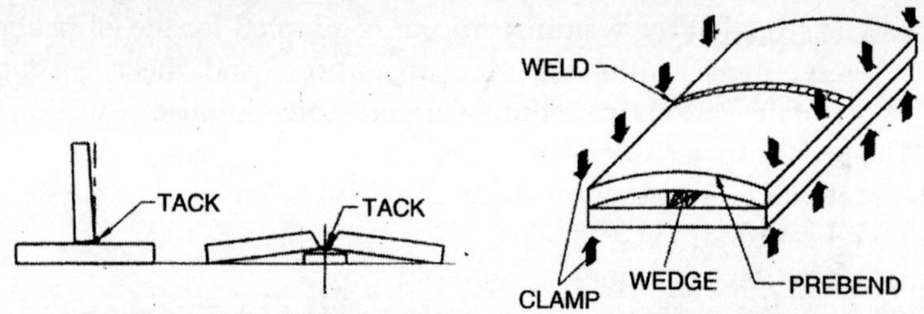

(a) Presetting for fillet and butt welds (b) Prebending

Fig. 12.4

required varies with the plate thickness, plate width and the welding procedure. Prebending requires the clamping at the edges. In the second method, the restraint may be applied by clamps or by the use of fixture. An example is shown in Fig. 12.5 (a) where the clamping arrangement is designed to prevent angular distortion in a V-butt joint while permitting transverse shrinkage. In Fig. 12.5 (b), the job is totally and heavily clamped against distortion of any nature.

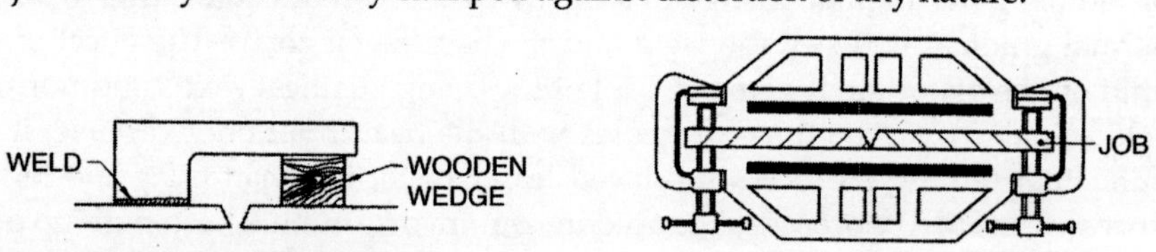

(a) Arrangement of clamp to prevent angular distortion (b) Fixturing with heavy clamps

Fig. 12.5

12.5 CONTROL OF DISTORTION

Control of distortion can be done in three stages. The stages are :

(i) Design stage.

(ii) Fabrication stage.

(iii) Correction Stage.

12.5.1 Design Stage

To prevent or control distortion, the following methods are used :

(i) Minimise the number of joints by use of standard, rolled and forged sections in the place of welded parts. This will reduce the amount of weld metal deposition and thus distortion. (Fig. 12.6)

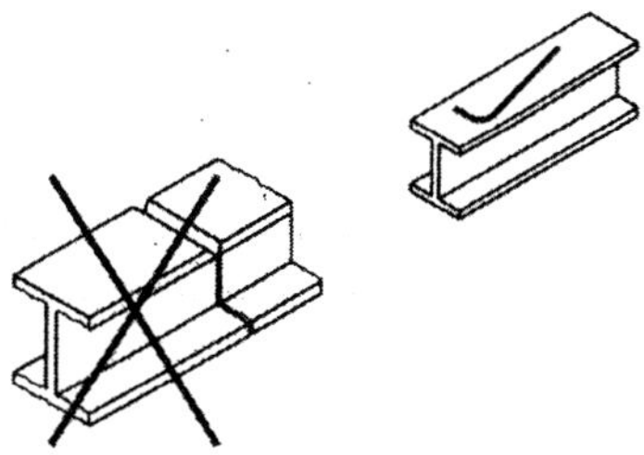

Fig. 12.6 Minimise number of joints.

(ii) Provide sufficient rigidity against bonding loads that may arise during welding. [Fig. 12.7(a)]

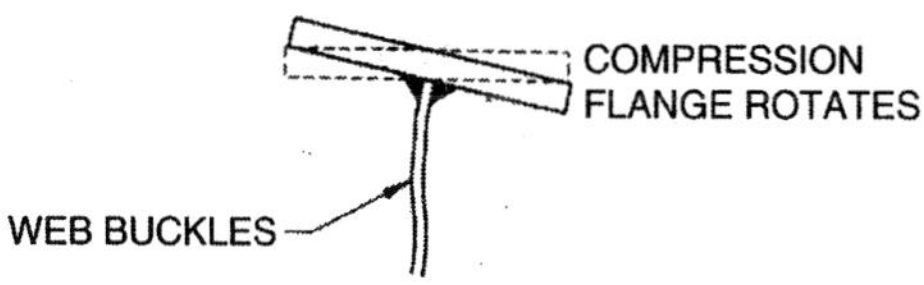

(a)

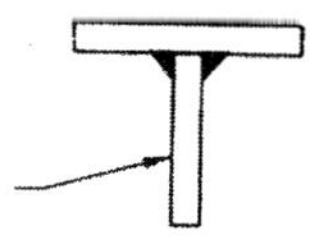

(b)

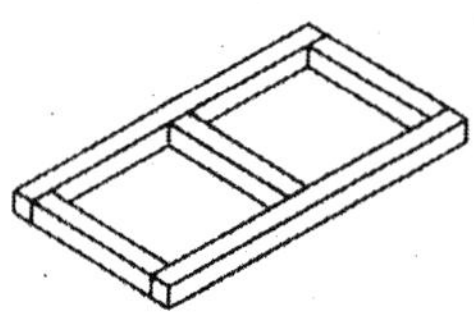

(c) Likely to buckle

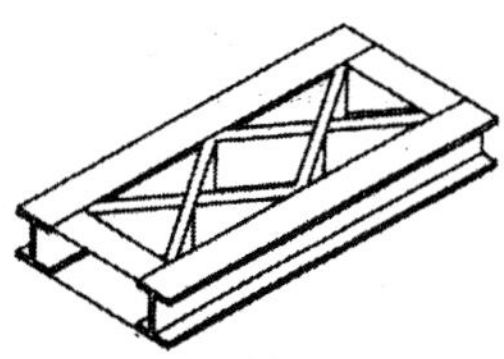

(d) Rigid

(a) Rigidity of welded assemblies

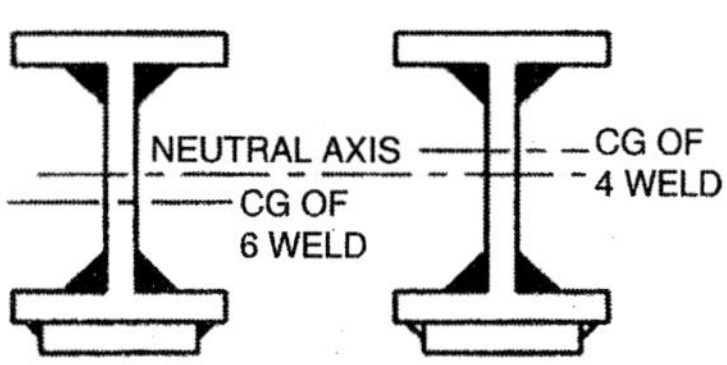

(a) (b)

(b) Welds located all around and near to the neutral axis

Fig. 12.7

(iii) Welds should be located all around and near to the neutral axis to minimise distortion. [Fig. 12.7(b)].

(iv) Groove-fillet joints should be used in the place of plain fillets to minimise distortion.

(v) Use unsymmetrical double 'V' or double 'U' joints to minimise distortion in butt joints.

12.5.2 Fabrication Stage

To minimise distortion at the fabrication stage, the following steps are recommended :

(1) Do Not Over Weld

The shrinkage forces will increase as the actual deposition is more. The amount of weld metal in a fillet can be minimised by employing a flat or slightly convex bead and in a butt joint, by proper edge-preparation and fit up. The excess weld metal in highly convex bead does not increase the strength but it promotes shrinkage stresses. Correct joint design and proper control of edge-preparation and fit up help to reduce the amount of weld metal. In relatively thicker plates, to minimise the weld metal deposition the joint design can be made to a 'J' or a 'U' preparation or the angle of bevel can be decreased if the root opening is increased.

(2) Sequence the Welding

Weld sequence defines the order and method of executing the different welds with the object of minimising distortion and residual stress while welding the assembly.

(i) Break the structure into such assemblies. For each assembly prepare suitable jigs and at each stage compensate for distortion.

(ii) Use symmetrical or approximately symmetrical arrangements.

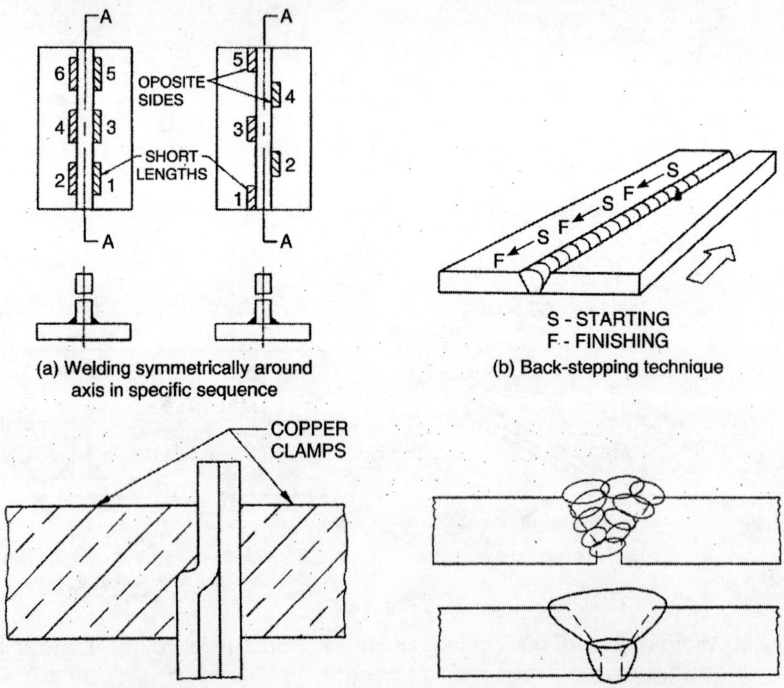

(a) Welding symmetrically around axis in specific sequence

(b) Back-stepping technique

(c) Copper heat sinks

(d) Stringer Vs weave bead welding

Fig. 12.8

The welding sequence should ensure dissipation of heat away from the weld. This can be achieved in the following ways :

(a) By welding symmetrically around the axis [Fig. 12.8(a)].
(b) By building up welds in short lengths [Fig. 12.8(a)].
(c) By back stepping : Making welds in short lengths. [Fig. 12.8(b)]
(d) By opposed blocks : Welding on opposite sides of the joint [Fig. 12.8(a)]
(e) By the use of copper heat sinks. [Fig. 12.8(c)]
(f) By the type of passes, stringer or weave. [Fig. 12.8(d)]

(3) Presetting and Fixturing

Assembly techniques can to a extent take care of distortion especially in the fabrication stage of large complex structures. One of the two assembly methods mentioned below can yield good results :

(i) Estimating the amount of distortion likely to occur during welding and then assembling the job with members. Pre-set or even pre-bend to compensate distortion, known as pre-setting or pre-bending.
(ii) Assembly welding of the components by employing some from of restraint to minimise. The distortion after welding, known as fixturing.

In the first method the parts have the full freedom to move about during welding and there is less residual stress than with fixturing. On the other hand it is a difficult method to apply correctly especially on final assemblies. Therefore, the individual component parts are welded as much as possible with pre-setting and pre-banding and the final assemblies are welded under the restraint of clamping fixtures.

12.5.3 Correction Stage

It is not always possible to control distortion within acceptable limits especially when dealing with a new design for the first time. Inspite of the most careful planning, excessive distortion may result. Correction methods can be broadly classified as below :

(1) Thermal Straightening.
(2) Mechanical Straightening.
(3) Other Methods.

(1) Thermal Straightening

The most common distortion removal technique is to use a flame and heat the part at selected spots or along certain lines then cool it in air. The area to be straightened is heated to between 600 to 650°C for plain carbon and low alloy steels and suddenly cooled in air, or, if necessary with a spray of water in low carbon steels. The methods of flame straightening are shown in Fig. 12.9. In line heating, heat from the torch is applied along a line or a set of parallel lines. This method is frequently used for removing the angular distortion produced by the fillet welds attaching a plate to its stiffener.

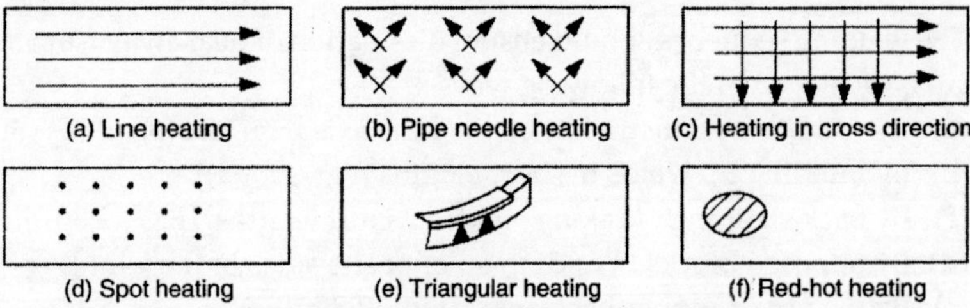

Fig. 12.9 Methods of flame-straightening.

In pine-needle heating, heat is applied along two short lines crossing each other. This method is half way between line heating and spot heating. Since the shrinkage and angular distortion occur in two directions. This method produces a uniform distortion removal effect. In checker board (cross-directions) heating, heat is applied along a pair of two lines crossing each other. This method is used to remove severe distortion.

In spot-heating, heat is applied on a number of spots. It is widely used for distortion removal, especially in this plate structures. In triangular heating, heat is applied on a wedge shaped area and this method is useful for the removal of bending distortion in frames. Red hot heating is thought of when severe distortion has occured in a localised area and it may be necessary to heat the area to a high temperature and beat it with a hammer. This method can cause metallurgical changes.

(2) Mechanical Straighening

Distorted members can be straightened in a press if the members can be moved to the press. Heat may or may not be required for straightening. Jacking is also adopted to remove distortion. Hammering of locally heated arears is sometimes resorted for straightening though it is not advisable.

(3) Other Methods

To reduce distortion, various thermal treatments are done. They include preheat and post weld thermal heatments.

(i) Preheating : Weld shrinkage is generally reduced by preheating. Actual measurements across welds during cooling have shown that less than 30% total contraction occured in joints preheated to 200°C, compared to non-preheated joints.

(ii) Stress Relief : In many cases thermal stress relief is necessary to prevent further distortion being developed before the weldment is brought to its finished state. Residual tensile stress in welds is always balanced by compressive residual stresses. If a considerable portion of the stressed material is machined out, a new balance of residual stress will result, causing new distortion. Weld stress-relieving prior to maching is thus very important for prolonged dimensional accuracy of sliding and rotating parts.

(iii) **Vibration Stress Relieving :** This technique reduces distortion by means of vibrating the weldments. The equipment consists of a variable spead vibrator which is clamped to the work-piece and an electronic amplifier. By varying the speed of the vibrating motor, the frequency can be varied until a resonant frequency has been reached for the work-piece. The piece is then allowed to vibrate for a period which varies in relation to the weight of the piece. Usually it ranges from 10 to 30 minutes. In this process even 50% of the residual stresses can be relieved. The component thus balances roughly its residual stresses and remains undistorted.

❑❑❑

13

CHAPTER

WELD DEFECTS

❑ INSIDE THIS CHAPTER

13.1 General classification, 13.2 Cracks, 13.3 Incomplete penetration (ICP), 13.4 Lack of fusion, 13.5 Lamellar tearing, 13.6 Slag inclusions, 13.7 Gas holes and porosity, 13.8 Tungsten inclusions, 13.9 Arc Strikes, 13.10 Craters, 13.11 Weld spatter, 13.12 Undercut, 13.13 Excess reinforcement, 13.14 Burn through and excess penetration, 13.15 Overlap, 13.16 Root cavity or buckback, 13.17 Mismatch, 13.18 Improper weld profile.

It is an established fact and agreed by one and all that defect free weld is a rarity. One has to necessarily accept this statement as the word 'defect' itself has got various meanings. A 'defect' unacceptable in one situation may be acceptable in another. They are no more 'defects' but acceptable discontinuities. Thus by the word 'defect' we mean any variation of metal structure, discontinuity, imperfection, flaw, inhomogenity etc. which will definitely influence the performance of the component.

13.1 GENERAL CLASSIFICATION

The discontinuities can be generally due to inherent processing and service conditions and accordingly they can be classified as :

(i) **inherent discontinuities** : related to the solidification and subsequent rolling of metalcast type and wrought type.

(ii) **processing discontinuities** : related to the various manufacturing processes such as machining, forming, welding, heat treating etc.

(iii) **service discontinuities** : related to various service conditions such as stress, corrosion, fatigue etc.

Weld defects can also be classified broadly into two types :

(i) **Planar defects :** Cracks, lack of fusion, lack of penetration, underent etc.

(ii) **Voluminar defects :** Slag inclusions, pores, cavities etc.

Besides these, there are geometric defects like excess reinforcement, root cavity etc.

13.2 CRACKS

These are grouped under metallurgical defects. They are caused due to processing operations like grinding, heat treating etc. Then they are called grinding cracks, heat treating cracks etc. Depending on the nature, orientation and direction, cracks are also called longitudinal cracks, transverse cracks and crater cracks. Cracks are with a short tip and linear ruptures of the metal under stress. (Fig. 13.1) Cracks are dangerous defects and most of the manufcturing codes do not accept cracks irrespective of their dimensions. Surface cracks are best detected by liquid penetrant and magnetic particle inspection techniques. Radiography and ultrasonics are capable of revealing internal cracks.

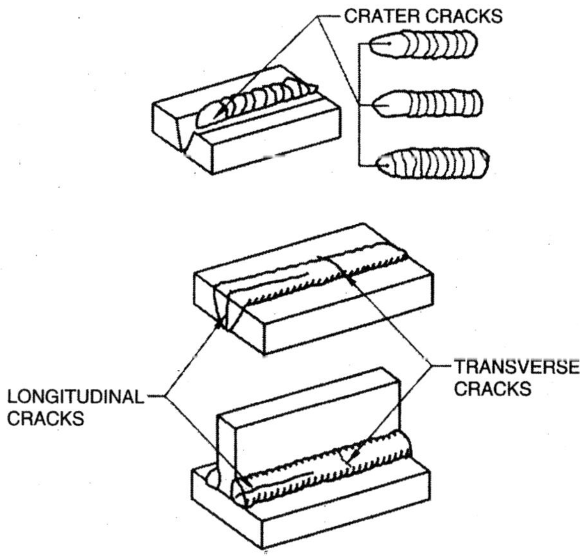

Fig. 13.1 Typical cracks in weld metal.

13.2.1 Hot Cracks

Hot cracks occur at elevated temperatures during cooling and solidfying from the molten stage. The presence of carbon and sulphur in higher percentage induce hot cracking. The restraint on the joints so motivates hot cracking. A proper cooling and a proper matching of electrodes will reduce the possibility of hot cracking.

13.2.2 Cold Cracks

Cold cracks otherwise known as 'delayed cracks', occur after even a few days of cooling. This is also called 'hydrogen induced cracking' as the effect of hydrogen in predominant. Preheating, post-heating/delayed cooling and usage of low hydrogen electrodes will reduce cold cracking.

13.2.3 Crater Cracks

These cracks occur in the crater of a weld bead. This is due to improper filling of the crater before raising the electrode away from the weld puddle or the arc suddenly. Grinding off the starts and stops will remove the crater cracks. Simple NDT checks like PT, MT will reveal crater cracks.

13.3 INCOMPLETE PENETRATION (ICP)

Incomplete penetration is considered as a serious defect in a weld. During welding, if there is no proper melting, fusion and penetration at the lip of root in the weld, a notch-like defect will be formed at the root of the weld and it is known as incomplete penetration (Fig. 13.2). It can initiate brittle fracture, and cracks are

Fig. 13.2 Incomplete penetration.

initiated from the unfused root. Hence most of the codes do not accept incomplete penetration. Proper joint design and fit up, correct electrode for root welding etc. will eliminate ICP. Radiography and ultrasonics detect these defects easily.

13.4 LACK OF FUSION

Lack of fusion refers to incomplete or partial melting and fusion of the weld metal. It may either occur between the base metal and weld metal or between weld passes. (Fig. 13.3) Improper welding techniques like improper manipulation, low welding current, too high welding speed, improper cleaning between passes and improper joint fit up, all these things will lead to lack of fusion. Radiography is capable of detecting lack of fusion depending on the orientation of the defect. The radiography test is also supplemented by ultrasonic tests.

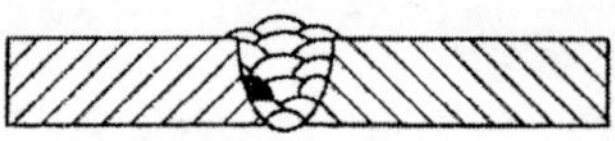

Fig. 13.3 Lack of fusion.

13.5 LAMELLAR TEARING

Normally this occurs in 'T' joints that are fillet welded on both sides. The stresses developed by this design result in separation within the base metal. (Fig. 13.4) Such

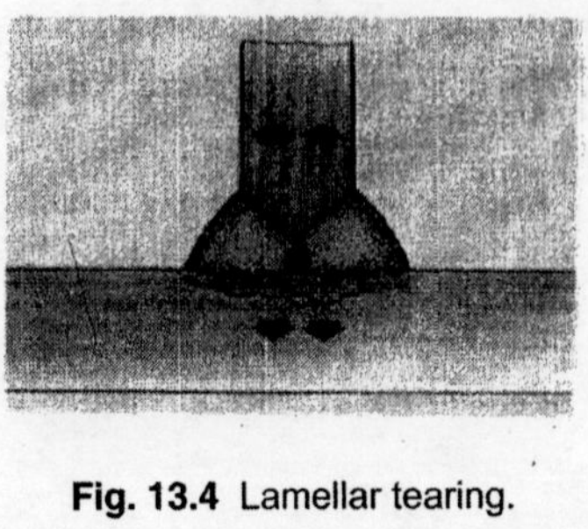

Fig. 13.4 Lamellar tearing.

a discontinuity is often associated with lamination or other planes of weaker regions in the base metal.

13.6 SLAG INCLUSIONS

It refers to the oxides and non-metallic solid materials that are entrapped in the weld metal or between the base metal and the weld metal. They occur as spot inclusions or interrupted or continuous lines. Slag inclusions can occur in MMAW with coated electrodes or in SAW where the protective slag over the weld puddle gets entrapped. Careful cleaning betwewen passes, suitable type of electrode coating or flux, proper electrode manipulation, proper preparation of the groove etc. willreduce the slag inclusions (Fig. 13.5). Radiography as well as ultrasonic techniques are equally good in detecting these discontinuities. Slag inclusions are, to some extent, permitted by the codes.

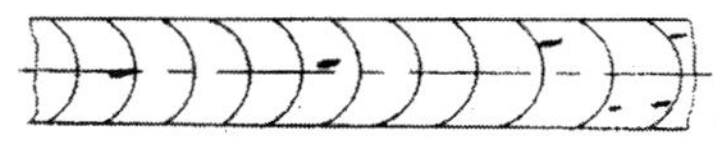

Fig. 13.5 Slag inclusion.

13.7 GAS HOLES AND POROSITY

This refers to the entrampment of the gases evolved during weld metal solidification, i.e., gases formed by chemical reactions in the weld also lead to formation of gas porosity. Porosity is usually spherical in shape. (Fig. 13.6)

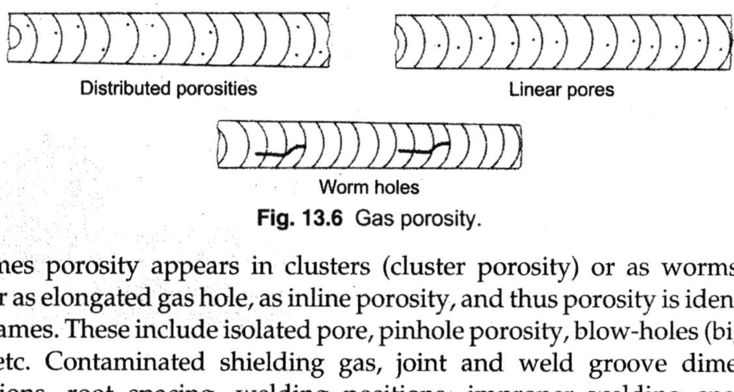

Fig. 13.6 Gas porosity.

Sometimes porosity appears in clusters (cluster porosity) or as worms (worm holes) or as elongated gas hole, as inline porosity, and thus porosity is identified by many names. These include isolated pore, pinhole porosity, blow-holes (bigger gas holes) etc. Contaminated shielding gas, joint and weld groove dimensions/ orientations, root spacing, welding positions, improper welding speed, low current, presence of rust, moisture etc. will encourage porosity formation and so by controlling these porosity can be minimised. Radiography technique will be more suitable to detect porosities even though the rounded indications are sometimes picked up by ultrasonics. Failures due to parasities in weld have not been reported frequently. Porosity charts are available in various codes for acceptance/non-acceptance of gas holes.

13.8 TUNGSTEN INCLUSIONS

These are tungsten particles entrapped in the weld deposit during occasional touching of the tungsten electrodes on to the molten weld pool or filler wires in TIG welding. Timely grinding or replacement of these tungsten electrodes and proper handling of these electrodes during welding will eliminate the occurance of these inclusions. Radiography is the best NDT technique to detect tungsten inclusion. Normally tungsten inclusions are not considered as dangerous defects as long as they do not cross a limited number and size. When present, they are evaluated at par with gas inclusions.

13.9 ARC STRIKES

This is one of the common surface defects which can be visually observed. This is due to stray arc-melting, resulting in a localised area of high hardening. It may sometimes lead to surface cracks also. It can also be produced by the heat generated by the passage of an electric current. Some welders have the bad habit of scratching the live electrode on the parent metal causing arc strike. Proper education of the welders on the significance of the arc strike, keeping the insulated electric wire and clean magnetic particle teting prod will help in avoiding arc strikes. The arc strike locations are to be ground smooth. (Fig. 13.7)

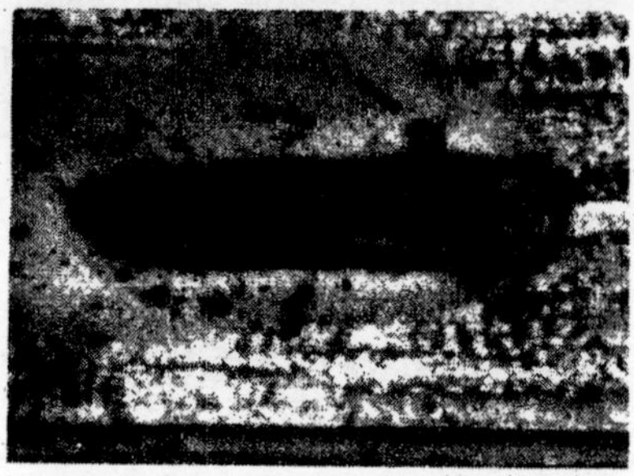

Fig. 13.7 Arc strike.

13.10 CRATERS

A crater/pit represents a circular surface cavity extending into the weld metal. Cracking sometimes starts from these craters and thus they become significant (Fig. 13.8). Avoiding the abrupt interruption in the welding arc or gradually reducing the welding current or increasing the travel speed may reduce crater formation.

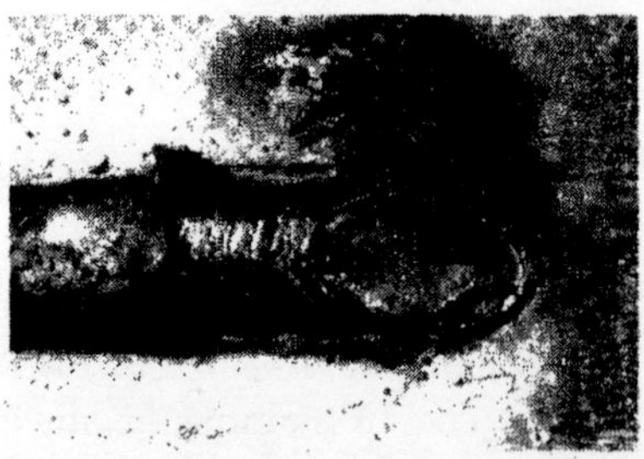

Fig. 13.8 Crater and spatter.

13.11 WELD SPATTER

These are particles of metal usually spherical in shape, getting ejected from the molten pool during welding and depositing on the surface of the base metal or weld deposit. While spatter is not considered as harmful, too much of spatter in and around the weld region may mask the quality of the weld X-ray picture. Hence, some of the standards insist for cleaning of spatter prior to radiography.

13.12 UNDERCUT

It represents a groove spot or continuous melted into the base metal along the edge of the weld and left unfilled by the weld metal. Simple visual inspection may be sufficient for identifying under cut. (Fig. 13.9) An undercut is considered as a serious defect as it weakens. The strength of the weld zone by reducing the effective thickness. Undercuts can reduce the fatigue strength significantly. Correct choice of current and good manipulation of the electrode during welding will prevent undercuts. Undercut regions can be ground smooth and accepted.

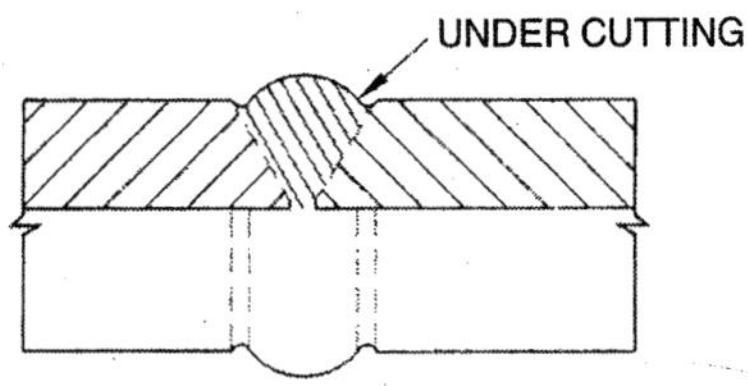

Fig. 13.9 Under cut.

13.13 EXCESS REINFORCEMENT

Excess of the weld metal more than the required weld size is called excess reinforcement and it is not desirable (Fig. 13.10). The reinforcement angle is the primary parameter affecting fatigue strength. So, the excess metal over the required size needs to be ground off/dressedup. Many codes have established limits on the maximum reinforcement permissible on the surface of butt welds. Some codes insist on flush grinding of the weld zone at par with the base metal. This is a simple defect which can be easily avoided.

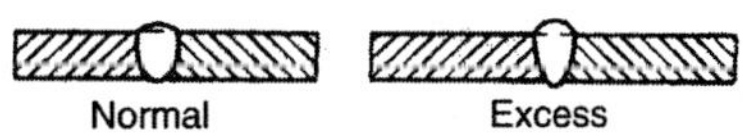

Fig. 13.10 Reinforcement.

13.14 BURN THROUGH AND EXCESS PENETRATION

This refers to the excessive melting of the root and undesirable heavy protrusion at the root, obstructing the flow of the medium through the tube or pipe. The tube's bore diameter will be sometimes drastically reduced by uniform excess penetration. By avoiding high current and low welding speed. This can be reduced (Fig. 13.11). Ball test and radiography are made use of for detection of these defects. Borescopes are also employed for viewing through small diameter tubes. Use of consumable inserts restricts excess penetration.

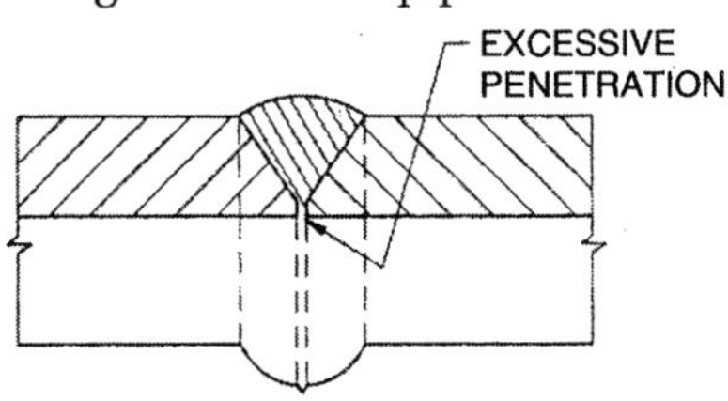

Fig. 13.11 Excess penetration.

13.15 OVERLAP

It represents an excess or overflow of unfused weld metal extending beyond the fusion limits over the surface of the base metal (Fig. 13.12). It is more common with fillet welds than the butt

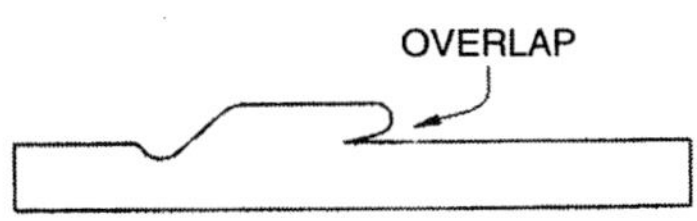

Fig. 13.12 Overlap.

welds. Proper welding techniques and proper welding input will avoid these kinds of defects. Overlap can be removed by grinding.

13.16 ROOT CAVITY OR SUCK BACK

Normally a root should have sufficient penetration and a convex shape. Sometimes due to excessive shrinkage of the weld deposit at the root, high current, slow welding speed, excessive joint gap, insufficient filler material etc., the root gets a concave shape and thus a suckback occurs (Fig. 13.13). Radiography can easily bring out this defect. Depending on the concavity and thickness loss, the Joint can be accepted provided sufficient reinforcement is avilable for making up the required thickness level.

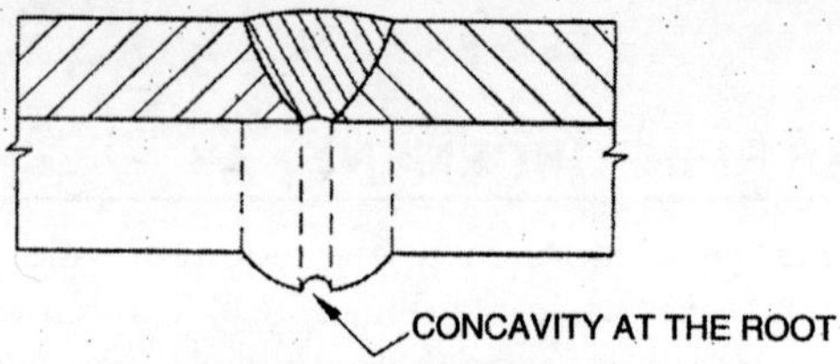

Fig. 13.13 Concavity.

13.17 MISMATCH

This refers to the centre line offset of two members in a welded but specimen (Fig. 13.14). It is a stress raiser and causes difficulty in welding. Hence many specifications stress the limit for acceptance of mismatches. When the joining pieces are having the equal thickness the mismatch can be easily measured at the surface itself.

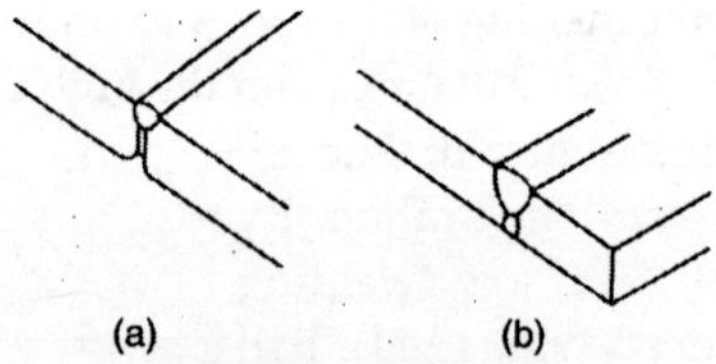

Fig. 13.14 Mismatch.

13.18 IMPROPER WELD PROFILE

Apart from the above illustrations, the improper weld profile especially is fillet welds will make the welds unacceptable. Excessive concave or convex beads come under this category. Sharp grooves between adjacent layers are also an example of this. Simple visual inspection aided sometimes by weld gauges, will be sufficient to evaluate these defects. Proper weld design, technique and parameter selection will eliminate these defects. Almost all discontinuities can be detected through NDT, their causes and effects can be established. All discontinuities are not defects to become unacceptable. Acceptance or rejection of discontinuities depends on many factors. A few of these factors are :

(i) Type of stresses and location of stresses applied during service.
(ii) The temperature and pressure that is applied on the component.
(iii) The material and thickness.
(iv) The safety, environment.
(v) Cost are reputation.

The decision of acceptance or rejection, therefore, should be taken considering the prime function and end use of component.

❑❑❑

14

CHAPTER

TESTING OF WELDS

❑ **INSIDE THIS CHAPTER**

14.1 Non-destructive testing of welds, 14.2 Destructive testing and metallography of welds.

14.1 NON-DESTRUCTIVE TESTING OF WELDS

A product without proven quality will not have any market. Every customer wants the product trouble free and work satisfactorily for a reasonable period. Thus the importance of the product quality, in the present day context, cannot be understimated. The quality of the component is assessed through testing and inspection. Non-destructive testing thus plays a vital role in industry for quality control.

As the name itself implies, a test done without damaging the specimen, without altering its property or affecting its future useful function is called non-destructive testing (NDT). NDT techniques are many and varied starting from simple visual testing to highly sophisticated and complicated. Opto Holography, Accoustic Emission testing etc. The four important non-destructive tests are :

(i) Liquid Penetrant Test (PT).

(ii) Magnetic Particle Test (MT).

(iii) Radiographic Test (RT).

(iv) Ultrasonic Test (UT).

14.1.1 Visual Test (VT)

The simplest and the important NDT technique is the visual examination test (UT) and precedes all other types of NDT. In fact, VT forms the basis of NDT examination. Visual test does not require costly and sophisticated instruments and accessories but simple aids like mirror, magnifying lens, borescopes etc. Of late, there is lot of awareness about this technique and guidelines have been prepared for the proper training qualification and certification of personnel doing VT. Not all the surface abnormalities and flows can be inspected and identified by VT but more information can be gathered by improving the visibility.

14.1.2 Liquid Penetrant Test (PT)

Liquid penetrant test is a simple and easily carried out NDT technique. It is applicable to almost all materials except porous materials. PT effectively detects surface and open surface discontinuities like cracks, seams, laps, cold shuts, laminations, porosity etc. The surface condition as well as surface preparation for this test is very important. The principle of this technique is based on capillary action. An active liquid penetrant dye is applied on the surface of the part under inspection. The penetrant seeps into the very small discontinuities and openings. After a sufficient period, the liquid is wiped off/removed and the surface of the part is given a developer coating. This acts as a blotter as well as a good contrasting background.

The developer sucks out the penetrant to the top surface from the discontinuities. Under the white developer contrast, the visibility of discontinuities is enhanced very much. Almost all the dye penetrants are red/pink in colour and are easily seen against the white background (Fig. 14.1). For meeting improved

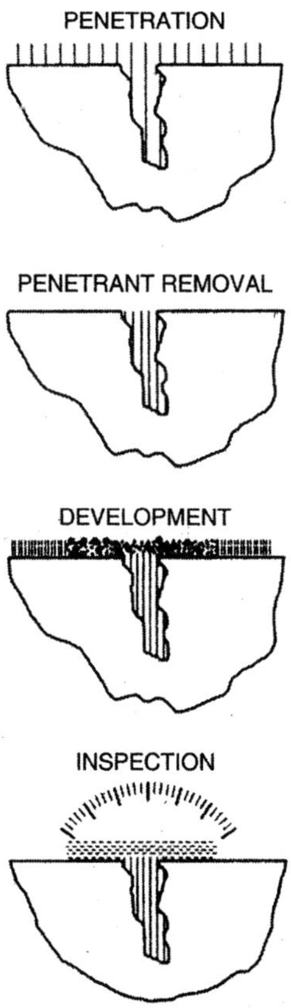

(a) Penetrant on surface seeps into crack

(b) Hand wipe with dry cloth removes penetrant

(c) Developer acts like a blotter to draw penetrant out of cracks

(d) Black light (UV) causes penetrant to glow in dark visible penetrant inspected under white light

Fig. 14.1 Liquid penetrant test.

senstivity and critical quality requirements, fluorescent penetrants are used. In this case the inspection and interpretation are done using an ultraviolet light. The basic steps involved in PT technique are as follows :

(1) Surface preparation.
(2) Application of penetrant.
(3) Excess penetrant removal after sufficient penetration time (Dwell time).
(4) Application of developer.
(5) Interpretation and evaluation.
(6) Post-cleaning.

The merits of the PT technique are :

(i) It is very effective and economical for surface inspection.
(ii) It is applicable to almost all materials.
(iii) It can be used even in fairly inaccessible areas.

The limitations are :

(i) Suitable only for open discontinuities and hence surface defects.
(ii) Process variables including penetration time and temperature are to be carefully controlled for better results.
(iii) Inflammable consumables–hence not applied for base metal temperature above 50°C.

14.1.3 Magnetic Particle Testing (MT)

Another versatile surface NDT technique is magnetic-particle testing applied on ferro-magnetic materials for the detection of surface and sub-surface discontinuities to a limited depth. It finds applications right from the raw materials to finished products. It also finds an extensive application in processing operations like machining, heat-treatment as well as inservice inspection programmes. The basic steps involved in this technique are as follows :

(i) The job (ferromagnetic material) is to be suitably magnetized.
(ii) The application of examination media. (Iron powder either in dry form or in wet ink form).
(iii) The interpretation and evaluation.
(iv) De-magnetization and post-cleaning.

The principle involved in this technique is that the magnetic lines of force run through the job, creating what is known as flux density. Any discontinuity in the specimen perpendicular to the flux lines disturbs the uniform flux density and cause flux leakage, i.e., formation of local tiny magnets which heavily attracts the iron powder, applied to form a pattern representing the discontinuity. Depending on the requirements, situation, facility available etc., different type of magnetisation techniques can be employed :

(a) Permanent magnetic yoke method [Fig. 14.2(a)].

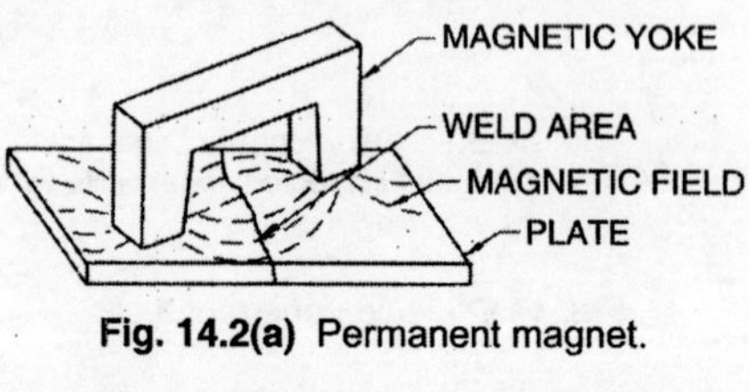

Fig. 14.2(a) Permanent magnet.

(b) Electromagnetic yoke method [Fig. 14.2(b)].

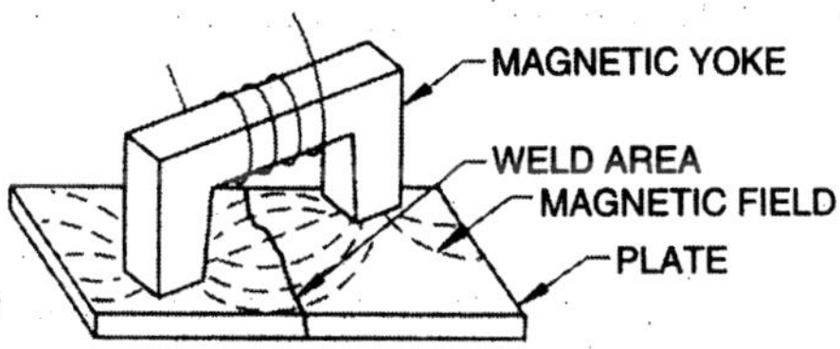

Fig. 14.2(b) Electro magnetic yoke.

(c) Coil method [Fig. 14.2(c)].

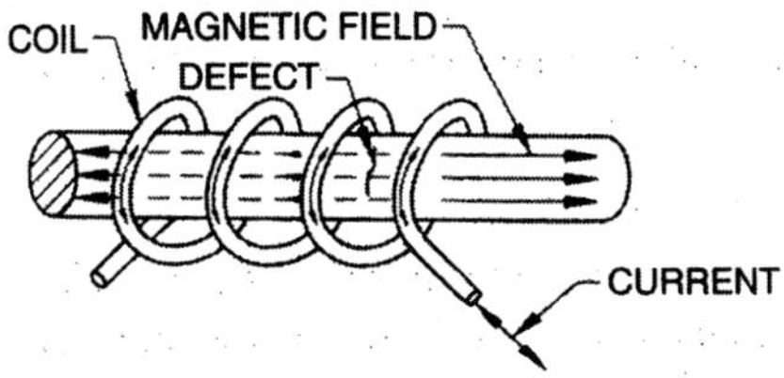

Fig. 14.2(c) Longitudinal magnetisation using coils.

(d) Head shot method [Fig. 14.2(d)].

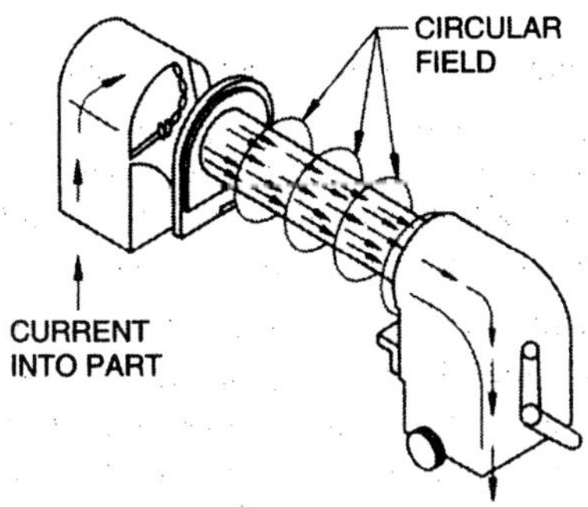

Fig. 14.2(d) Direct circular field magnetization by head shot method.

(e) Prod method [Fig. 14.2(e)].

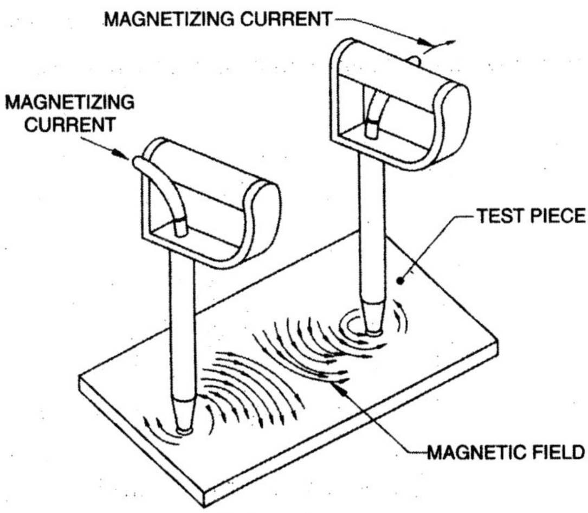

Single prod contacts

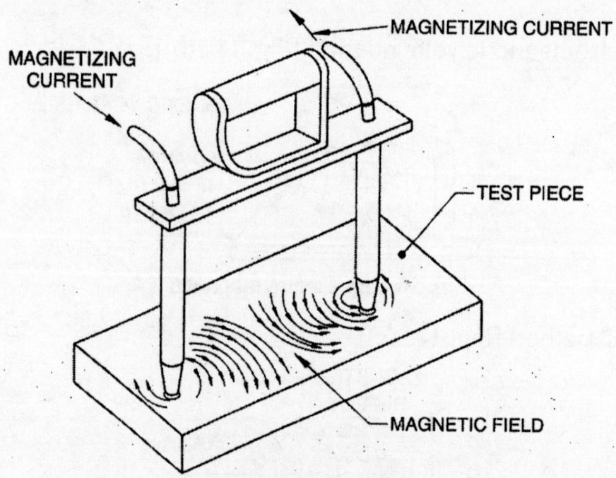

Double prod contacts

Fig. 14.2(e) Prod method.

(f) Central conductor method [Fig. 14.2(f)].

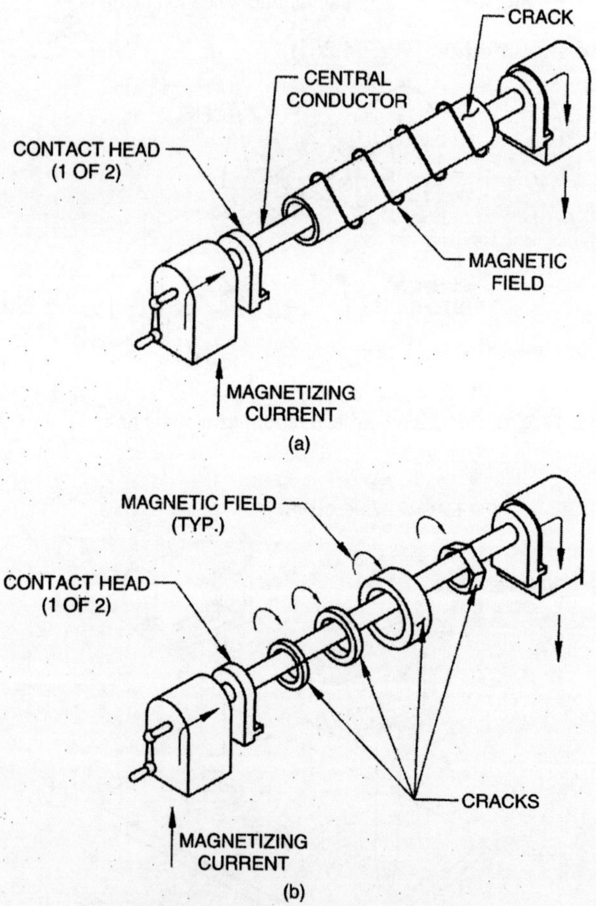

Fig. 14.2(f) Central conductor method.

The merits of MT techniques are :

(i) Effective for surface and subsurface inspection.
(ii) Easy, fast and simple.
(iii) Can be used in areas where accessibility is limited.

The limitations are :

(i) Applicable only for ferromagnetic materials.
(ii) Some components may require demagnetisation after testing.

14.1.4 Radiographic Testing (RT)

Radiographic testing is an important, versatile NDT technique. It gives a permanent visual records of results in the form of a radiographic picture. In the radiographic technique, highly energetic and penetrating electromagnetic radiations (X-rays and gamma rays) are used for making a picture of the component. X-rays and gamma rays are made to penetrate through the specimen. As they pass through the specimen, they undergo what is known as "differential absorption." Different portions of the object will have different absorption factors, and, hence, the final emerging radiation intensity from the object will be different. This variation in intensity is recorded in a film, which after proper chemical processing is known as radiograph. The radiograph will be able to reveal the inner soundness of the material, (Fig. 14.3), through different darkness levels. If the specimen is homogeneous and uniform, the emerging intensity will also be uniform and have uniform influence in darkening the film. If lighter or denser inclusion is there in the material, radiation passing through the same will have less absorption or more absorption respectively. So the energy passing through the lighter areas will affect the film more and darken it deeper than the energy passing through the denser areas. So discontinuities like gas holes, slag inclusion, cracks, lack of fusion, incomplete penetration etc. will be appearing as 'darker' indications

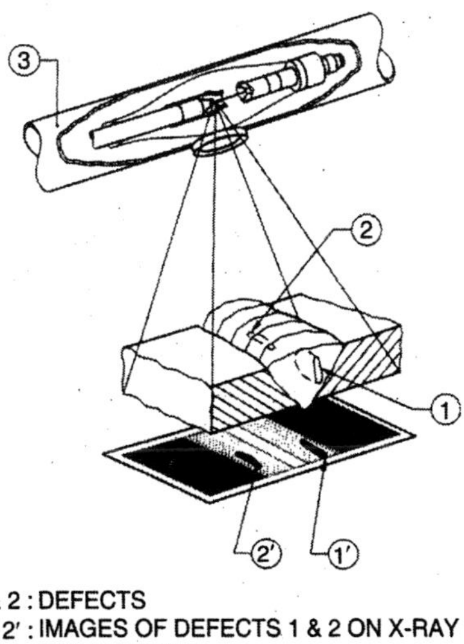

1 & 2 : DEFECTS
1' & 2' : IMAGES OF DEFECTS 1 & 2 ON X-RAY FILM
3 : RAY SOURCE

Fig. 14.3 Radiographic testing.

in the radiograph and excess penetration, tungsten inclusions, spatters etc. will appear comparatively lighter.

The radiographic technique is applied on all materials which may vary from a miniature electronic component to a mammoth sized component including those for space, military use etc. This technique is normally best suited for the detection of voluminar kind of defects. For planar type of defects, ultrasonic testing is preferred.

This technique has following advantages :

(i) Gives a permanent record.
(ii) Covers a wide variety of components.

Following are the limitation of this technique :

(i) Radiation hazard.
(ii) High recurring cost of films and chemicals.

14.1.5 Ultrasonic Testing (UT)

An ultrasonic testing is a fast growing popular technique in NDT. It is basically the usage of high frequency sound or mechanical vibration with a frequency above 20,000 cycles per second (cps or Hertz). A sound of frequency between 20 and 20000 cps is capable of recognising or being heard by the human ear. This is referred to as 'audible sound'. Sound with frequencies in excess of 20000 Hertz is referred to as ultrasonic frequency. These frequencies are used in ultrasonic testing. Most of the weld testings are done in the frequency range of 1 to 10 MHz. The same phenomenon is used in UT but with very high frequency sound waves with a great degree of sophistication in order to detect even a very small flow.

The simple steps involved in an ultrasonic testing are briefly stated as below :

(i) Generation of ultrasonic waves produced by energising a piezo electric crystal. (Piezo is a Greek word to mean 'press'). The crystals may be natural crystals like quartz or polarised ceramics like barium titanate, lead zirconate/titanate or chemically synthesised compounds like lithium sulphate. These crystals on receiving electrical energy, convert it into ultrasonic energy known as Piezo-electric effect.
(ii) Transmission of the ultrasonic waves into the specimen under test through a couplant media, such as water, oil, grease, glycerine etc.
(iii) Study of the reflected energy from the mismatch and inhomogenity. Usually the reflected energy will be converted back into electrical energy by the piezo electric crystal (i.e., probe) and would be shown as pips on a CRT screen.

The most popular ultrasonic technique is the pulse-echo technique, and it has got wide application in the industrial field. In this techniques, pulses of ultrasonic sound energy are sent into the specimen and the reflection from the impurities, if any, or reflections from any boundary are studied on the CRT screen [Fig. 14.4(a)]. The other techniques like through transmission technique [Fig. 14.4(b)] and the Resonance technique [Fig. 14.4(c)] are also used for some application.

The advantages of this system are :

(i) High sensitivity.
(ii) High penetrating power.

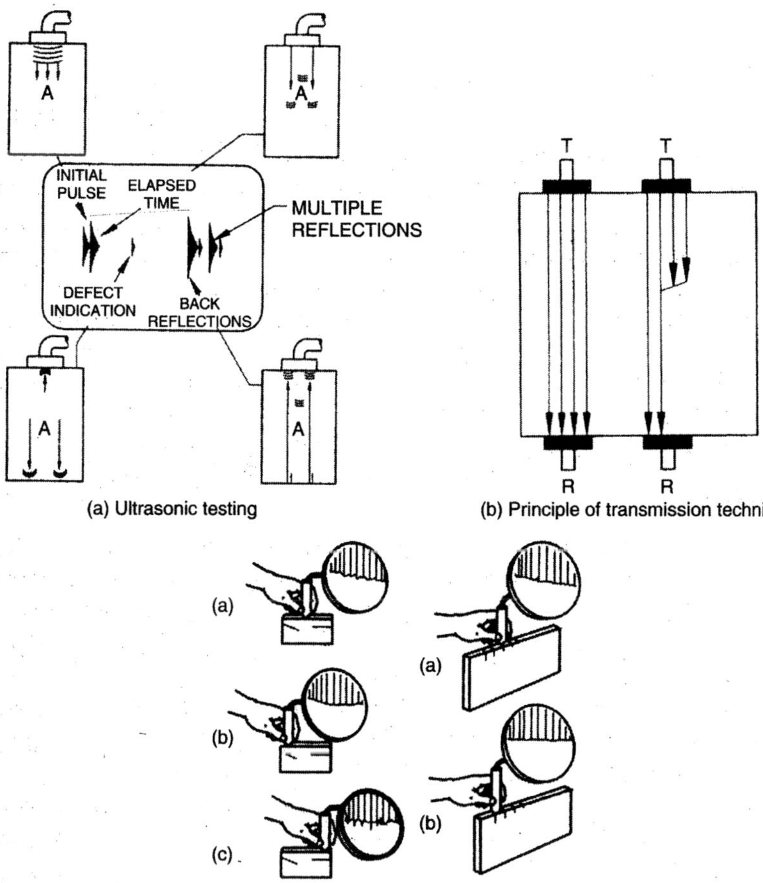

(a) Ultrasonic testing

(b) Principle of transmission technique

(c) The resonance technique

Fig. 14.4

(iii) Fast process and portable equipment.
(iv) One side accessibility is enough.
(v) Free from hazards.

The limitations are :

(i) Cannot be used for unfavourable geometry or shape of the component.
(ii) Cannot be used for undesirable internal structure.
(iii) Requirements of numerous reference and caliberation standards blocks.
(iv) Interpretation problems.
(v) Requirement of trained personnel.

With the advancements taking place in every sphere of life, newer and improved NDT techiques are coming into force. The significance of non-destructive testing is increasing day-by-day. The following factors will be of help to decide on the choice of NDT technique :

(i) Materials to be tested.
(ii) Method of fabrication.
(iii) Geometry of the material.
(iv) Defects possible or expected.

(v) Sensitivity desired.
(vi) Economic aspects.

NDT, if applied properly, taking into consideration of the above factors, brings immense benefit both to the user and the producer. Non-destructive testing ensures the following :

(i) Product reliability, performance and improvement.
(ii) Aids in better design.
(iii) Lowers manufacturing cost.
(iv) Maintains quality and controls in manufacturing processes.
(v) Ensures customer satisfaction.
(vi) Saves human life.

14.2 DESTRUCTIVE TESTING AND METALLOGRAPHY OF WELDS

Welding is widely used for fabrication of components like bridges, pressure vessels, high temperature parts of a jet engine etc. These components are designed to meet a specific service requirements. To ensure that these components perform satisfactorily, the quality has to be assesed and this is done by conducting non-destructive and destructive testing of weldments. Non-destructive testing is done without any damage to the weld while in destructive testing weld is broken. Destructive testing is done during welding procedure establishment and also it is carried out during actual production on test plates to ensure quality. These tests are employed in order to determine physical properties like tensile, impact strength, hardness and bend ductility of weldments and materials in general, while metallographic tests are employed to assess the soundness of the weld, disposition of weld passes, depth of penetration, width of HAZ, metallurgical structure of weld metal and HAZ. Following are the destructive tests :

(i) Tensile test.
(ii) Bend test.
(iii) Impact test.
(iv) Hardness test.

The metallographic examinations are :

(i) Macro-examination.
(ii) Micro-examination.

Destructive and metallographic tests for weldments are used for welder's performance, weld procedure and filler metal qualification and in general for quality control purposes.

14.2.1 Destructive Tests

(a) Tensile Test

To determine the strength and ductility the tensile test is used. Load is applied uniaxially on a machined sample and observations of the load and sample extension are made. The test is used to assess the strength of both the base metal and of the weld. The tensile testing machine used for conducting the test is shown in Fig. 14.5(a).

(a) Universal testing machine

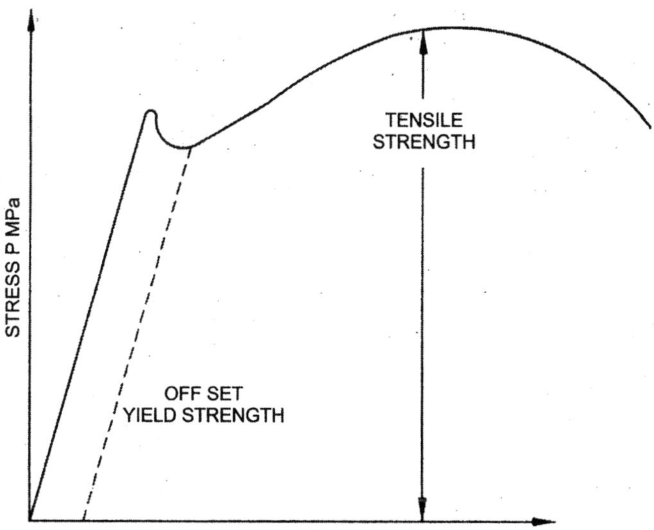

(b) Engineering stress—strain diagram

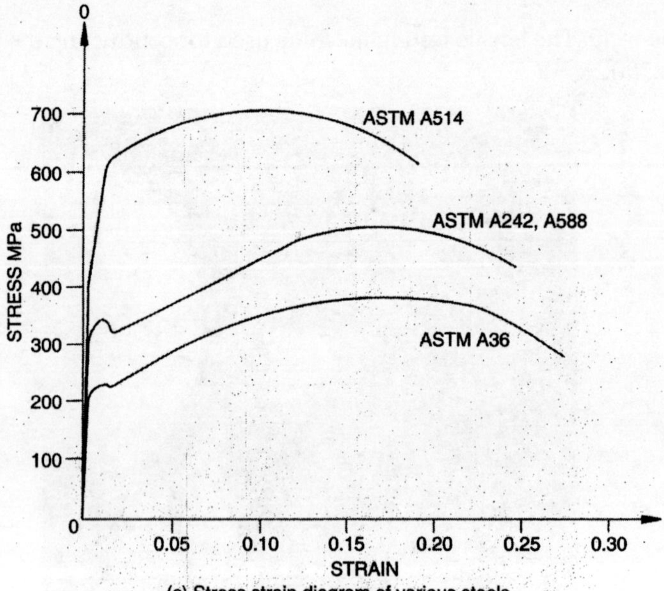

(c) Stress strain diagram of various steels

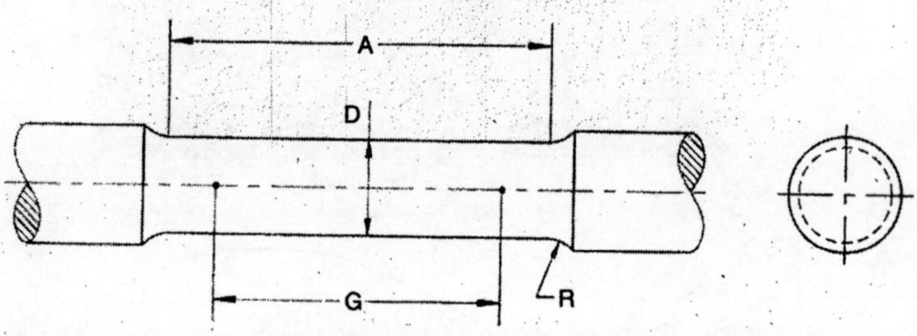

Nominal diameter	Standard specimen 12.5 mm	Small size specimen 6.25 mm	8.75 mm
Gas length **G**	50.0 ± 0.10	25.0 ± 0.10	35.0 ± 0.10
Diameter **D**	12.5 ± 0.25	6.25 ± 0.12	8.75 ± 0.18
Radius of fillet **R**	10	5	6
Length of reduced section **A**	60	32	45

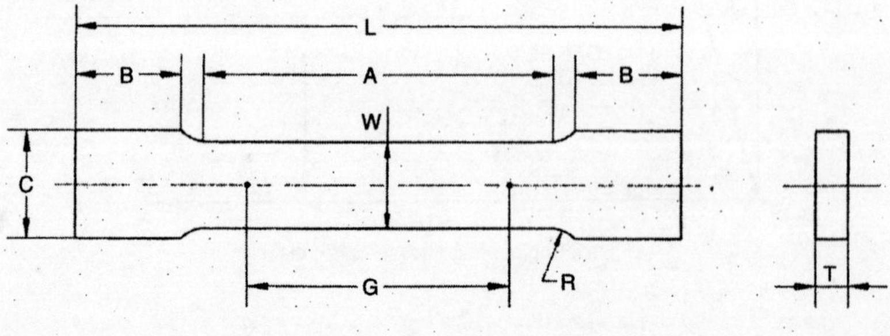

	Standard specimens	
	mm	mm
Gauge length **G**	200 ± 0.25	50 ± 0.10
Width **W**	40 ± 2	12.5 ± 0.25
Thickness **T**	Thickness of material	
Radius of fillet **R**	25	13
Overall length **L**	450	200
Length of reduced section **A**	225	60
Length of grip section **B**	75	50
Width of grip section **C**	50	20

(d) Standard tensile sample

Fig. 14.5

(b) Base Metal Tensile Test

The strength and ductility of materials are determined from the tensile test. Samples for testing are taken parallel to the rolling direction of the plate and 90° to the rolling direction, and these types of tests are called longitudinal and transverse tensile tests respectively. The plot of load versus elongation obtained from a tensile test is shown in Fig. 14.5 (b) and this is known as engineering stress/strain diagram and sometimes it is also called load-elogtion diagram. The ultimate tensile strength (UTS) is calculated by dividing the load (max) by original area and it is the maximum stress in the stress-strain curve. The yield strength (YS) shown by the stress/strain diagram is normally the strength at some point of extension under load or a permanent plastic strain. Typical stress-strain diagrams for some of the steels are shown in Fig. 14.5 (c). The ductility of the materials is evaluated by percentage elongation (% E) and percentage reduction in area (% RA). A standard gauge length (GL) is marked on the sample before application of load and after testing increase in GL is measured. The %E is the ratio of increase in GL by the original GL (dl/L). The % RA is the ratio of decrease in area by the original area. The standard specimen used for this test are shown in Fig. 14.5 (d).

(c) Weld Tensile Test

Tensile testing of welds is more complicated because the weld test section is composite in nature, the section is composed of the deposited weld metal, the heat affected zone and the unaffected base metal. To evaluation the ductility and the strength of the weld, samples in different orientations and locations are used for testing. Figure 14.6 (a) and 14.6 (b) shows the samples used for evaluation of welds.

(d) All Weld-Metal Test

The sample is machined from the undiluted weld metal deposit and the orientation is along the axis of the weld. The test is used for the qualification of filler metal.

(e) Transverse Weld Test

To assess the strength of the weld joint, a transverse weld test is used. A section is cut from the welded plate and a flat tensile specimen is machined. The reduced section of the specimen contains weld metal, heat affected zone and base

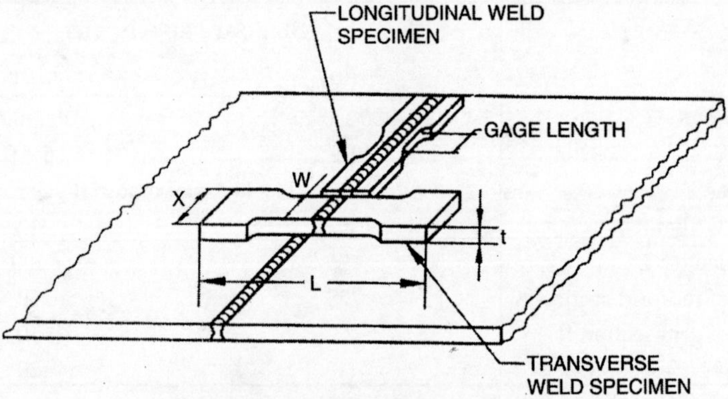

(a) Transverse and longitudinal weld tensile samples

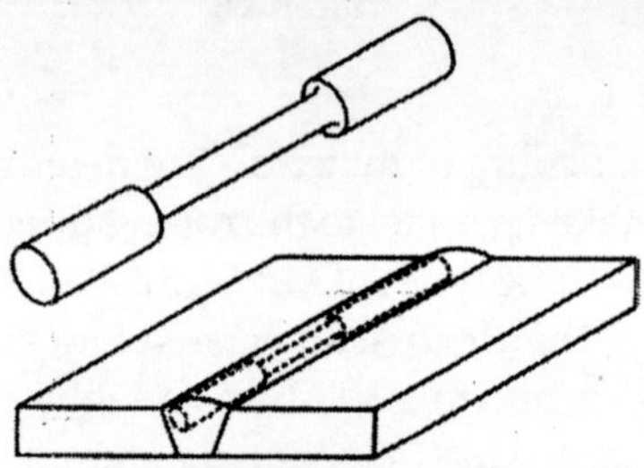

(b) All weld tensile sample

Fig. 14.6

metal. On application of uniaxial load all these regions are subjected to stress and the weak region is assessed. This test is useful for qualification of weld procedure. In this test, the direction of loading is along the weld axis, but this test differs from the all weld-metal secimen, the areas across the specimen contain weld metal, heat affected zone and base metal. All these zones are under strain equally during testing. As the strain is applied equally the weld metal and the base metal regions simultaneously elongate until fracture occurs. This test is useful for assesing the behaviour of a composite joint like welds whose strength differs significantly from that of the base metal.

14.2.2 Bend Test

The bend test measures the ductility and the soundness of material. There are various types of bend tests to evaluate the soundness of the welds. The guided bend test is generally used for weldments. A typical guided bend fixture is shown in the Fig. 14.7 (a). Bend test samples may be longitudinal or transverse to the weld. Face bend tests are conducted with the weld face in tension and root bend tests are done with the root in tension. A side bend test may be used when thick samples are tested. Illustrations of various bend tests are shown in Fig. 14.7 (b). Bend tests are also used in a weld procedure or a welder's performance. The strain on the outer fibre is calculated by the formula :

$$e = \frac{100\,t}{2\,R + t}$$

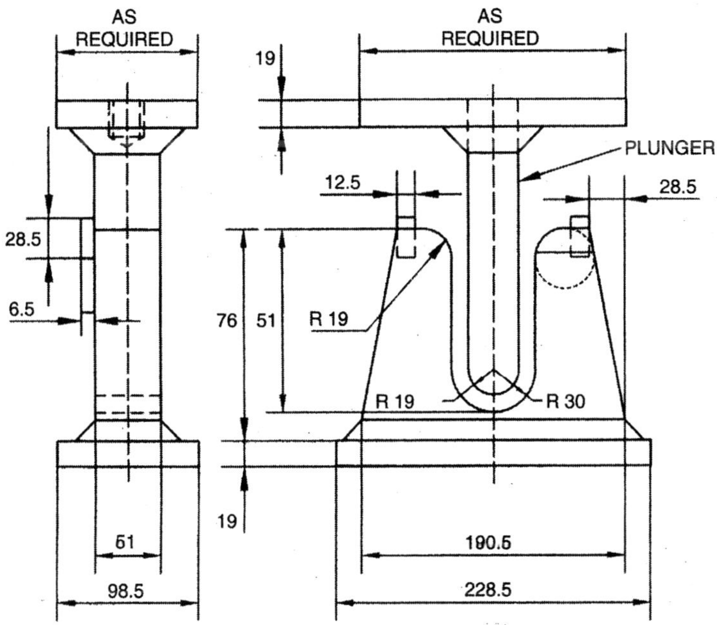

(a) Fixtures for conducting guided bend test

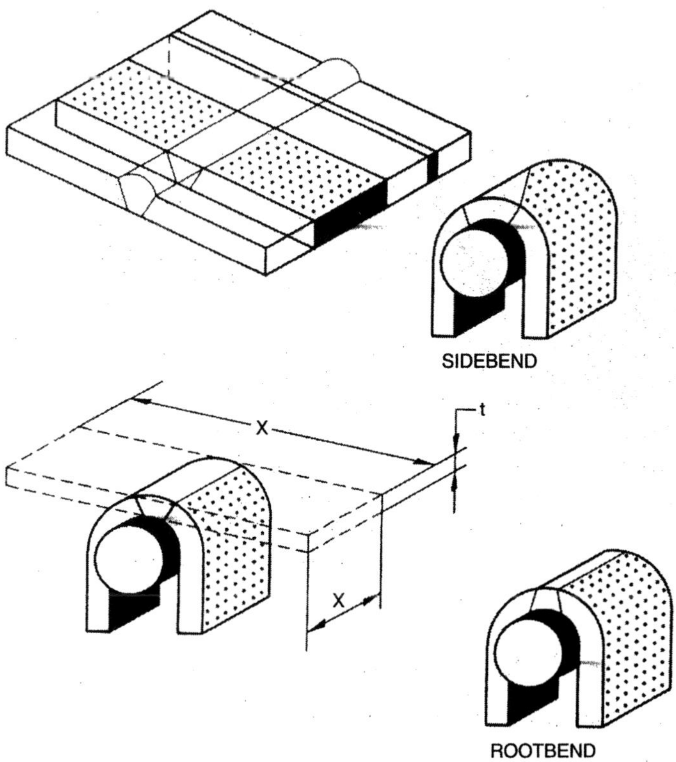

(b) Types of bend tests for a weld joint

Fig. 14.7

where, e = strain %

t = bend specimen thickness in mms.

R = bend radius in mms.

The specimen thickness and bend radius are chosen according to the ductility of the material to be tested. Bend tests for carbon and alloy steels are conducted with a mandrel diameter (D) equal to three/four times. The sample thickness (t) and these testing details are given in applicable codes. After bending the samples are visually examined and observations on the defects are made depending on the size of defect the sample is termed as failure. The acceptance criteria are also given in the applicable codes.

14.2.3 Impact Test

Charpy V-notch impact test is commonly used to assess the steel's toughness property and an impact tester of 30 kgf capacity is shown in Fig. 14.8 (a). A typical

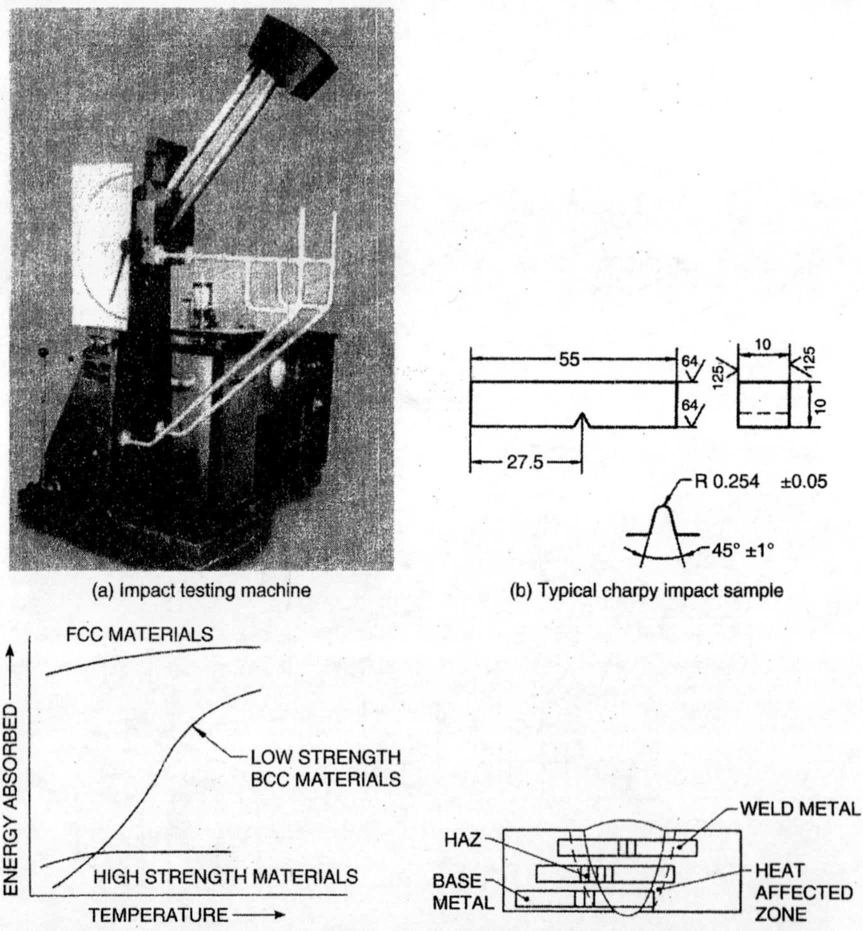

(a) Impact testing machine

(b) Typical charpy impact sample

(c) Toughness characteristics of materials

(d) Typical impact samples for a weld joint

Fig. 14.8

charpy specimen is shown in Fig. 14.8 (b). The toughness behaviour of the ferrous materials with respect to temperature is shown in Fig. 14.8 (c). As can be seen that the toughness depends upon or the structure of the material as well as the temperature. Face-centered cubic structure material shows excellent toughness property while the body-centered cubic structure material shows transition from ductile to brittle as the temperature is lowered. Materials with high charpy V-notch absorbed energy are resistant to brittle fracture in service. The units for absorbed energy are ft-lbs or J (Joules) of kgf.m. The energy requirements for a particular service is fixed for a particular temperature and these are given in applicable codes. In the case of components for critical service, the weld procedure qualification may require charpy V-notch determination for weld metal and heat affected zone. Figure 14.8 (d) illustrates the charpy impact samples for the weld metal and HAZ.

14.2.4 Hardness Test

Hardness is a measure of material's resistance to plastic deformation. It i s usually measured by making indentations. Hardness of weld is measured by Brinell, Vickers and Rockwell tests and these use the area of indentation under load for expressing hardness values. Welding is associated with rapid heating and cooling of metals and due to this metallurigcal changes take place. If cooling is quite rapid, hard microstructure like martensite may be formed in the HAZ. To detect such metallurgical changes the hardness test can be used. In the case of carbon and alloy steels, there is an approximate relationship between Brinell, Vickers and Rockwell numbers which is given in Table 14.1.

Table 14.1 : Relationship between various hardness and tensile strength Hardened steel and steel alloys

Rockwell standard			Vickers DPH 10 kgf	Brinell 3000 kgf 10 mm	Tensile strength PSI
C 150 kgf	A 60 kfgf	B 100 kgf			
Diamond Penetrator					
60	81.2	70.7	697	614	314
59	80.7	69.9	674	600	306
58	80.1	69.2	653	587	299
57	79.6	68.5	633	573	291
56	79.0	67.7	613	560	284
55	78.5	66.9	595	547	277
54	78.0	66.1	577	534	270
53	77.4	65.4	560	522	263
52	76.8	64.6	544	509	256
51	76.3	63.8	528	496	250
50	75.9	63.1	513	484	243
49	75.2	62.1	498	472	236

48	74.7	61.4	484	460	230
47	74.1	60.8	471	448	223
46	73.6	60.0	458	437	217
45	73.1	59.2	446	426	211
44	72.5	58.5	434	415	205
43	72.0	57.7	423	404	199
42	71.5	56.9	412	393	194
41	70.9	56.2	402	382	188
40	70.4	55.4	392	372	182
39	69.9	54.6	382	362	177
38	69.4	53.8	372	352	171
37	68.9	53.1	363	342	166
36	68.4	52.3	354	332	162
35	67.9	51.5	345	322	157
34	67.4	50.8	336	313	133
33	66.8	50.0	327	205	148
32	66.3	49.3	318	297	144
31	65.8	48.4	310	290	140
30	65.3	47.7	302	283	136
29	64.6	47.0	294	276	132
28	64.3	46.1	286	270	129
27	63.8	45.2	279	265	126
26	63.3	44.6	272	260	123
25	62.8	43.8	266	255	120
24	62.4	43.1	260	250	117
23	62.0	42.1	254	245	115
22	61.5	41.6	248	240	112
21	61.0	40.9	243	235	110
20	60.5	40.1	238	230	108

Also, the hardness numbers can be related to the tensile strength and this is also given in Table 14.1. To measure hardness of welds accurately, grinding, polishing and etching the samples should be done. Hardness of HAZ regions with etched samples is easier to measure accurately as the HAZ will be clearly identified and hardness measurement of any secific area can be done. Figure 14.9 (a) shows the hardness survey of a typical weld.

(a) Brinell Hardness Test

A steel ball is used for making an indentation on the surface of the sample. The indentation diameter is measured by means of a microscope to obtain the Brinell hardness number (BHN), the load is divided by the area of the indentation. A Brinell hardness tester is shown in Fig. 14.9 (b) and the ball indenter is shown in Fig. 14.9 (c). Hardness measurements of the hard material is done by choosing a load of 3000 kgf and for soft materials a load of 500-1500 kgf may be used.

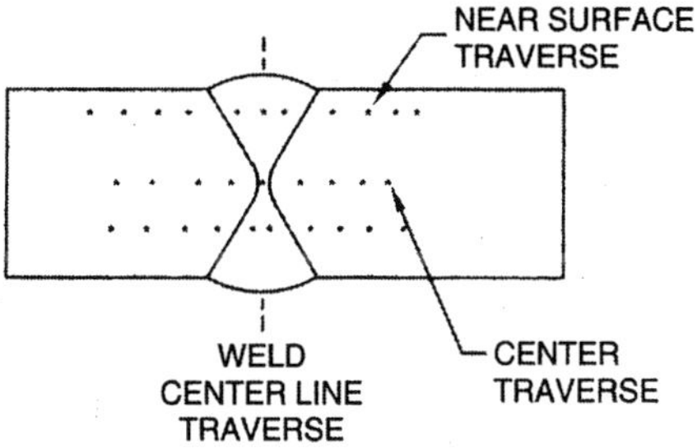

(a) Hardness survey for a weld

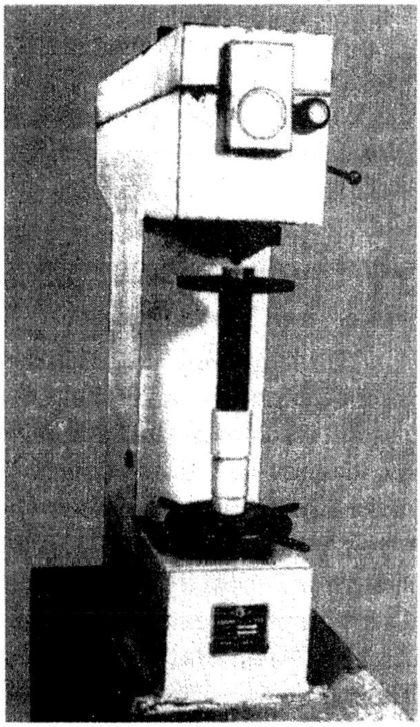

(b) Brinell hardness tester

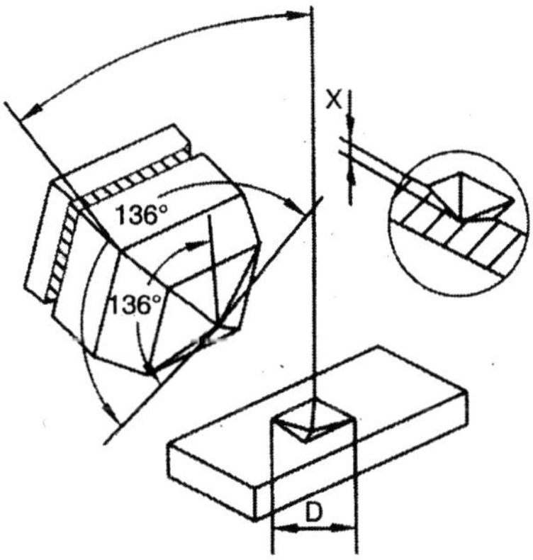

(c) Indenter for Vicker's hardness test-square based diamond pyramid

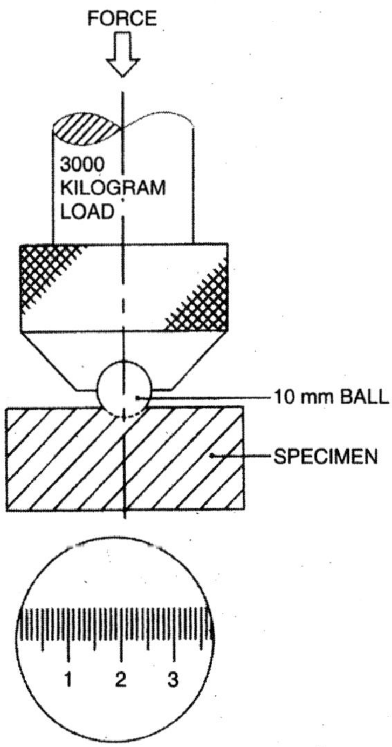

(d) Brinell indenter steel ball ϕ 10 mm

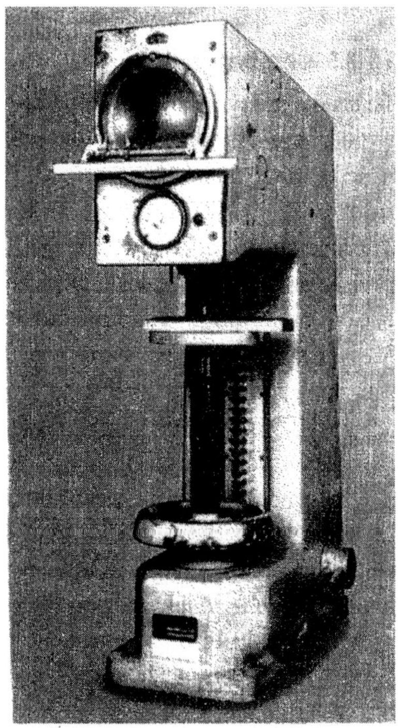

(e) Universal hardness tester with Rockwell, Brinell and Vickers testing facilities

Fig. 14.9

(b) Vickers Hardness Test

A square based diamond pyramid with an apex angle of 136° indenter is used which is shown in Fig. 14.9 (c). The hardness number (VHN or DPH) is defined as the load divided by the surface area of the indentation. The diagnosals of the indentation are measured by means of a microscope to obtain the hardness and the indentation made by vickers is shown in Fig. 14.9 (d). This test has received fairly wide acceptance for hardness measurements. The load of 5 kgf is selected for soft materials and 10 to 30 kgf is selected for harder materials.

(c) Rockwell Hardness Test

The hardness is related to the depth of penetration made by an indentor and the indenter may be a steel ball of 1/16 of an inch diameter or a 120° diamond cone. The test is conducted by the application of a minor load of 10 kgf initially to secure the indenter onto the specimen. The dial is adjusted to set the point and then a major load is gradually applied. The pointer of the dial moves and gradually comes to rest, the major load is released and the Rockwell hardness is read. Figure 14.9 (e) shows the Rockwell hardness tester with the dial for reading the hardness values. There are three scales of Rockwell hardness measurements. The scales are Rockwell A, B and C with major loads of 60 kgf, 100 kgf and 150 kgf respectively. The indenter used for A and C scales is a diamond cone while for B scale it is a steel ball. The hardness numbers are followed by a suffix HRB/HRC, e.g., 22 HRC for measurement in C-scale.

14.2.5 Metallography

To study the weld metal, HAZ or base metal microstructure or to know number of weld passes, macro/micro examination is employed. Macro examination is done upto a magnification of 10X and micro-examination is done at a magnification from 100X to 1500X. Figure 14.10 (a) shows a metallurgical microscope used for micro examination of materials. Metallographic examination of welds require prepa- ration of samples. There are three stages of preparation of samples :

(i) Lapping of sample with a series of silicon carbide emery paper.

(ii) Polishing on a rotating disc with alumina or diamond paste to obtain a scratch free surface.

(iii) Etching of the polished samples with a reagent to reveal the microstructure.

A metallographic specimen in the as polished/unetched condition will reveal inclusions, porosity, cracks etc. The etching process in defined as the process to reveal microstructural details by prefential attack of a metal surface with an acid/basic chemical solution. The commonly used etchants for ferrous metals are given in Table 14.2.

Table 14.2 : Etching Reagents

Etching Reagent	Composition	Remarks	Uses
General reagents for irons and steels (carbon, low and medium-alloy steel)			
Nital	2 ml HNO_3, 98 ml ethyl or methyl alcohol. (95% or absolute) also amyl alcohol.	Not as good as picral for high resolution work with heat treated structures. Excellent for outlining ferrite grain boundaries. Etching time a few seconds to a minute.	For carbon steels : Gives maximum contrast between pearlite and ferrite or cementite network; reveals ferrite boundaries; differentiates ferrite from martensite.
Picral	4 g picric acid, 100 ml ethyl or methyl alcohbol (95% or obsolute) Use absolute alcohol only when acid contains 10% or more moisture), and 4 or 5 drops zephiran chloride (17%) wetting agent.	Not as good as nital for revealing ferrite grain boundaries.Gives suerior resolution with fine pearlite, martensite, tempered martensite and bainite structure. Detects carbides. Etching time a few seconds to 1 minute or more.	For all grades of carbon steels, annealed, normalized, quenhed, quenched and tempered, spheriodized, austempered.
Ville's reagent	5 ml HCl, 1 g picric acid, 100 ml ethyl or methyl alcohol (95% or obsolute).	Best results obtained when martensite is tempered.	For revealing austentitc grain size in quenched and quenched, and tempered steels.
General reagents for alloy steels (high alloy, stainless and tool steels)			
Ferric chloride and hydrochloric acid	5 g $FeCl_3$, 50 ml HCl, 100 ml distilled water.	Immerse until structure is revealed.	Reveals the structure of austenitenickel and stainless steels.
Mixed acids in glycerine	(A) 10 ml HNO 20 ml HCl, 30 ml, glycerine (B) 10 ml HNO, 20 ml HCl, 20 ml glycerin, 10 ml H_2O_2.	Mix HCl and glycerine thoroughly before adding HNO_3. Before etching, heat the specimen in hot water. Best results are obtained with alternate polishing and etching. Use hood. Do not store. Action can be modified by varying the proportion of HCl.	Reveals the structures of CrNi and Cr Mn steels, and all Fe-Cr austenite alloys.

Cupric chloride and hydrochloric acid	5 g $CuCl_2$, 100 ml HCl, 100 ml ethyl alcohol, 100 ml distilled water.	Use cold.	For austenite and ferrite steels, the ferrite being most easily attacked. (Carbodies and austenite are not attacked.)
Ville's reagent	5 ml HCl, 1 ml picric acid, 100 ml ethyl or methyl alcohol (95% or obsolute).	Immerse to reveal structure.	Can etch numerous types of Fe-Cr, Fe-Cr-Ni and Fe-Cr-Mn steels. Also attacks the grain boundaries in Cr-Ni austenitic steels.
Cupric sulphate and per-chloric acid	10 g $CuSO_4$, 45 ml per-chloric acid (70%), 55 ml distilled water.	Boil 15 min. Do not use in presence of organic materials. Use hood. Do not concentrate acid. Highly explosive.	Etches stainless steels, and shows chromium segreation by revealing areas poor in chromium.

Macro-etching of samples is a process used to reveal the quality of a material by subjecting it to a corrosive action of an etchant. Examinations are limited to usual observation or upto a maximum magnification of 10X.

Fig. 14.10(a) Metallurgical microscope with maximum magnification 1000x.

Micro-etching is used to reveal the microstructure of the metals and also it is used for observing other fine details like cracks, inclusion shapes etc. While micro-etching will provide information on variation of structure grain size, flow lines, calumnar structures, dentrites, segregation etc. macro-etching of welds will reveal weld structure, weld penetration, dilution, slag inclusion, porosity, cracks, disposition of beads etc.

(a) Macro Examination

Macro-examination involves visual examination of polished samples or etched samples by unaided eye or with a mgnifying glass upto 10X. Fig. 14.10 (b) shows some of the weld defects.

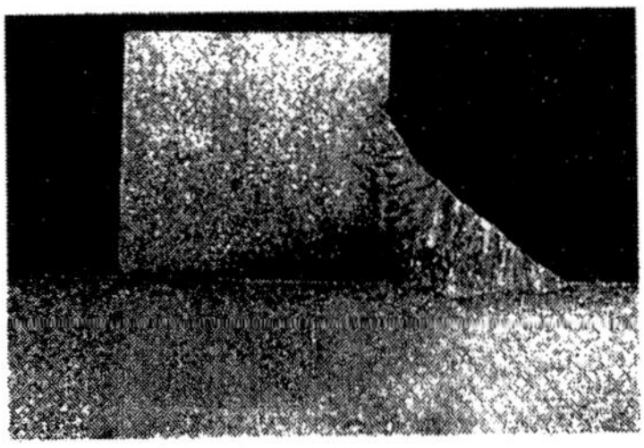

Fillet weld

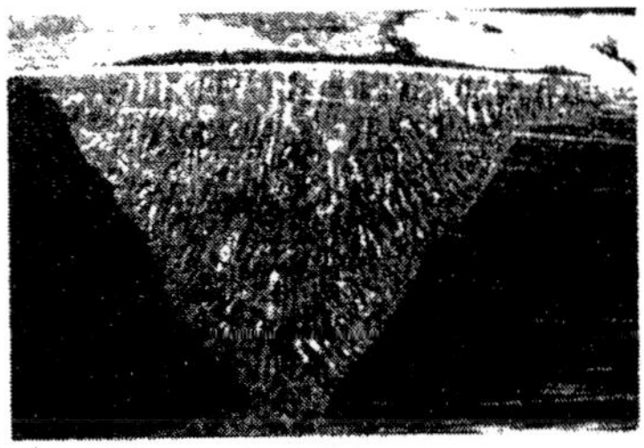

Butt weld

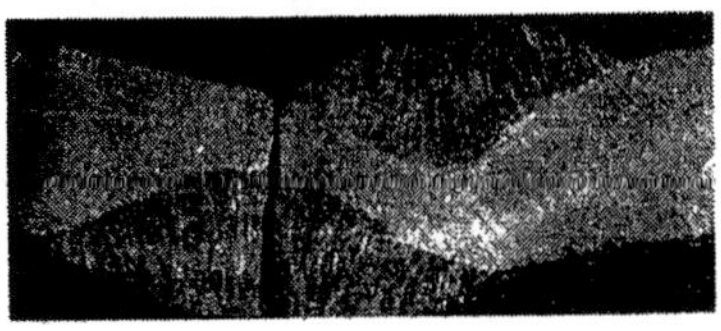

Off-line weld

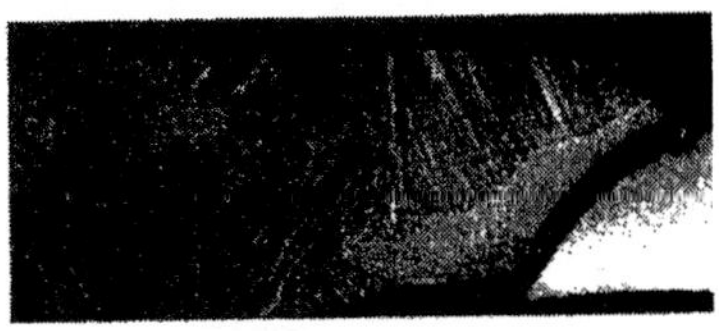

Lack of fusion

Fig. 14.10(b) Macrographs of weld sections.

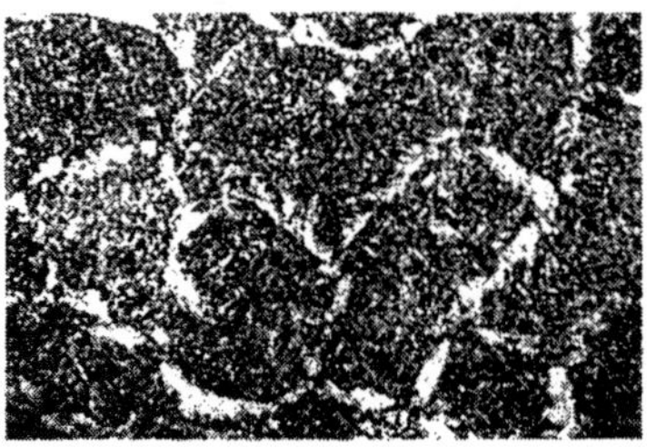

Weld metal

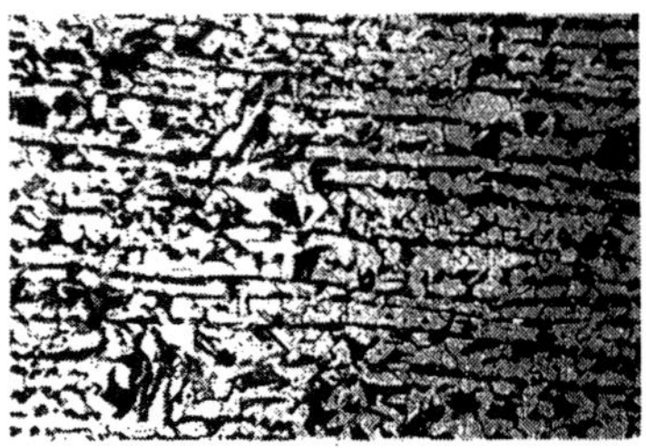

Carbon steel showing banded structure

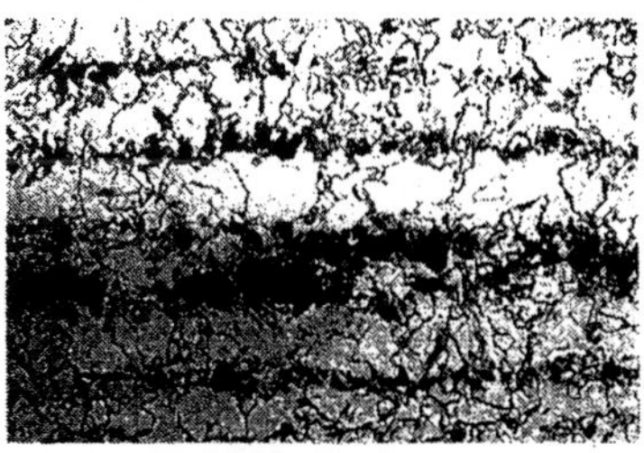

Carbon steel with inclusions

Austentic stainless steel

Fig. 14.10(c) Microstructure of various steels and weld metal.

(b) Micro Examination

The etched sample is observed under a metallurgical microscope for identifying the microstructure and other details. Typical micro-structures of carbon, low alloy and stainless steels are shown in Fig. 14.10 (c).

❑❑❑

15

CHAPTER

SAFETY IN WELDING AND CUTTING

INSIDE THIS CHAPTER

15.1 Possible hazards in welding and cutting, 15.2 Causes and remedies.

In any manufacturing industry welding and cutting are important processes. For avoiding the hazards in welding and cutting, certain safety precautions are very much needed to prevent serious injuries and accidents. Accidents occur mainly due to lack of knowledge about the hazards in an operation and due to wrong operation. The hazards in welding and cutting in fabrication industries are described in this chapter.

15.1 POSSIBLE HAZARDS IN WELDING AND CUTTING

(i) Burn injuries.
(ii) Electrical shock.
(iii) Eye injuries.
(iv) Air Pollution.
(v) Fire and explosion hazards.
(vi) Heat effects.
(vii) Back fire and flash back.
(viii) Other hazards like fall from heights, noise, cuts etc.

15.2 CAUSES AND REMEDIES

15.2.1 Burn Injuries

These can be caused by :

(i) Hot spatters from the welding.

(ii) Touching hot jobs or welding holders.
(iii) Burns due to flash from short-circuit or exposed cables.
(iv) Burns due to excessive heat.

The following are remedial measures :

(i) Use personal protective equipment like mask, gloves, safety shoes and protective clothing.
(ii) Display warning boards for preheating and post heating operations.

15.2.2 Electrical Shock

This can be caused by :

(i) Poor earth connections.
(ii) Equipment fault.
(iii) Wrong connection.

Welding voltage is low, though any operating voltage (upto 100 volts) more than 24 V is not considered as safe voltage. Electrical shock depends on the following :

(i) The amount of current flowing through the human body.
(ii) The resistance offered by the body and the condition of the body.
(iii) The path of the current flowing through the body.
(iv) The duration of the current flow through the body.

The following are remedial measures :

(i) There must be double earth in electric are welding. One earth from the welding set and the other earth from the job as shown in Fig. 15.1.

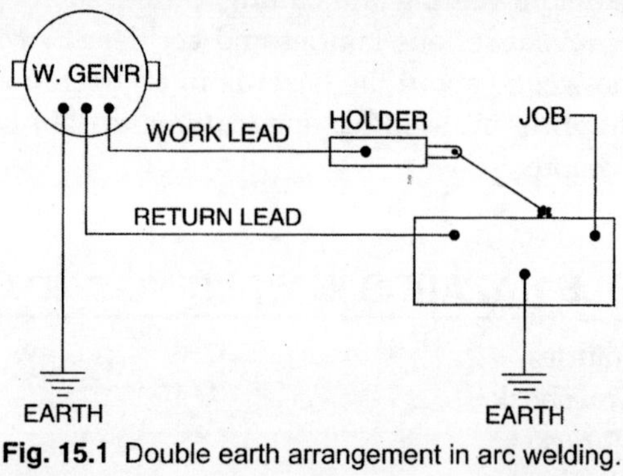

Fig. 15.1 Double earth arrangement in arc welding.

(ii) The welding earth should be capable of carrying the full welding current.
(iii) Only trained electricians should be allowed to repair welding sets.
(iv) Only well insulated electrode holders should be used.
(v) Proper welding shoes and hand gloves should be used.

15.2.3 Eye Injuries

The causes are :

(i) Seeing the arc directly or indirectly.
(ii) Seeing the gas welding and cutting flame.
(iii) Inhailing the fumes emanating while welding.
(iv) Fall of flying objects while removing the excess weld metal and slag.

(A) Effect of Ultraviolet Rays

Arc welding emits ultraviolet radiation which will sun-burn the eyes. The ultraviolet rays damage the interior layer of the eye and ratina. The most common eye disease due to this 'welders conjunctivitis'. The eyes will become reddish and there will be irritation as if sand is rubbed into the eyes. In the case of continuous exposure to ultraviolet rays it may lead to total blindness.

Remedial Measures :

(i) Should not see the arc welding rays.

(ii) Welder should use welder's helmet with opaque glass of 10 to 14 shade filter lenses.

(iii) The helper/workers assisting the welders should use welding goggles fitted with 3 mm thick shatter-proof glasses.

(B) Effect of Infrared Rays

Gas welding emits infrared rays. The infrared rays may cause industrial cataract.

Remedial Measures : For gas welding and cutting operations, gas welding goggles fitted with shades 4 and 5 should be used. Filter lenses for various welding and cutting operations are given in Table 15.1.

Table 15.1 : Filter lens shade numbers for various welding and cutting operations

Types of operation	Recommended shade number
Resistance weldings and for protection against stray light from nearby welding and cutting.	Clear or filter up to No. 2
Torch brazing or soldering.	3 or 4
Light oxygen cutting and gas welding (to 1/8″)	4 or 5
Oxygen cutting, medium gas welding (1/8″ to 1/2″) and arc welding upto 30 amps	5 or 6
Heavy gas welding (over 1/2″) and arc welding and cutting from 30 to 75 amps.	6 to 8
Arc welding and cutting from 75 to 200 amps.	10
Arc welding and cutting from 200 to 400 amps.	12
Arc welding and cutting exceeding 400 amps.	14

(C) Effect of Fumes

The fumes emanating from welding and cutting operation cause eye irritation.

Remedial Measures : The hazard from fumes can be avoided by wearing leather mask and goggles by both the welder and his helper. In addition the helper should turn his face away when the work is in progress.

(D) Effect of Flying Objects

The flying particles are as a result of the process of removal of slag and excess metal which is inevitable. They can cause injuries to the worker.

Remedial Measures :

(i) This can be avoided by wearing shatter-proof glass goggles.

(ii) By practising safe work procedure like holding and striking the chip removing chiesel away from the worker's body, one can avoid injury to oneself.

15.2.4 Air Pollution

Air pollution in welding depends upon :

(i) The type of electrode used.
(ii) The type of flux used.
(iii) The type of coating on the electrode.

The formation of contaminants and a proper assessment of fumes concentration should be carefully made. The level of concentration should be within the threshold limit value for a particular pollutants. The work space, exhaust arrangements and forced air circulation are to be provided as per requirements. Table 15.2 lists the possible contaminants, their harmful effects and their respective threshold limit values.

Table 15.2 : Air pollution in welding

Contaminants	Chromium	Manganese	Molybdenum	Nitrous fumes	Carbon monoxide	Ozone	Flourides
Harmful effects	Irritation and deep ulceration of nose/skin	Affects central nervous system. Irritates mucous membranes of the respiratory tract.	Toxin irritates mucous membranes of respiratory tract.	Intensely corrosive. Irritates respiratory tract, causes dizziness headache and weakness.	Asphyxia anoxemia, giddiness and swooning	Headache and dryness of the upper respiratory tract. Pulmonary edema	Irritates mucous membranes of eyes, nose and respiratory tract. Cause ulcerations, conjunctivities, coughing and eosinophilia
Tlv	mg/m^3	5 mg/m^3	15 mg/m^3	5 ppm	50 ppm	0.1 ppm	2.5 mg/m^3
Remedial measures	Adequate exhaust ventilation		Wearing of gas masks	Use of barrier cream.		Medical check up.	

15.2.5 Fire and Explosion Hazards

The causes are :

(i) The fire can be due to sparking.
(ii) The loose joints in the welding cables may heat the wire and cause a fire.
(iii) Flammable materials such as oily scrap, cotton wastes etc. may cause a fire.

The remedial measures are :

(i) Remove flammable materials, oily scraps, cotton wastes from the welding spot.

(ii) The joints in the welding cables should be properly insulated.
(iii) Oil drums, paints and thinner tins should be kept away from the welding area.

15.2.6 Effects of Heat

Welding is a process of heating and joining metals with or without the application of pressure. Preheating and post heating of jobs release radiant heat. The effects of such heat are as follows :

(A) Heat Cramp

It is a mild form of heat stress. The symptoms are painful muscle spasms, sweaty skin but normal body temperature.

The remedial actions are :

(i) Sit or lie down in a shade.
(ii) Drink plenty of cold water.
(iii) Stretching muscles may help but do not massage.

(B) Heat Exhaustion

It is the second stage of heat stress. The symptoms are profuse sweating, clammy or pale skin, dizziness, nausea, vomiting, headache, muscle ache, rapid pulse, thrist and slightly elevated body temperature.

The remedial actions are :

(i) If it is mild, treat as for heat cramps.
(ii) If persistent headache and vomiting occur, gently apply wet towels and call the emergency help.

(C) Heat Stroke

It is a serious condition in which the temperature regulating center in the brain 'shuts down', body temperature soars, which will lead to severe even fatal brain, lever or kidney damage. The symptoms are agitated mental status of the person, confused, staggered walk, hot dry skin, rapid pulse and body temperature high.

The remedial actions are :

(i) Move the affected person to a reclining position in the shade.
(ii) Call for emergency help.
(iii) If the person is conscious, offer him sips of cool water.
(iv) Fan the person and sponge with cold water.
(v) An ice bath under the supervision of a doctor may be necessary.

15.2.7 Back Fire and Flash Back

(A) Back Fire

It is a small explosion of the flame at the torch tip which may be the result of several avoidable conditions.

The causes are :

(i) Pre-ignition of the gases.
(ii) Too low a pressure of gases at the tip.
(iii) Overheated tip.
(iv) Hot metal particles and carbon deposits inside the orifice.

The remedial action are :

(i) Set slightly higher pressure for both the oxygen and acetylene flames.
(ii) Cool the tip. Do not keep the tip very close to the hot molten metal.
(iii) Clean the tip by using a tip-orifice cleaner.

(B) Flash Back

A flash back is the burning of gases in the mixing chamber or beyond, towards the regulator. It is an extremely dangerous occurance which will cause hose-fires, regulator ruptures and fuel cylinder fire.

The causes are :

(i) Failure to purge before lighting the torch.
(ii) Overheated tip.

The remedial actions are :

(i) Purge the systems properly.
(ii) Cool the tip.
(iii) Use a flash back arrester in the line and maintain a proper water level in the arrester.

15.2.8 Other Hazards

(A) Fall from Heights

During structural erection works, welders warking at a height may fall from elevated floors, plateforms and temporary structural stands. The causes are :

(i) Improper support to platform.
(ii) Non-use of safety belts.
(iii) Electric shocks.
(iv) Touching hot metal surfaces.

The remedial measures are :

(i) The support to platform should be rigid.
(ii) The welders and workers working at heights should use safety belts with life line connected to permanent structures.
(iii) The welders and helpers should use proper personal protective equipments like safety gloves, safety shoes, goggles, welders' helmets and shields.

(B) Cuts and Injuries

This is due to handling of material. The causes are :

(i) While fitting jobs for welding edge preparation, burrs may cause injuries.
(ii) While handling the material during welding it may cause injuries.

The remedial measure is to use proper personal protective equipments like safety shoes, gloves and aprom.

(C) Noise Hazard

The noise level in certain welding processes is hazardous. Mainly in plasma spraying the noise level is measured to be as high as 120 db. Most of the noise produced in plasma spraying is in the upper octave band, in the dangerous range and is due to the design of the nozzle. A similar problem may occur in ultrasonic welding.

The remedial actions are :

(i) Ear plugs/ear muffs must be worn while working in plasma spraying/ultrasonic welding areas.

(ii) Closed containers that have held flammable liquids or other combustibles should be thoroughly cleaned before welding or cutting.

(iii) For stationary containers of inflammable materials, purging with inert gas/steam is recommended before welding/cutting with the vent valve kept open. After purging/cleaning check the level of contaminants and if it is within the threshold limit value, then proceed with welding/cutting work.

The following precautions are necessary is preparing the containers, tanks and drums for welding.

(i) Remove all the sources of ignition like open flames, unguarded electric lights etc. from the interior of drums to be welded.

(ii) Ensure free draining by removing debris and wastes.

(iii) Use a portable electric hand lamp which is marked for using in hazardous locations.

(iv) Ensure strictly the work permit system.

(v) Experienced welders under qualified supervisors should perform the welding or cutting operation in confined spaces and storage tanks that contain explosive material.

Safety aspects during welding and cutting operations in confined spaces :

(i) Means of escape for the welder in an emergency should be provided.

(ii) Attach the safety belt and life line to the welder keeping the other end of life line outside of the confined space.

(iii) A capable person should be kept outside the confined space to observe the welder all the time with planned rescue procedure.

(iv) Suitable fire extinguishers should be kept near the confined space.

15.2.9 First-Aid for Welding Accidents

(A) First-aid for Burn Injury

(i) Avoid handlingof the affected area more than it is required.

(ii) Do not apply lotions of any kind.

(iii) Cover the area with clean lint, free cloth or freshly laundered materials.

(iv) Contact a doctor for further treatment.

(B) First-aid for Electric Shocks

(i) Switch off the electric supply.

(ii) Remove the victim from contact with the live line.

(iii) Administer artificial respiration till the victim breaths normally.

(iv) Treat for shock and burns.

(v) Transfer the victim to hospital and seek medical aid.

(C) First-aid and for Eye Injury

Since eyes are the most precious part of the human body utmost care must be exercised in treating the eye injuries.

(i) Trust only a doctor for treating the eyes.

(ii) Take the victim to the hospital.

In general, Table 15.3 gives an easy reference of the Do's and Don'ts from the recommended safety rules for Arc welding operations by The American Welding Society.

Table 15.3 : Safety rules for arc welding operations

Do's	Dont's
The electrode holder must be effectively insulated and properly maintained.	Do not drop the electrode since the insulating material will break.
If the skin has become wet owing to perspiration or some other cause, its resistance is reduced and purge of current may be serious.	Never carry the electrode holder in the arm pits since they are often somewhat damp. Avoid standing on damp floors.
Make sure that the welding mchine is properly earthed so that in case of possible short-circuit the current is directed to earth.	Do not touch the workpiece with hand or foot at the same time in contact with the electrode holder.
Make sure that the welding helmet or face shield is not defective. Arc rays are dangerous to the eyes.	Never strike an arc if other workers are nearby without eye protection.
Make sure that the exhaust system functions effectively, esecially in confined spaces where fumes are dangerous and may cause unconsciousness.	Do not forget to take additional safety precautions before working in a confined space.
Maintain the electrical cable in good condition and make sure the connections are tight. Cable connectors should be used for joints. Keep the cables as short as possible.	Do not allow the cables to lie where they can be burnt, pinched, run over, frayed by sharp edges or where it may be hazardous.
Have all electrical equipment inspected periodically by a qualified electrician, and the defects should be eliminated before they become serious.	Do not undertake any maintenance on welding machines unless they are disconnected from the main supply.
Use approved good quality goggles and head shields which provide proper eye protection. Use cover glasses over coloured glasses.	Do not use incorrect shade or tint filter lens, because that will cause eye strain and headache.
Good housekeeping reduces fire risk.	Do not allow rubbish, old hose, paint, grease, oil tins and other inflammable materials to accumulate. Do not throw the electrode bits in the work spot.
To avoid fire, remove either the inflammable material or the job to another location where there will be no possibility of starting a fire causing explosion.	Do not cut or weld in rooms containing inflammable materials, gases, vapours or liquids or flammable paint.

❑❑❑

CHAPTER 16

HEAT TREATMENT OF WELDS

❑ INSIDE THIS CHAPTER

16.1 ANNEALING

The general purpose of annealing treatment is to reduce the hardness of steel or to produce a structure that facilitates the progress of subsequent manufacturing operations. The term annealing by itself usually implies full annealing which in most cases involves raising the temperature to get the austenite phase, followed by slow cooling. Annealing is carried out to improve machinability and cold formability, to restore ductility, to reduce or eliminate structural non-homogenity, to refine grain size, to relieve internal stresses and to prepare the structure of the steel for subsequent heat treatment. Annealing is not applied to improve mechanical properties of alloy steels. The mechanical properties of structural steels are improved by hardening, followed by tempering. There are different types of annealing for different purposes.

16.1.1 Full Annealing

Full annealing consists of austenitization of the steel followed by slow cooling in the furnace (Fig. 16.1). The temperature to be adopted depends on the carbon content of the steel. Annealing is prescribed for castings of carbon and alloy steels and high carbon hot worked steels. Full annealing improves the machinability and also increases the strength because of a fine grain structure.

16.1.2 Spheroidized Annealing

This is done by heating the steel just above or slightly below AC_1 critical temperature (Fig. 16.2) for a prolonged time followed by slow cooling in order to soften the steel as much as possible. Spheroidized annealing is done to transform

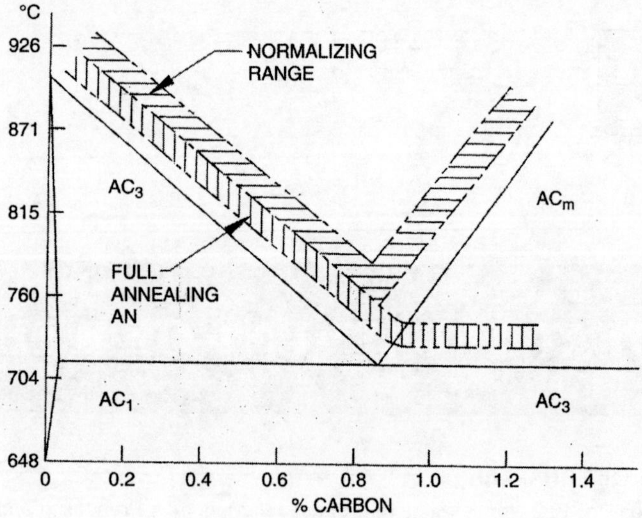

Fig. 16.1 Full annealing.

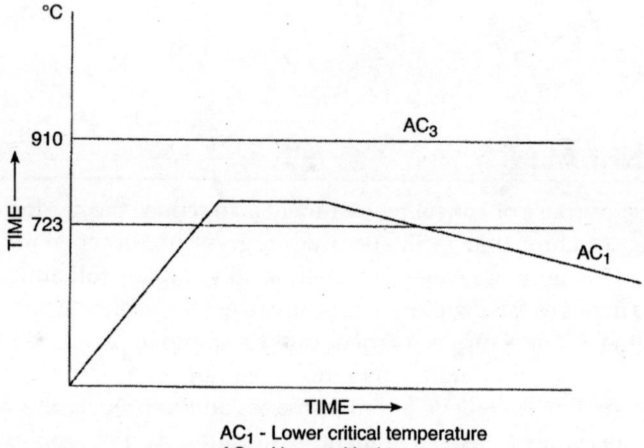

Fig. 16.2 Spheroidized annealing.

such as martensite that are produced by quenching and welding. This is to improve machinability and in some cases to improve cold-formability.

16.1.3 Isothermal Annealing

This is useful for softening steels for the subsequent maching operations. Figure 16.3 shows a typical isothermal annealing given to a ball bearing steel for softening and machining.This consists of austenitizing the steel at the normal annealing temperature and then cooling rapidly to the appropriate temperature. This temperature is held for a predetermined time waiting for the complete austenite decomposition to take place. After this the steel is cooled in a furnace or air-coded or rapidly cooled. Case hardening alloy steels are normally subjected to isothermal annealing.

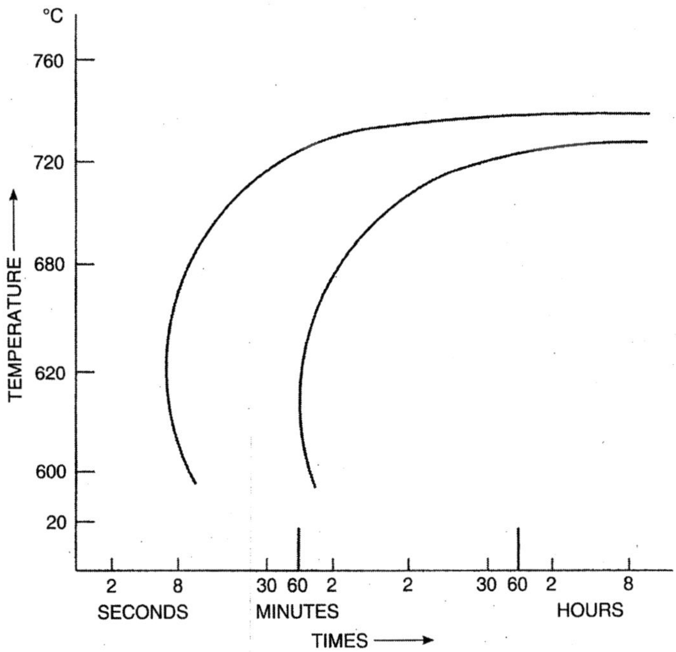

Fig. 16.3 Isothermal annealing.

16.1.4 Homogenizing Annealing

This process is adopted to alloy steel ingot which, when solidified after pouring, have a non-homogeneous structure (Fig. 16.4). This lack of homogenity in structure is generally removed by hot working. Where it is not possible, diffusion or homogenizing annealing treatment is done. The phenomenon of diffusion brings out a uniform concentration of alloying elements across. The matrix structure. The homogenizing is carried out for several hours at temperatures between 1150°C to 1200°C and the further cooled in air. This process is used only for special cases since the cost is high.

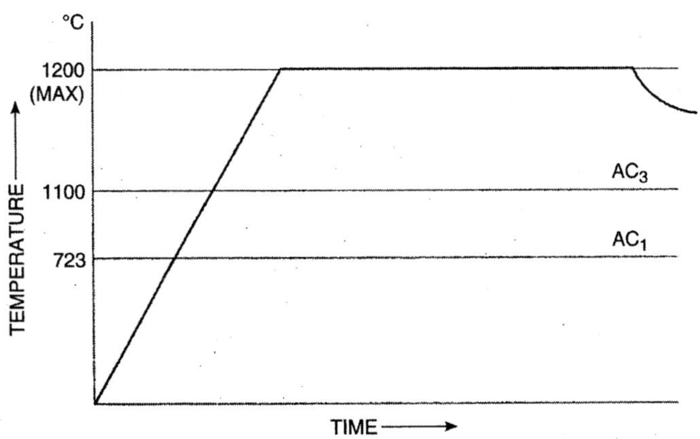

Fig. 16.4 Homogenizing annealing.

16.1.5 Intermediate Annealing

This is carried out after case hardening in order to carry out further machining. The components are held at 630°C to 680°C for 4 to 6 hours followed by slow cooling.

16.1.6 Bright Annealing

This is done to obtain a bright surface free from oxides. The charge is protected from oxidation with suitable gas atmosphere such as pure nitrogen, pure hydrogen, cracked ammonia etc. Bright annealing is done for both ferrous and non-ferrous metals, in the form of wire, strip sheet, tube etc.

16.1.7 Solution Annealing

Following welding or thermal processing of stainless steel, a subsequent re-annealing may be required for optimum corrosion resistance, softness and ductility. During annealing, chromium carbides, which markedly decrease resistance inter-granular corrosion, are dissolved (Fig. 16.5). Because carbide precipitation can occur at temperatures between 425°C to 900°C it obviously is desirable that the annealing temperature should be safely above this limit. Moreover, because all carbides should be in solution before cooling begins, and because the chronium carbides dissolves slowly, the highest practical temperature consistent with limited grain growth is selected. This temperature range is 1010-1120°C. Cooling from the annealing temperature carbides remain in solution. Time temperature and the method of cooling depend on the thickness of material. Thicker sections are water quenched.

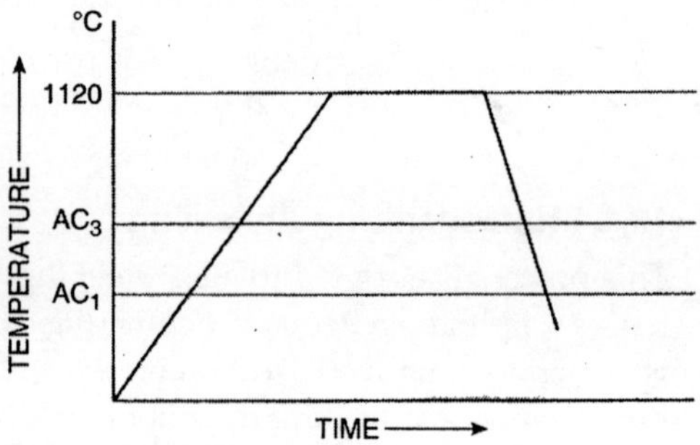

Fig. 16.5 Solution annealing.

16.2 NORMALIZING

This treatment is also called normalised annealing. This is done to refine the grain structure, improve hardenability, relieve internal stresses and improve the mechanical properties of carbon and low alloy steels etc.

Normalising consists of heating the steel to its hardening temperature and holding at this temperature for a short time depending on the type of steel, around 20-40 minutes. After the holding time has elapsed, the components are cooled in air. If the initial structure is coarse grained or irregular, normalising leads to a considerable improvement in fineness of the structure together with improvement in its mechanical properties. The mechanical properties obtained after normalising depend on the rate of cooling in air. A faster rate of cooling will give higher strength and greater hardness. Normalising is a heat treatment somewhat similar to annealing and is frequently employed in treating the steel both the base metal prior to welding and the finished weldment. Normalising involves heating the steel aproximately 30°C above its critical range, AC_1 in Fig. 16.1 to transform the structure to austenite, followed by cooling in still air.

A normalising treatment is used to :

(i) reduce stresses from cold working or welding.

(ii) remove hardened zones adjacent to the weld.

(iii) create a more uniform and desirable microstructure in both the weld metal and base metal.

(iv) refine, by recrystallization, any coarse structure which may have been developed in the steel through hot working or welding.

16.3 STRESS RELIEVING

Stress relievingis a heat tretment operation designed to relieve the stresses induced in the component, minimise the distortion during subsequent heat treatment and to avoid formation of cracks in extreme cases. This treatment consists of heating the components to a temperature of 550°C to 700°C and holding them for sufficient time at that temperature to achieve uniformity throughout the components and subsequent cooling them to an ambient temperature in a furnace. This treatment will not cause any phase changes but recrystallisation may take place. Stresses may be developed due to machining operations, forming operations, heat treatment, casting, welding etc. Internal stresses induced by welding can be relieved by heating the components to 600-650°C and holding at this temperature for a predetermined time. Normally the holding time is about 3 to 4 minutes per mm of the maximum thickness of the plate. It is subsequently coded at a rate of 50-100°C per hour to a temperature of 300°C. This slow and uniform cooling is necessary to prevent the addition of new internal stresses on cooling and to avoid formation of cracks in stress-relieved welded components (Fig. 16.6). Heating decreases the

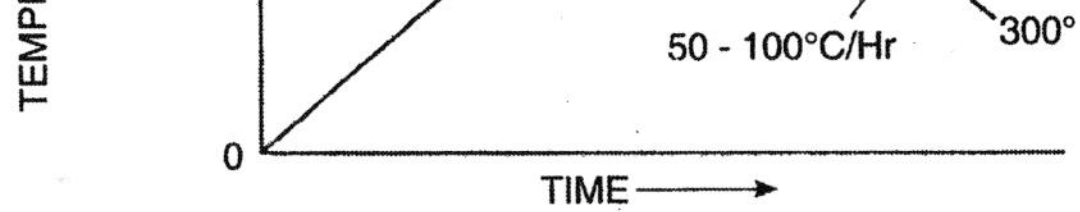

Fig. 16.6 Stress relieving.

yield strength of the material. When it is warm and plastic, the stresses are relieved. Components to be heat-treated should be provided with sufficient machining allowance to compensate for any distortion resulting from stress-relieving. Mostly electric furnaces are used for stress-relieving.

16.4 PREHEATING AND POST HEATING

The application of supplementary heat to a joint before or after welding can be a very important part of welding process/procedure. Elevating the temperature of the work-pieces immediately before a welding or cutting operation is called preheating. The application of supplementary heat to a welded joint or to a weldment after welding is called post heating.

Both preheating and post heating may be applied locally to the joint area or to a entire weldment. These treatments are employed in cutting and particularly in welding operations for number of reasons. Following are some :

(i) To retard the cooling rates and thus to avoid cold cracking in the heat affected zones of hardenable steels.
(ii) To increase the toughness of the weld joint and improve its ability to withstand adverse service conditions involving impact loading or low temperatures.
(iii) To alleviate the effects of hydrogen which enters the weld metal and heat affected zone of the base metal.
(iv) To alleviate the effects of hydrogen which enters the weld metal and heat affected zone of the base metal.
(v) To minimise shrinkage and distortion.
(vi) To produce particular mechanical or physical proportion in the steel of which the weldment is constructed.

16.5 HEAT TREATMENT OF WELDS

16.5.1 Heat Treating in Furnaces

Generally, in all the furnaces electrical power is used for heating. In places where electrical power is not available direct gas fired heat treatment furnaces are used. The furnaces for weld heat-treatment may be classified based on temperatures as follows :

(i) 700°C stress relieving, tempering.
(ii) 1000°C Hardening, grain refining, homogenizing annealing.
(iii) Over 1000°C Hardening, Solution annealing.

(A) Gas Fired Heating Systems

Gaseous fuel such as natural or producer gas is fired along with air through appropriate burners. Heat is developed by the circulation of the fuel gases which are at high temperatures. The flow of gases is controlled to alter the heating rate in furnace.

(B) Electrically Heated Furnaces

An electric furnace is of either an open elument type or indirect type. In the later the heating elements are protected from furnace atmosphere. A radiant tube furnace is an example of the indirectly heated furnace. Electrical furnaces can be made to develop a very vast range of temperatures, i.e., 150°C to 1500°C. The temperature control is very accurate in electric furnaces. Shop manufacture is always preferred with electric furnaces.

16.5.2 Heating Burners

Burners are used for preheating for maintaining interpass temperature during welding and for post heating. Burners are readily available in the market or they can be fabricated in different sizes and shapes to suit the welding products of different sizes and shapes. Long tubes fitted with serial nozzles are used for long seam welding of shells (Fig. 16.7). A tube, bent in the form of a ring fitted with nozzles is used for circumferential welding of pipes and drums. Sometimes box type burners are used. This consists of a mild steel chamber inside which some inconel wire mesh and ceramic honey comb structure etc. are kept. In the rear of the chamber, a gas inlet is provided which can be connected directly to the gas line hose. The inner face reaches 1000°C in 5 to 10 seconds. There is no flame production and the heat is concentrated. The box can be handled while preheating, since the outer casing will be at room temperature. A thermocouple can be attached, the readings of which can be fed back to a controller which opens or closes the gas flow to the burner, thereby varying the furnace temperature.

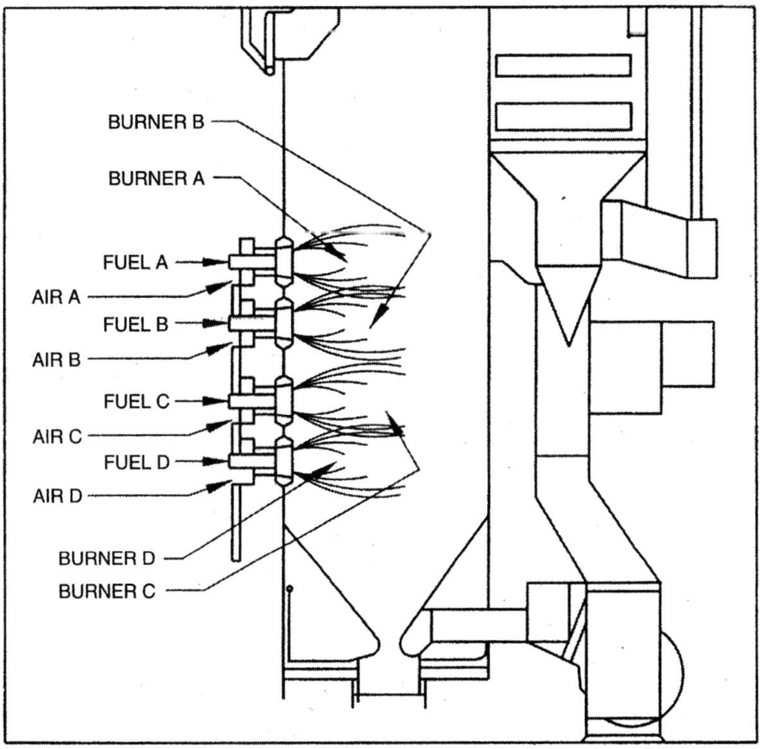

Fig. 16.7 Heating burners.

(A) High Velocity Gas-Fired Burners

Systems are now available for post-weld heat treatment of pressure vessels. Fuel gases circulated at very high velocity inside the shells impart temperatures to the order of 600-700°C. Mostly site welds are stress-relieved with this technique.

16.5.3 Resistance Heating Elements

In welding, the most commonly used method of heat treatment is resistance heating pads. The three different types of pads are used :

(i) nichrome wire with beads.
(ii) finger elements.
(iii) ceramic pads.

(A) Nichrome Wire with Beads

In welding, an extensively used heating element consists of lengths of 80/20 nickel-chromium wire with insulating beads. These elements are wrapped around the pipe and connected to the welding transformer and the current adjusted manually. This method has many disadvantages :

(i) During heating, the nichrome wire expands causing gaps between the insulating beads. Also the insulating beads often break during handling. These can cause short-circuiting with the pipe and result in stray arcing.
(ii) Controlling of the temperature difference at the top and bottom cannot be practised.
(iii) High rate of element failures at high temperatures.

(B) Finger Elements

This consists of a twin 9-SWG nichrome wire, bent in the form of fingers. Each finger is 100 mm long and insulated by sintered alumina ceramics. The number of fingers in these elements can be varied and made to suit any diameter of pipe. They use heavy gauge wires and result in low voltage drop across each element. The voltage drop is 1.3 V across each finger. The limiting factor in fixing a number of fingers is the maximum open circuit voltage, available on the heating transformer. The following are advantages :

(i) The sintered alumina ceramics has greater resistance to breaking, while handling.
(ii) Electrical insulation and heat conduction are very good.
(iii) Larger gauge nichrome wire will last for many cycles of heat-treatment without breaking.
(iv) This can be used upto 1050°C.

(C) Flexible Ceramic Pads (FCP)

FCP consists of insulated alumina beads with two holes interlocked to each other through which a multi-strand 80/20 nichrome wire is threaded. They are available in standard electrical rating (165 V, 50 Amps, 3.25 kW) and in various dimensions. Figures 16.8 and 16.9 (a) and (b) shows the ceramic pads setup for preheat in a pipe at site. The advantages are :

(i) This is a voltage controlled heater rather than a current controlled heater. This feature makes it suitable for accurate control of heat-treatment and allows it to be used with less costly on/off controllers.

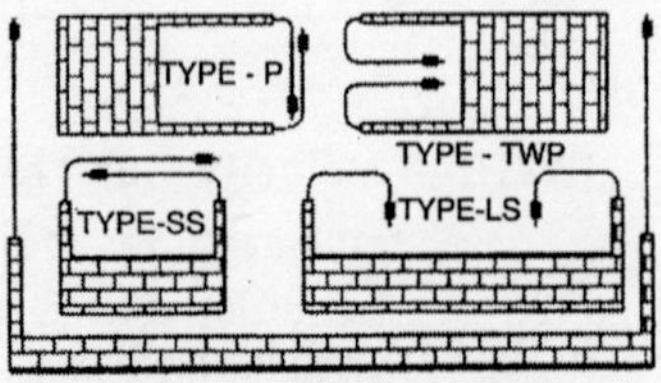

Fig. 16.8 Schematic diagram of flexible ceramic pads.

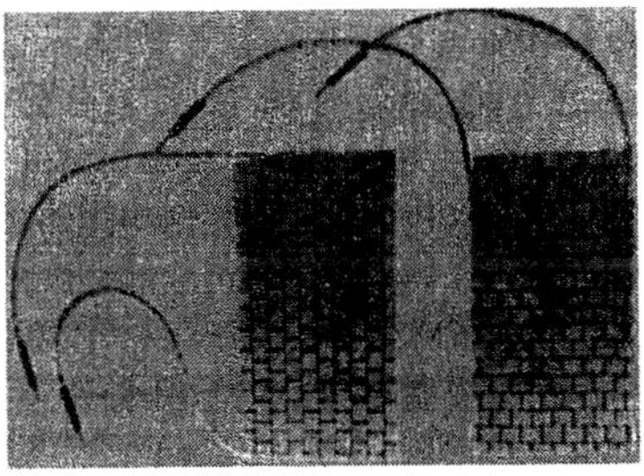

(a) Flexible ceramic pads

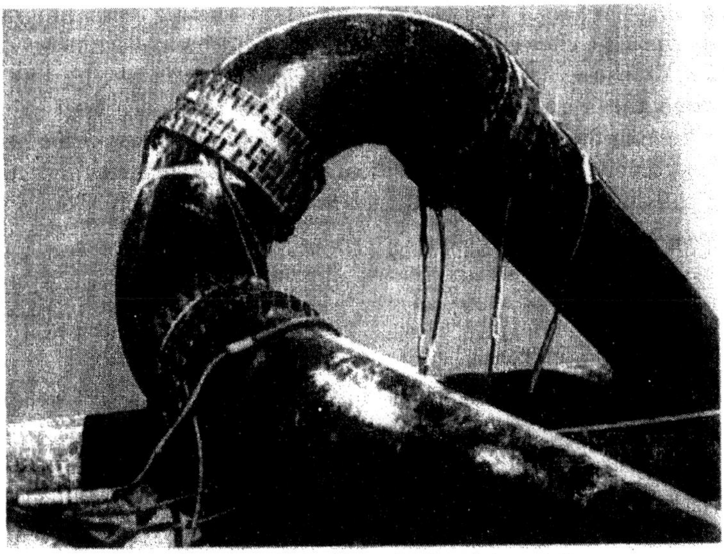

(b) Ceramic pads used for preheating of pipes at site

Fig. 16.9

(ii) These elements are extremely flexible and can be wrapped around any size of pipe.
(iii) The elements are of modular type and require less inventory.

16.5.3.1 Merits of Resistance Heating

(i) Continuous uniform heat can be maintained throughout the welding operation.
(ii) Temperature can be adjusted quickly and accurately.
(iii) Heating and welding can go on simultaneously.
(iv) Differential heat input can be given easily, as is required for the top and bottom of the pipe or where pipes are welded to higher wall thickness valves and fittings.

16.5.3.2 Demerits of Resistance Heating

(i) Element burnont may take place during heat treatment causing interruption.
(ii) It is possible that the resistance heating element can short the pipe causing stray arcs.

16.5.4 Induction Heating Coils

Induction heating is done at frequencies of 1000 Hz or higher for obtaining the above frequency range. Either a motor generator set or a solid state electronic-controlled induction heaters are to be employed. The various induction heating schemes are illustrated in Fig. 16.10 (a). The equipments are available at different ratings varying from 5 kV to 2000 kW. Depending on the job thickness and size, the frequency, power rating and cycle time for heating is determined. The equipment can then be specified using the above parameters. For welds which are subjected to continuous weld heat treatment cycles beginning from preheat to post-weld stress relief at one stretch, induction heating with programmable controls is the best choice. Different types of inductor coils can be designed. Figure 16.10 (b) shows a typical one for pipe welds. In induction heating the pipe walls receive the induced current and develops resistance heat whereas in resistance heating the resistance coil has to develop and transmit the heat to the pipe wall extremely. High power densities can be realised in a very short time with induction coils which can heat up the job wall instantaneously. The response to controls is also excellent.

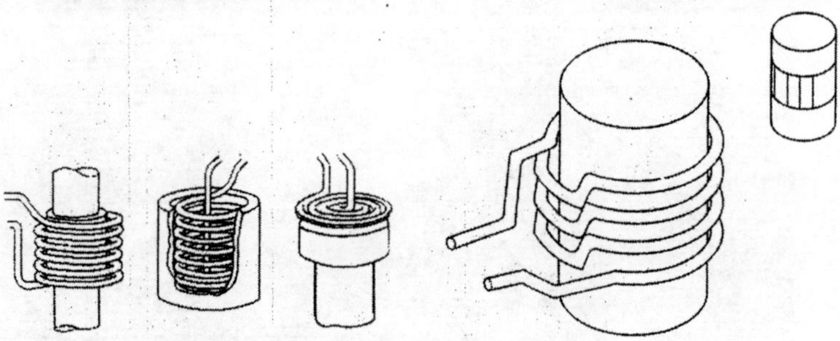

(a) Different types of induction coils (b) OD heating multi turn induction coil

Fig. 16.10

16.5.5 Temperature Measurement

During the heat treatment cycle the temperature is measured and recorded with the help of thermocouples and recorders. Further, the voltage produced by thermocouple can be used to control a contactor for starting and stopping of the heating. The principle of the thermocouples that when two dissimilar metal wires are in contact with each other, and if, there is a temperature difference at the junction of both the ends, an electromotive force (emf) is developed. The difference in temperature between them changes. By proper calibration, the voltage readings are translated into temperature readings. For temperature measurements upto 1250°C, chrome-alumel thermocouple is widely used. Chromel wire is non-magnetic and connected to the positive terminal of the recorder. The alumel wire is magnetic and connected to the negative terminal of the recorder. The thermocouple wires are normally short in length. To connect them to the recorder, kept at a distance, a compensating cable is used which has emf characteristics similar to the thermocouple.

16.5.6 Thermocouple Fixing

Fixing of a thermocouple on the spot is very important. If the thermocouple are not properly fixed, serious error in measurement of temperature can occur. If the thermocouple wires touch each other or the pipe away from the hot junction then the recorder may not read the temperature of the spot where the thermocouple is attached. Hence, the thermocouple wires must be electrically insulated from one another and also from the pipe surface. If the hot junction of the thermocouple wires is not insulated from the direct heat, radiated towards it by the heating elements, it will read higher temperature than the pipe surface. A fool-proof method of fixing the thermocouple is to weld directly 21 SWG thermocouple wire to the pipe using a capacitor discharge unit. This will eliminate the error in measurements. Figure 16.11 (a) and (b) shows a typical arrangement as required by welding codes in preheating and post weld stress relieving. A typical comparison of temperature errors due to the method of thermocouple attachment is given below.

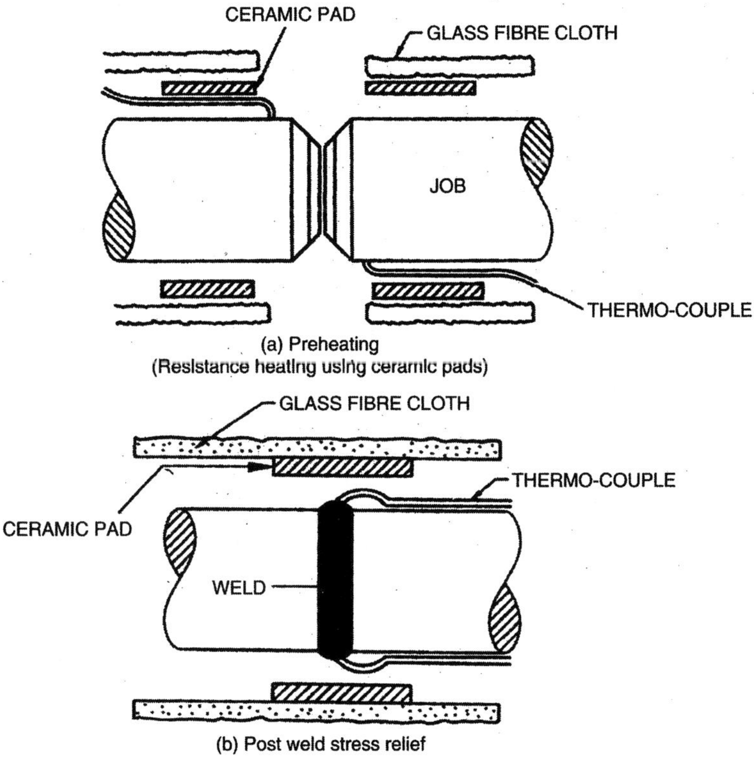

(a) Preheating
(Resistance heating using ceramic pads)

(b) Post weld stress relief

Fig. 16.11

A minimum of two diametrically opposite thermocouples are recommended to monitor the temperature of weldments on pipes above 150 mm in diameter and 20 mm wall thickness. On larger diameter pipes additional thermocouples must be fixed so as to indicate the temperature at significant points on the weld.

The recommended thermocouple location for the weld :

(i) At top and bottom of weld.

(ii) At a distance 1.5 times the thickness on either side from the centre line of the weld.

(iii) At 2.5 RT from the centre of the weld to achieve at least half the soaking temperature, where R is the radius and T is the thickness of pipe.

16.5.7 Temperature Recording

Single channel and multi-channel (upto six channels) temperature recorders are available in the market. If the thermocouple output voltages are connected to the input of the recorder, the recorder will produce a temperature recording against time. The recorder can give a recorded output from a few minutes to several hours, upto 72 hours. The temperature recording is generally done at a slow speed, throughout the welding cycle. The output is printed out continuously which can be form off after each weld is over.

17

CHAPTER

MECHANISATION AND AUTOMATION IN WELDING

❑ INSIDE THIS CHAPTER

Machanisation is a process of substituting mechanical means for mundane human work. When partial or complete replacement of manual work is achieved by a piece of equipment or machine, the process is often referred to as mechanisation. In welding, the three basic operations one encounters are :

(i) Positioning of the work with respect to the welding device or head.
(ii) Guiding the welding device along the weld line (seam tracking).
(iii) Control of the welding parameters.

The welding processes fall into two main types as regards their application. There are processes in which a local heat source or a single point of welding is moved along the joint to complete the weld seam. Most of the arc welding processes as well as electron beam, resistance seam, induction and high frequency resistance and cold pressure seam welding fall into this category. Then there are processes in which a whole joint is made in one operation, for example, friction welding, reesistance butt, flash butt, diffusion bonding, spot and projection welding, explosive and cold pressure welding. The sequence of operations in making a weld will be different for local heat source type processes. The steps in such a case are :

(i) assemble the parts.
(ii) present the work-piece to the machine or vice versa.
(iii) initiate welding.
(iv) create relative movement between the welding device and the work-piece.

(v) control the welding parameters.

(vi) stop welding and remove the work-piece.

For welding processes where a whole joint is made in one operation, steps (iv) and (v) are greatly simplified and therefore, a high degree of mechanisation is those stages is achieved compared to mechanisation of the process using local heat source. The degree of mechanisation extends from the performance of an isolated function up to full mechanisation of welding, carried out entirely by machine. The later is better known as automatic welding. In addition, if the machine is also responsible for bringing in and taking out the parts to be manufactured, and skills are to be provided to the machine by incorporating various adoptive controls and sensors, then the degree of automation also increases.

17.1 DEFINITIONS

Considering the factors as enumerated above the definitions of the concepts especially for arc welding can be restated as below :

17.1.1 Manual Welding

Any welding operation that is carried out in a manner in which the welding torch, electrode or gun are manually held and controlled can be termed as manual welding.

17.1.2 Semi-automatic (Partially Mechanised) Welding

Any welding operation of continuous nature, after the welding parameters have been set, where the movement of the welding gun along the joint is carried out and controlled manually is known as semi-automatic welding or partially mechanised welding, e.g., SAW, flux cored arc welding etc.

17.1.3 Fully Mechanised Welding

Any welding operation in which the welding gun is supported and driven mechanically but is manually supervised comes under fully mechanised welding, e.g., SAW, ESW etc.

17.1.4 Automatic Welding

This is a welding operation in which the welding gun is supported mechanically and it may be controlled by pre-set instructions and utilise in-process information to make corrections during welding to maintain consistency, by employing appropriate adaptive and sequential controls. For example, automatic tube to tube welding.

17.1.5 Automated Welding Systems

It is a fully mechanised process where in all loading operations from assembly, loading and transfer of the parts to the machine, their welding in one or more stages and the final ejection of the completed product, are carried out mechanically without manual intervention.

17.2 WELD SEAM

The line of weld joint which the welder or welding gun has to follow is termed as weld seam. Various types of weld seams are :

(i) Linear
(ii) Curved
(iii) Circular
(iv) Combination of linear, curved and circular seams
(v) Complex : two or three dimensional welds in different planes.

17.3 MAJOR CATEGORIES OF WELD TYPES

(i) Long, linear single pass welds, e.g., fillet welds.
(ii) Long, linear multi-pass welds, e.g., pressure vessel longitudinal and circumferential joints.
(iii) Small, circular, single or multi-pass welds, e.g., tube or pipe joints.
(iv) Short complex three dimensional welds, e.g., Bourdon to Bourdon tube in temperature and pressure gauges.

17.4 MECHANISATION OF THE WELDING OPERATION

In mechaical welding, the operation is performed under the constant observation and control of a welding operator. The equipment may or may not perform the loading and unloading operations. The equipment should control the following variables during welding :

(i) Initiation and maintainence of the welding arc.
(ii) Feeding of the filler metal into the joint.
(iii) Travel speed along the joint or the welding speed.

The operator must be near the point of welding to closely observe the operation. The operator interacts continuously with the equipment to ensure proper placement and quality of the weld metal. Mechanised welding should provide the welding operator with sufficient time to monitor and control the guidance aspects of the operation as well as the welding process variables. Weld quality can be enhanced as a result of proper control of the process variables.

17.5 MECHANISING RELATIVE MOTION

The provision of relative motion between the welding head and the work-piece is essential to all arc welding processes and the mechanisation of this motion is the first step to automatic welding. There are only two choices : Moving the head or moving the work. However it is possible to employ both together if a complicated shape has to be welded. The choice depends on the shape, size and weight of the work-piece on which the welding is carried out and the form of the joint. Large, heavy and awkwardly shaped works with long weld seams clearly require a moving head. Circumferential welds, however, in very large objects such as pressure vessels can be made with relative ease by turning the work under a stationary welding head. Longitudinal seams in pipes, even if the pipe is of considerable size, can often be made by moving the work or by moving the head.

Equipment for relative motion may be secially designed to suit the application as would be done for a machine tool or use of standard maniulators may be made. Manipulators also known as positioners are mechanical devices used to turn a job to place the joint in the most suitable position for welding. This is usually a flat position for arc welding since this enables highest possible current to be used and therefore the highest productivity to be obtained. The simplest manipulators are merely holding devices with possibly the facility for turning by hand. More eleborate types can mechanically turn work-pieces weighing many tonnes, continuously at different welding speeds. As essential requirement of the mechanism for relative motion is that it should provide constant, reproducible and jerk-free motion with smooth starting and stopping.

(A) Moving Head Systems

(i) Linear carriage for all positions and tractors running directly on the work or on rails supported by the work.
(ii) Stationary rail gantries positioned over the work with self-propelled or independently driven heads.
(iii) Column and boom manipulators carrying welding heads giving horizontal or vertical movement.
(iv) Moving gantries carrying one or more heads which span the work.
(v) Orbital pipe welders.
(vi) Tube to tube plate welding heads working from spigots inside the tube.
(vii)Industrial welding robots and NC machines.

(B) Moving Work Systems

(i) Rotating tables.
(ii) Turning rolls : roller positioners or rotators.
(iii) Welding lathe : head stock and tail stock positioners.
(iv) Continuous forming machines for longitudinally folded tubes, sector finned tube and helically wound pipe.
(v) Continuous movement of preformed componants like cans under fixed welding head.

17.6 MOVING HEAD SYSTEMS

(A) Machine Carriages

A welding mchine carriage provides relatively inexpensive means for providing arc motion. A typical carriage rides on a linear or curved track of the same contour as the joint to be welded. Some carriages are specifically designed to ride on the surface of the material being welded in the flat position and some use the material for weld joint guidance. The travel speed of the carriage is a welding variable and uniformity is important for weld quality. Tractor type carriages are used primarily for welding in the flat or horizontal position (Fig. 17.1). Welding

Fig. 17.1 Machine carriage.

carriages may be employed for welding in the horizontal, vertical or overhead position. Others may be designed to follow irregular joint contours. They employ a special track upon which the welding carriage is mounted. Since the welding operator must continually interact with the controls, the carriage controls and welding controls are typically placed within easy reach of operator. This type of equipment is most productive in producing long flat position groove and fillet welds such as those encountered in ships and barge buildings. It is also useful in field work such as bridge or storage tank fabrication (Fig. 17.2). The rigidity with

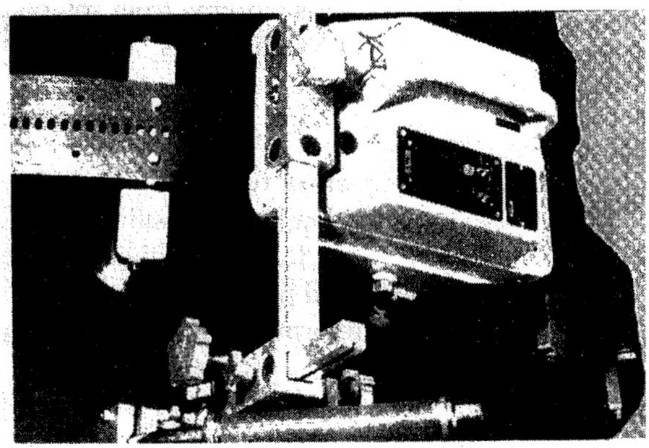

Fig. 17.2 Carriage controls.

which the welding carriage is held to the track and the uniformity of travel are critical to good performance. A side beam carriage consists of a welding carriage mounted on a horizontal beam as shown in Fig. 17.3. It provides powered linear travel for the welding head. The welding carriage supports the welding head, filler wire, wire feeder and sometimes the operator's' controls. The welding head on the carriage is adjustable for vertical height and transverse position. During operation, the welding process is monitored by the welding operator. Welding position and travel speed of the side beam carriage are adjusted by the operator to accommodate different welding procedures and also variations in part fit up.

Fig. 17.3 Side beam carriage.

(B) Stationary Gantry

Rail systems can also be supported independently of the work. Fixed overhead rails along which the welding head is carried by lead screw or similar mechanism are favoured when welds are to be made in relatively small work-pieces which can be easily moved into the position under the rail. This will apply to the longitudinal welds in small pressure vessels or drums (Fig. 17.4).

Fig. 17.4 Stationary gantry.

(C) Column and Boom

If more flexibility is required, a boom carrying the tractor or welding head may be mounted on a column, which can be moved up and down mechanically to the position of welding. Boom and column type manipulators are also used for positioning the welding head for circumferential welding and are found in most of the heavy pressure vessels fabrication shops (Fig. 17.5).

Fig. 17.5 Boom carrying welding head.

(D) Moving Gantry

A more eleborate form of rail system employs a travelling gantry on which one or more welding/cutting heads may be carried over the work. Each welding/cutting head has a limited individual transverse and vertical adjustment and may also have facilities for seam tracking and automatic height adjustment. Gantry type manipulators are much used in ship building particularly for panel lines (Fig. 17.6).

Fig. 17.6 Moving gantry.

(E) Orbital Welders

Different tye of equipments known as orbital welders have been devised for welding tubes and pipes which cannot be rotated. One form of machine for TIG Welding of thin wall tubes is shown in Fig. 17.7. In this system the motion of the

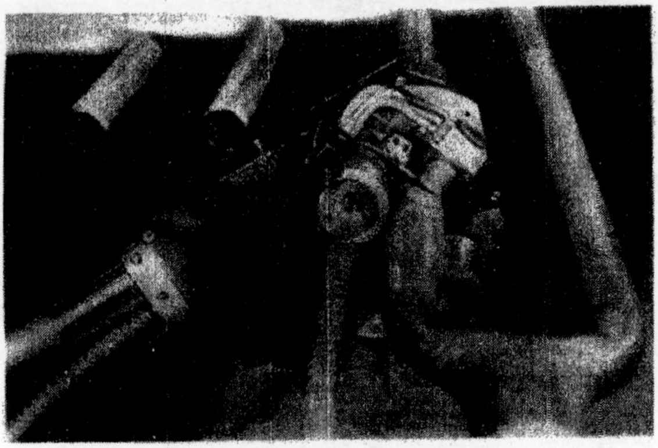

Fig. 17.7 TIG welding thin tubes.

head would wrap power cables and shielding gas and filler wire feed tubes around the pipe. For welding thick wall larger diameter pipes such as those used for gas and oil transmission, gas metal ard welding heads are used with suitable shielding gases. A number of different methods for moving such heads around the pipe are employed. The simplest system shown in Fig. 17.8 uses a chain which clamps the wheels of the machine to the pipe and provides a positive drive for the sprockets on the welding head, alignment of the welding head, while welding conditions are under continuous supervision of the operator. Other systems employ flexible straps or split rings clamped to the pipe, along with the tractor wheels run independently of the pipe. This gives more accurate tracking of the joint.

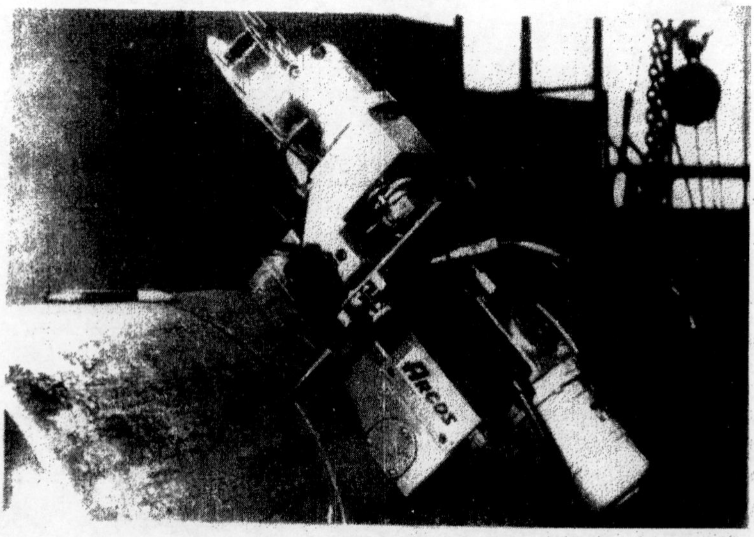

Fig. 17.8 Gas metal arc welding head-mover using chain.

(F) Tube-to-Tube Plate Welders

Tube-to-tube sheet joints as found in heat exchangers have been made for many years by devices in which a welding head or welding gun is rotated around a spigot clamped in the tube (Fig. 17.9). The weld is made between the surface of the plate and the end of the tube. The joint can be seen and hence the mechanism is relatively simple. Techniques have also been developed for producing butt type welds between the tubes and the back faced of the tube plate. For this type of joint the welding torch must be inserted inside the tube and the equipment is usually referred to as a bore welder. The welding is blind.

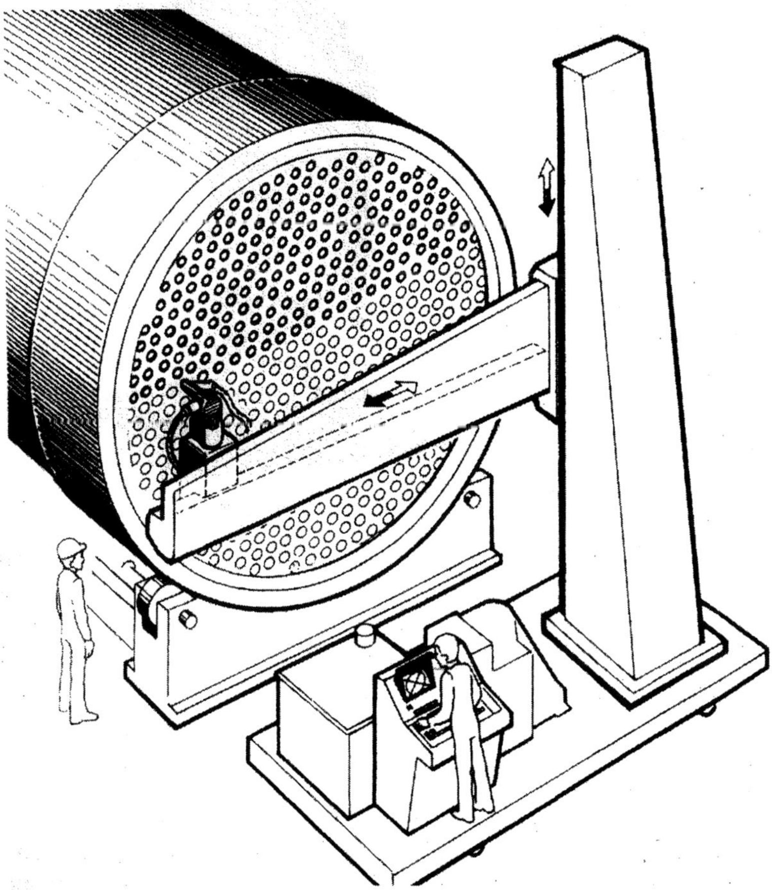

Fig. 17.9 Welding tube with tube plates of heat exchangers.

(G) Industrial Welding Robots and N.C. Machines

In flexible automated welding, a numerical or a computer control program replaces complex fixturing. The sequecing devices of fixed automated welding programs can be easily changed to accomodate small lots of production of several designs that require the same welding process, while providing productivity and quality equivalent to automatic welding. Small volume production of similar parts can be produced automatically in a short time. A flexible system can be easily reprogrammed for a new part. An industrial robot is the most flexible of the

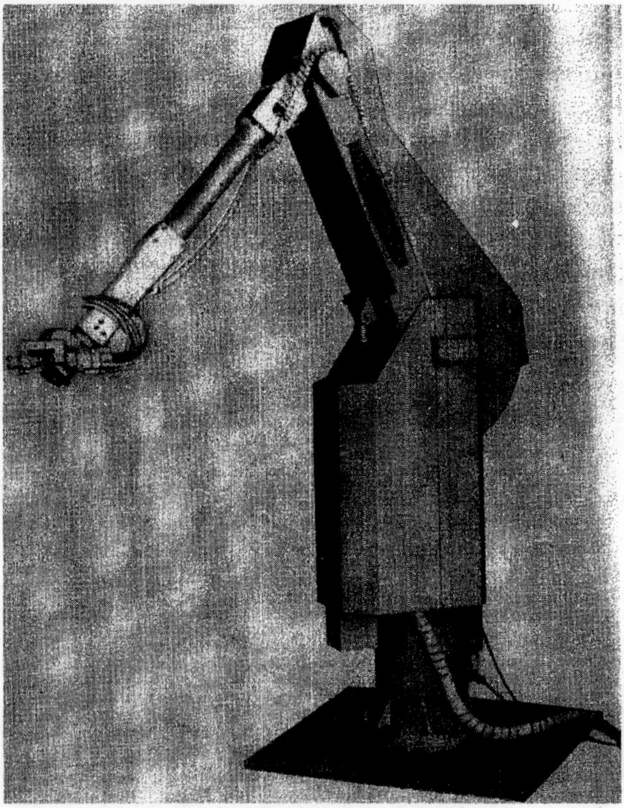

Fig. 17.10 Robotised welding plant.

automated systems used in manufacturing operations. A robot is essentially a mechanical device that can be programmed to perform some task of manipulator under automatic control. A welding gun or a torch can be mounted on a 'wrist' at the end of robot's arm. Robots may be designed with various motor systems depending upto the applications for which robot is to be used. An automated work station is composed of various pieces of equipment, totally integrated by a single control. A typical station incorporates a robot arm, a robotic work positioner, a welding process package and the required fixturing to hold the parts for welding. Figure 17.10 shows a robotised welding plant. A welding robot can be programmed to follow a predetermined path, duplicating a seam in an assembly to welded. With some equipment this is done by moving the welding torch along the weld joint. Signals are sent back to the robot controller so that it can memorize the taught motion. A replay of the programme enables the robot to weld. After the job is completed certain robots provide for copying of the welding programme for future production of the weldment. In addition to controlling the welding torch path, some robots can also set and change welding variables.

If distortion is a significant factor or the components parts are not accurately prepared or located, the weld joint may be outside of the programmed location. To compensate for this dislocation, the flexible or robotic welding system may require sensory feedback to correct the program. Sensors can provide information to the robot controller to make real time adaptive corrections. A limited number of sensors and adaptive controls are available for use with certain welding processes. Flexible automation for welding can eliminate the need for complicated custom built fixtures and it is finding increased acceptance in low volume production. The inherent flexibility of robotic systems can accommodate several different type of weldments at the same work station. Further more, flexible welding automation can be ionterfaced with other automatic manufacturing operations.

17.7 MOVING WORK SYSTEMS

(A) Turntable Positioner

A turntable positioner also known as turntable is a single axis positioning device that provides rotation of a horizontal table or platform about a vertical axis. They may range in size from a small bench unit to large floor models. Turntable positioners are available in capacities from 20 kg to over 500 Tonnes. They are used extensively for machine welding, cutting, cladding, grinding, assembly and testing. Turntables can be built in such way as to index several assemblies on a common table. Several smaller turntables are mounted on the main table together with an indexing mechanism. A turntable positioner can be equipped with a number of optional features and accessories. The following are some popular adaptations :

(i) Constant or variable speed drive.
(ii) Manual rotation of the table.
(iii) Indexing mechanism.
(iv) Tilting base.
(v) Travel carriage and track.
(vi) Through hole in the centre of the table.

(B) Tilting Rotating Positioner

Where versatility in positioning is required for shop operations a tilting rotating positioner is recommended. In most of the cases, the positioner table can be tilted through an angle of 135°, including the horizontal and vertical positions. The turntable mechanism is mounted on an assembly that is pivoted in the main frame of the machine to provide the tilt axis (Fig. 17.11). The basic capacity rating of the positioner is based on :

(i) the total load.
(ii) the location of centre of gravity of the total load.

(C) Turning Rolls or Rotators

Turning rolls are used in sets. Each set consists of one powered roll and one idler roll. Each roll has two or more wheels. Turning roll design is quite simple. A set of rolls normally consists of a fabricated steel frame, wheels, drive chain, drive motor and controls (Fig. 17.12). Simlicity of design offers low initial cost as well as low maintainence and repair costs. Standard models can be used for many applications. Turning rolls can be manufactured to specific requirements such as

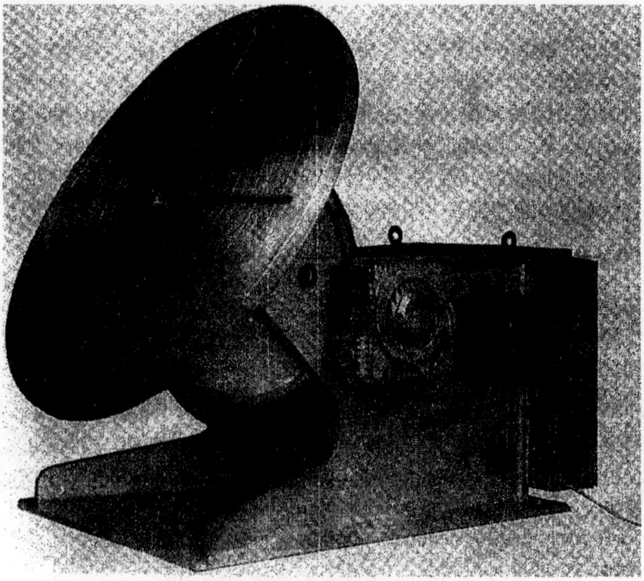

Fig. 17.11 Tilting arrangement.

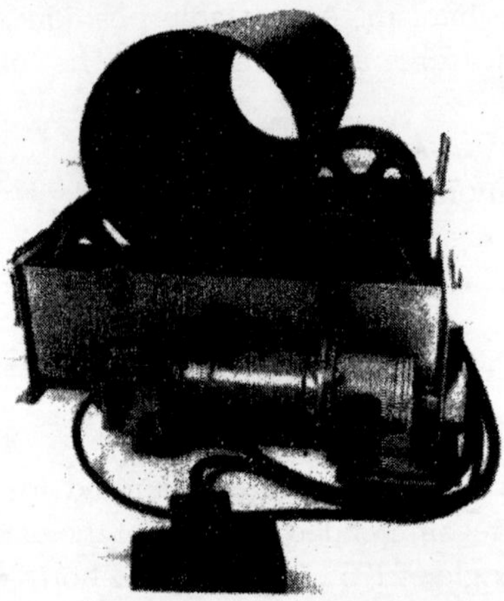

Fig. 17.12 Turning rolls.

wheel construction, weldment weight and diameter, special motions and modular frames.

Usage of turning rolls is normally limited to the rotation of a cylindrical shaped weldment about its horizontal axis. Non-cylindircal assemblies can also be rotated on turning rolls using special round fixturing to hold the assembly. The fixtures rest on the turning rolls. Turning rolls are built even in capacities upto 1000 tonnes. Basically rotators comprise of two types :

(i) Self-aligning rotators.

(ii) Conventional rotators.

In self-aligning rotators, the rubber tyred wheels automatically align themselves to the change from one diameter of vessel to another. In the case of conventional rotators, the rolls will have to be manually positioned across the frames to suit each diameter of the vessels. A few companies have developed anti-creep mechanism for turning rolls, to eliminate vessel 'drift' which arises during the fabrication of thick walled vessels. The anti-creep mechanism is provided to ensure that any vessel 'drift' is monitored and maintained within ± 1 mm. Turning rolls can position a seam for manual, semi-automatic or machine welding, in the flat position. With machine welding, circumferential welds may be rotated under a fixed welding head. Longitudinal wells can be made with a welding head mounted on a travelling carriage or manipulator.

(D) Welding Lathe

A welding lathe consists of a head stock and a tail stock positioner. The concept and application of the head stock-tail stock positioner are much the same as lathes used in machine shop operations. It is used for positioning large bulky assemblies. This positioner is extremely flexible because the head stock and tail stock are separate units (Fig. 17.13). The head stock is a single axis positioning device, providing a complete rotation of a vertical table about the horizontal axis. The tail stock usually have the same configuration as the head stock but it is not powered. A simple trunnion, an outboard roller support or any other free wheeling support structure best suited to the configuration of the weldment, may be used in place of a tail stock.

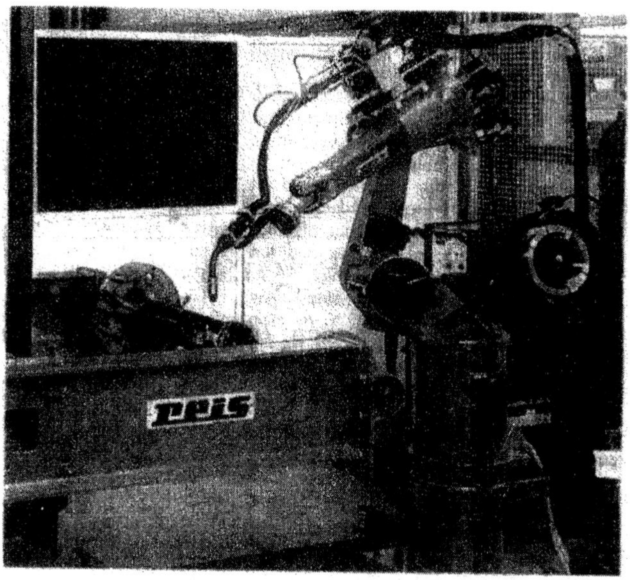

Fig. 17.13 Head stock and tail stock positions.

In all instances, precise installation and alignment of the head stock tailstock positioner is essential so that the axis of rotation is aligned as is done in the case of a lathe machine, otherwise the equipment or the weldment may get damaged.

(E) Special Welding Machines

Specialised custom-built welding machines are available with work holding devices, torch travel mechanisms and other specific features. They are

Fig. 17.14 Special machine for welding pipes longitudinally.

economically employed only in high volume production where a large number of identical parts are produced on a continuous basis. Applications include making circumferential and longitudinal welds on tanks and cylinders, welding piping and tubing, fabricating flanged beams, welding studs or bosses to plates and for maintainence work such as the rebuilding of track pads for crawler type tractors. Figure 17.14 shows a specialised custom-built machine wherein longitudinally folded pipes are further aligned for proper butt joint fit up to be welded continuously.

17.8 AUTOMATIC WELDING

Automatic welding is done with equipment that performs an entire welding operation without adjustment of the controls by a welding operator. An automatic welding system may or may not perform the loading and unloading of the components. The equipments and techniques currently used in automatic welding are not capable of controlling the operation completely. A welding operator is required to over see the process. An important aspect of automatic welding is that the operator need not continuously monitor the operation. Compared to machine welding, this results in increased productivity, improved quality and reduce operator fatigue. All the elements required for machine welding are utilised in the automatic welding system. Continuous filler wire feeding (if used), arc movement and work-piece positioning, are all part of automatic welding. Automatic welding also requires that the parts to be joined be prepared in a consistent and uniform manner. Parts must be uniform if the resultant welds arc to be uniform. Therefore, part preparation for automatic welding is more expensive than for machine welding. Also, automatic welding requires a complex welding cycle controller, that must provide for control by the welding operator as well as the material handling equipment and component fixturing. The controller must precisely time a complex sequence of operations. Advantages and limitations of automatic welding are :

Advantages :

(i) Consistent weld quality.

(ii) Improved duty cycle.

(iii) Increased productivity as a result of higher welding currents and speeds.

(iv) Less distortion.

(v) Predictable welding production rate.

(vi) Integration with other automatic operations on the weldment.

Limitations :

(i) Capital investment requirements are much higher.

(ii) Dedicated fixturing is required for accurate part location and orientation due to reduced tolerance to variations in fit up.

(iii) Elaborate arc movement are control devices with predetermined welding sequence may be required depending on the complexity of the part and geometry of the weld seam.

(iv) Production requirements must be large enough to justify the costs of equipment and installation, the training of personnel and the maintainence of the equipment.

❑❑❑

18

CHAPTER

ARC WELDING POWER SOURCES

❑ INSIDE THIS CHAPTER

An electric arc is formed by discharge of current passing through an ionised gaseous medium. It is an intense source of heat. The electric arc is used for many application like arc furnaces for steel melting, lamps for cinema projections and also for welding. In order to produce and maintain the electric arc we require electric power. This is obtained from special systems called Arc welding power sources. The electric supply available in a factory is 415 V, 3 phase, 50 Hz A.C. which cannot be used directly for arc welding for the following reasons :

(i) High Voltage is hazardous : In shileded metal arc welding (SMAW) the welder has to change electrodes, clean the slag etc. There is a possibility of electric shock if the welder comes into contact with live parts. This shock may prove to be fatal if the voltage is high. Therefore, the open circuit voltage should be limited to lower values usually less than 100 V.

(ii) Longer arc is difficult to manipulate : The arc length is proportional to the arc voltage. Therefore, if we use the high voltage the arc will be long, making it difficult for the welder to guide and manipulate the arc in the groove. Moreover, the melting efficiency of longer arc will be low. Atmospheric interference will also be high.

(iii) Melting requires high current : Consequently, a separate system is required to reduce the voltage and to increase the current level. This is done by arc welding power source.

For many of the arc welding processes direct current (DC) is used whereas the industrial power supply outlet as mentioned earlier is AC. The power source is mainly used to reduce the line power supply to a low voltage which is fairly safe.

When we reduce the line voltage of single phase 230 V or three phase 415 V, generally using a transformer, to a safe voltage of 65 to 100 V, the output current increases proportionately. From such a transformer we get low voltage high current output.

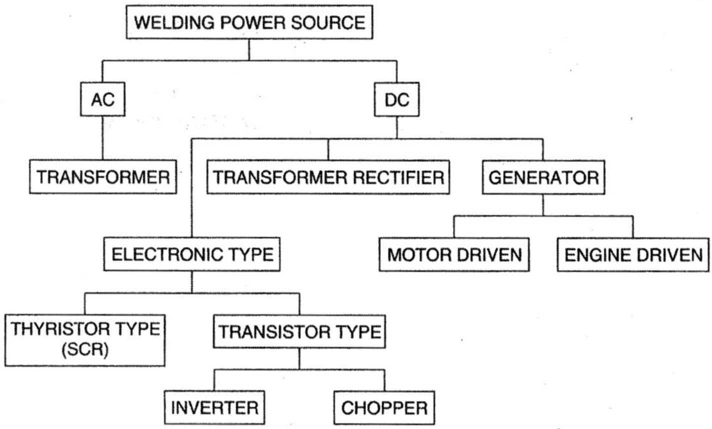

Fig. 18.1

The D.C. power for welding can be obtained by employing a transformer-rectifier or a DC generator or an electronically controlled power source. The family tree of arc welding power sources is given in Fig. 18.1.

18.1 CHARACTERISTICS OF POWER SOURCES

The variations of the output voltage with respect to lead (arc) current is called the output characteristics of welding power sources, each of which decides the welding performance in different ways. They are :

(i) Drooping characteristics.

(ii) Flat characteristics.

Dropping chracteristics can be further subclassified into :

(i) Constant current characterstics.

(ii) Constant power characteristics.

The output characteristics can be readily measured by conventional test procedures. A set of output voltage verses output current characteristics curves (Volt/ampere) curves are usually known as static characteristics. Figure 18.2 shows three possibilities of output characteristics of a power source.

18.1.1 Constant Current Characterstics

Figures 18.2 (a) and 18.2 (b) shows drooping characteristics and Fig. 18.2 (c) shows flat characterstics. Constant power [Fig. 18.2(a)] and constant current [Fig. 18.2(b)] types are also illustrated. Figure 18.3 shows the characteristic curves of constant power type of power source. It is also called drooping characteristic because of the substantial downward slope of the curve. The power source might have an open circuit voltage adjustment in addition to output current control. A change in either will change the slope of the voltampere curve. In these types of power sources, a large change in the arc length (arc

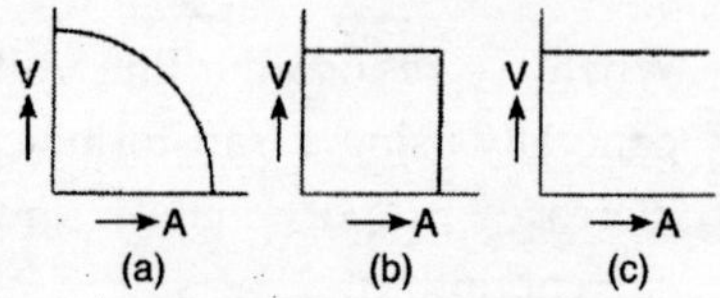

Fig. 18.2 The 3 possibilities of output characteristics of a power source.

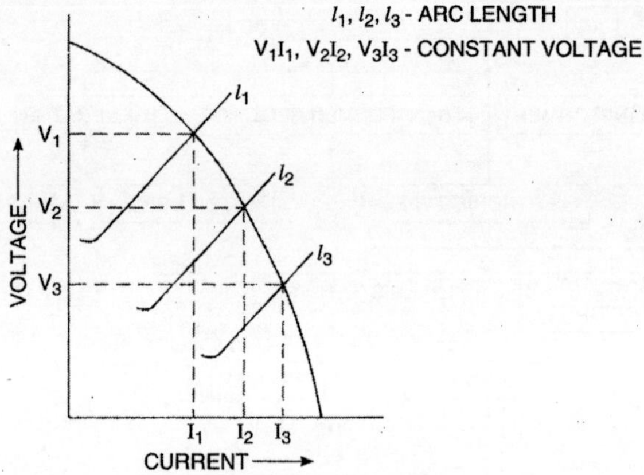

Fig. 18.3 Constant power.

voltage) will result in a small change in the welding power. Therefore, in manual metal arc welding electrode melting rate will remain fairly constant. The constant current type of drooping characteristic is obtained from the electronic type of power sources.

18.1.2 Constant Voltage Characteristics

Figure 18.4 shows the output characteristic of a constant voltage type power source. This type of power source is suitable for constant electrode feed processes

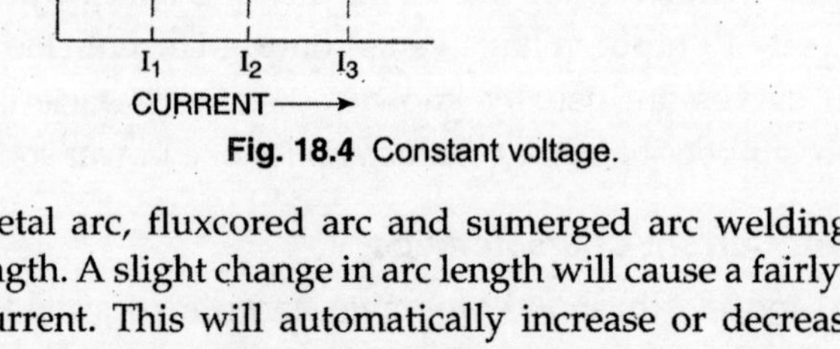

Fig. 18.4 Constant voltage.

such as gas metal arc, fluxcored arc and sumerged arc welding to maintain a constant arc length. A slight change in arc length will cause a fairly large change in the welding current. This will automatically increase or decrease the electrode melting rate to regain the desired arc length.

18.2 TRANSFORMER POWER SOURCES (AC)

The methods commonly used to control the welding circuit output of a transformer are given below :

18.2.1 Movable Coil Control

A movable of transformer consists of an elongated core on which the primary and secondary coils are located. One of the coils is fixed and the other is movable. Generally, the secondary coil is fixed and the primary coil is attached to a lead screw, and as the screw is turned the coil moves closer to or further away from the secondary coil. The varying distance between the two coils regulates the inductive coupling of the magnetic lines of force between them. When the two coils are away from each other, minimum welding current flows. Figure 18.5 shows the schematic diagram of the moving coil transformer and the minimum and maximum power outputs.

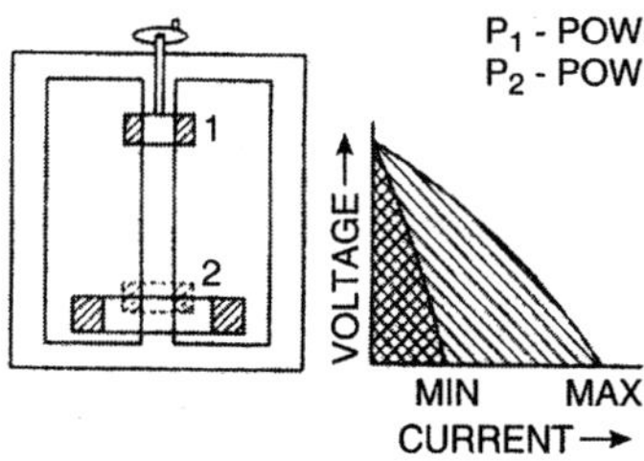

Fig. 18.5 Moving coil type constant power arc welding transformer.

18.2.2 Moving Shunt Type

Also called moving iron type. In this design both the primary coils and secondary coils are fixed in a position. A laminated iron core shunt is moved between the primary and secondary coils. It is made of the same material as that of the transformer core. The shunt core acts to divert magnetic flux around the coils. The magnetic flux is obstructed when the iron shunt is in between the primary and secondary coils (Fig. 18.6). The flux is diverted through the shunt. Accordingly, the welding output power is minimum as shown in curve-1. If the shunt is not inserted the power output will be as represented in curve-2.

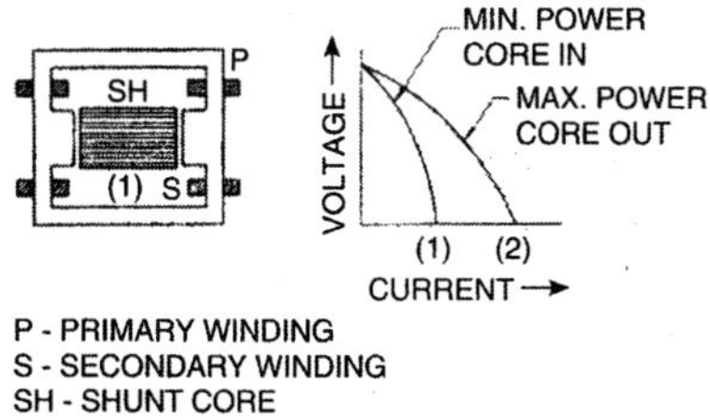

Fig. 18.6 Moving shunt code type control.

18.2.3 Series Inductance Control

The movable core inductance (reactor) of an arc welding machine consists of a constant voltage transformer and a reactor (Fig. 18.7). The inductance of the reactor is varied by mechanically moving a section of its iron core. When the iron core is moved, the inductive reactance of the coil is varied from the minimum

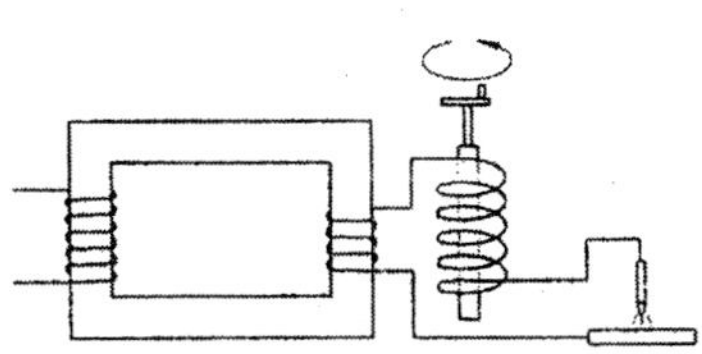

Fig. 18.7 Series inductance (Reactor) constant voltage power source.

to the maximum value, which controls the welding power from the minimum to maximum values.

18.2.4 Tapped Secondary Coil Control

A tapped secondary coil may be used for the control of the volt ampere output of a constant voltage type transformer. In this transformer, the secondary coil is tapped to permit the adjustment of the number of turns (Fig. 18.8). Decreasing the secondary turns reduces the output voltage. This type of power source is used for AC type MIG and SAW processes.

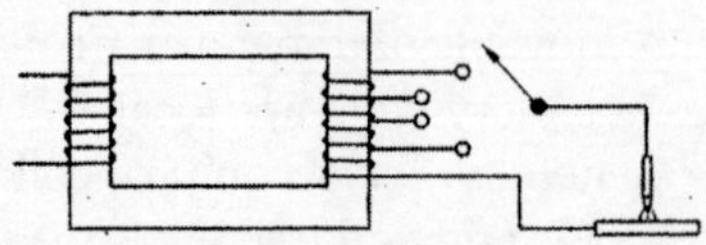

Fig. 18.8 Taped secondary constant voltage power source.

18.2.5 Magnetic Amplifier Control

In this design the main transformer has no moving parts. The equipment has a transformer and a saturable reactor. By adding a DC control circuit to the reactor system, it is possible to adjust output welding current from minimum to maximum. A large magnitude of alternating current is controlled by a relatively small magnitude of direct current. Figure 18.9 shows a magnetic amplifier circuit. By using a different connection for the welding current and rectifying diodes in series with the coils, the load ampere turns are used to assist the control the ampere turns in the magnetizing core. A small amount of current control will cause a large welding current variation, hence it is called magnetic amplifier.

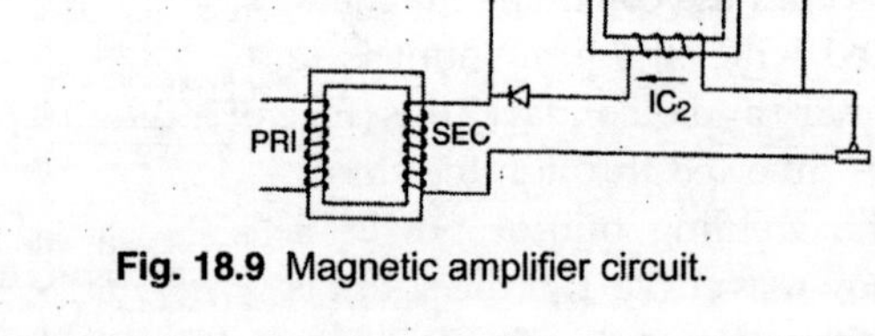

Fig. 18.9 Magnetic amplifier circuit.

18.3 RECTIFIERS (DC)

Rectifiers are used to convert the alternating current to direct current. Figures 18.10 and 18.11 show the single phase and three-phase transformer-rectifier sets. The arrangement of rectifier diodes is shown in these figures. An inductance is used in the D.C. welding circuit to smoothen the load current. A three-phase rectifier has relatively low ripple and hence the inductance required is small. Some D.C. power sources have optional current pulsing circuits also incorporated in them. Pulsing power sources are capable of switching from high to low welding current with desired repeat rate. Normally, low and high current values, pulse duration, pulse

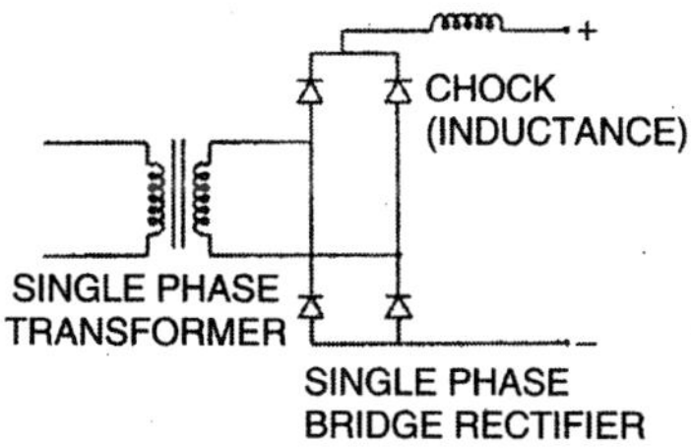

Fig. 18.10 Single phase transformer rectifier circuit.

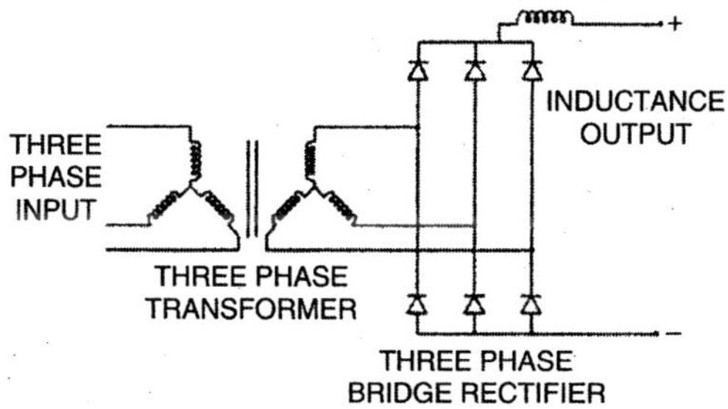

Fig. 18.11 Three phase transformer rectifier circuit.

repeatition rates are independently adjustable. Figure 18.12 compares the direct current with the pulsed output current. The pulse current is useful for out of position welding and thin foil welding especially using GTAW.

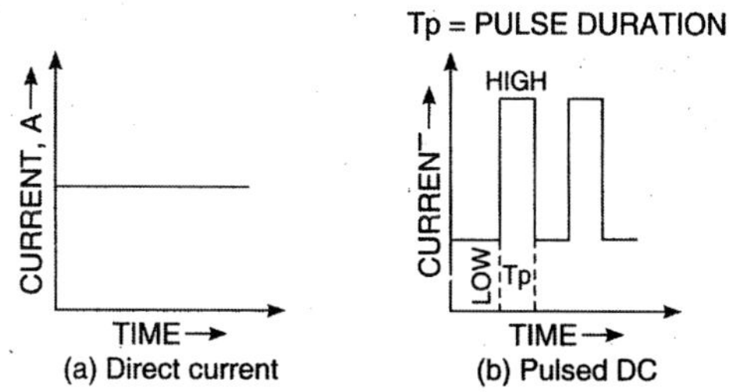

Fig. 18.12 Direct current vs pulsed current.

18.4 WELDING GENERATORS (DC)

Generator type power sources convert mechanical energy into electrical power suitable for arc welding. The mechanical power is obtained from an internal combustion engine or an electric motor. A DC generator has a rotor on an armature and a stator. The basic principle of the generator is that when electrical conductors

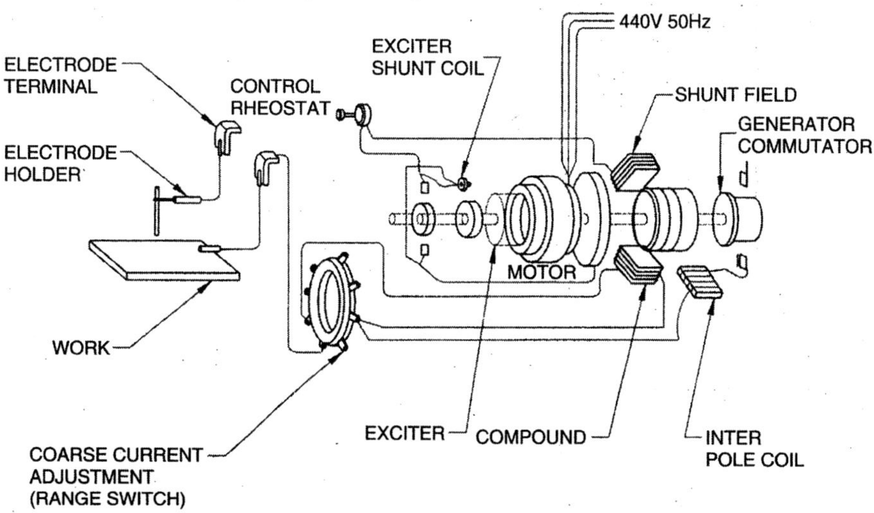

Fig. 18.13 DC motor — generator welding machine with electric motor drive.

are moved through a magnetic field, electric power is induced in the conductors as explained earlier in this chapter. A welding generator has a stator carrying coils that produce the magnetic field. The field will have two poles or four poles or two sets of field coils. The second set can be connected to the generator output such that it either aids or opposes the main field, changing the output current. Figure 18.13 shows the schematic diagram of a welding generator. The armature conductors of a welding generator are relatively heavy as they carry welding current.

The commutator is located at one end of the armature. It is a group of conducting bars arranged parallel to the rotating shaft to make switching contact with a set of stationary carbon brushes. These bars are connected to the armature conductors. When the armature rotates, the commutator performs the function of mechanical rectification. By adjusting the field current, using the current control rheostat (Fig. 18.13), the output current changes from minimum to maximum. There is also a selection of two ranges (maximum and minimum) possibly by changing. The series field as illustrated in Fig. 18.14.

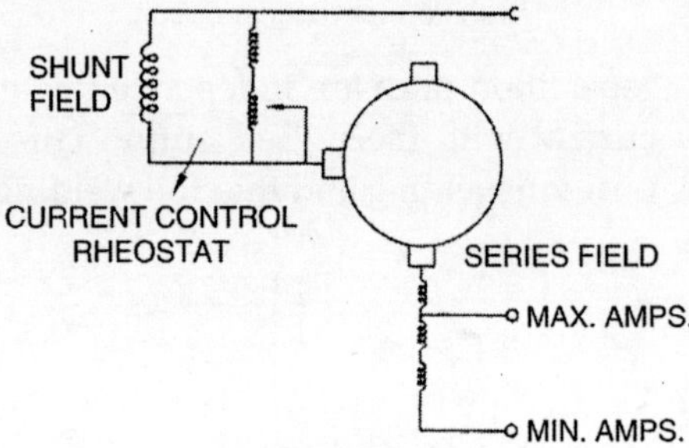

Fig. 18.14 Current control in welding generator.

18.5 THYRISTORISED WELDING POWER SOURCES

These are electronic type of power sources. Thyristorised power sources are usually D.C. type power sources. The input voltage is first stepped down to a low voltage by a transformer. This low voltage is rectified by a thyristor/rectifier. When the operator starts welding, a feedback of voltage and the current is fed into the electronic controller unit. Figure 18.15 shows the basic circuit of thyristorised power source. The controller compares the output parameters with the reference and gives out an error signal. Depending on this error signal value, the output voltage and current are modified to maintain the required output parameters. In

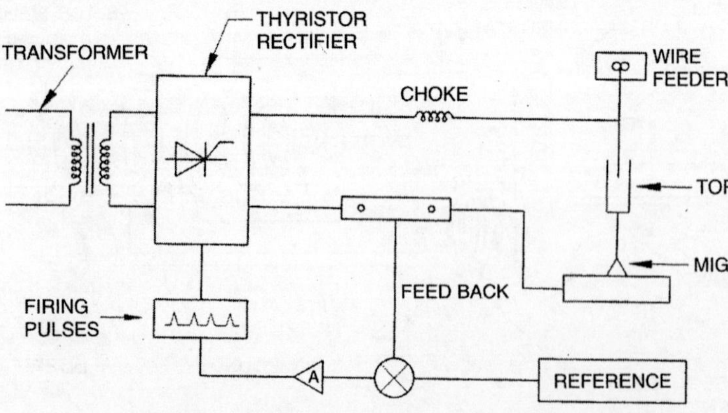

Fig. 18.15 Thyristor controlled welding power source.

this type of power source the output characteristics (constant current, constant power or constant voltage) can be selected.

18.6 TRANSISTORISED POWER SOURCES

These are solid state power sources where, after a transformer rectifier, instead of thyristor, we use power transistors. Here the feedback signal modifies the base current input to the transistors so that the output voltage/current is maintained constant. The output will also have a provision for variation of the pulse frequency parameters. The main advantage of a transistor power source is that in addition to output voltage control and current control, we can have frequency control as well as waveform control for the output current. These features help GMAW and GTAW. Figure 18.16 shows a transistorised power source circuitry. Table 18.1, shows the guidelines for power source selection and Table 18.2 gives the comparison of different welding power sources.

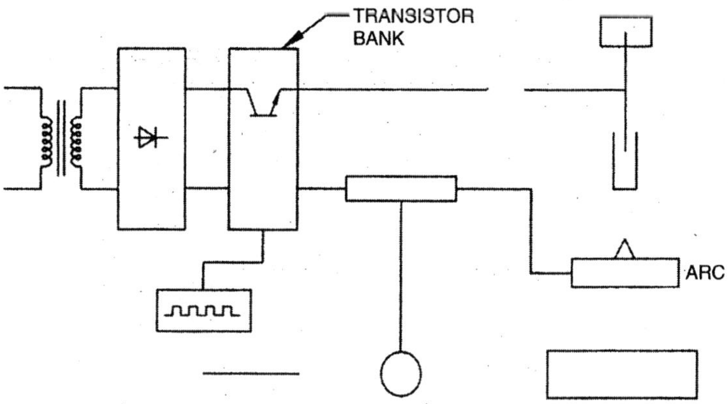

Fig. 18.16 Transistor controlled welding power source.

Table 18.1 : Selection of welding power sources

S. No.	Type	Output	Characteristics	Process
1.	Generator	DC	Constant power and constant current	MMAW/TIG
2.	Transformer	AC	Constant power and constant current	MMAW/TIG High current SAW
3.	Rectifier	DC	Constant power	MMAW/TIG High current SAW
		DC	Constant voltage	MIG/SAW
4.	Electronic type	DC	Constant current	TIG
			Constant power	MMAW
			Constant voltage	SAW/MIG
		DC pulsed	Constant current	TIG
			Constant voltage	MIG
		AC pulsed	Constant current	TIG

Table 18.2 : Comparison of power sources

S. No.	Power sources	Advantages	Disadvantages
1.	Transformer	1. Low price 2. Maintenance-free	1. Limited aplication 2. Poor efficienty (40%)
2.	Rectifier	1. Medium price 2. Maintenance-free 3. Good efficiency (80%)	1. Better weld quality than transformer 2. Less precise control of current
3.	Generator	1. Medium price 2. Good weld quality	1. More maintenance 2. Medium efficiency (60%)
4.	Electronic type	1. High weld quality 2. Selectable output characteristics 3. High efficiency (90%)	1. High price 2. Skilled maintenance personnel required.

18.7 ACCESSORIES FOR WELDING POWER SOURCES

(A) Remote Regulator

This is current control rheostat which forms a part of a magnetic amplifier circuit of arc welding power sources (Fig. 18.17). The advantage is that the welder can keep the rheostat with him at the work spot to achieve variation of the welding current while welding. He need not go to the welding machine every time to change the welding current. Normal remote regulators are used with extension cables which are connected to the power sources. There are also some remote regulators which do not have cable connection to the power sources. They send a signal through the welding cable itself to the power source for current variation which is achieved by a motorised control.

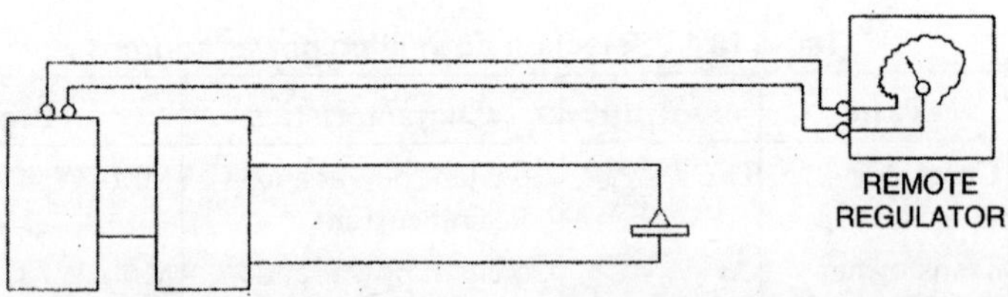

Fig. 18.17 Connection of remote regulator for current control.

(B) D.C. Suppressor

In the welding of aluminium, magnesium and their alloys, the refactory oxides on the surface of these metals hinders fusion. To weld them, we use alternating current. While using alternating current, when the tungsten electrode and the weld metal have different abilities to emit electrons, the currents during each half cycle will be unequal. Tungsten emits electrons more readily because it attains a much higher temperature than the metal being welded. Hence, due to the

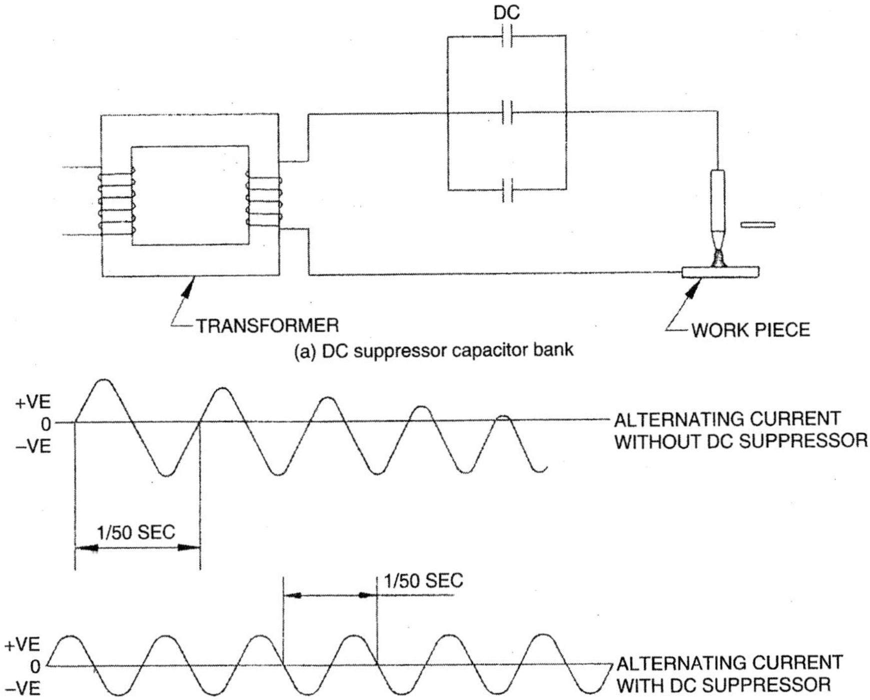

(b) Effect of DC suppressor

Fig. 18.18

difference in resistance to current flow, an unbalanced A.C. is produced. The current during (+) ve half cycle is more than the current during (–) ve half cycle. Sometimes rectification of AC to DC may also occur. To obtain a balanced current flow, the DC is to be removed. This is done by connecting a capacitor bank across the power source as in Fig. 18.18 (a). This will suppress the DC component and balance the alternating current [Fig. 18.18(b)].

18.8 HIGH FREQUENCY UNIT

A high frequency unit is used for arc stability in the case of TIG welding with DC or AC power sources for both manual and automatic applications. The high frequency unit superimposes a high voltage upon the welding circuit. This unit generally consists of a spark gap oscillator that delivers a high voltage output at radio frequencies 2 to 3 MHz. The HF output is transferred from the spark gap circuit by means of an air core transformer (Fig. 18.19). The HF unit produces a series of closely spaced bursts of high voltage energy. This high voltage ionises the gas between the electrode and the work. The ionised gas will then conduct welding current initiates the welding arc. The HF unit also stabilises the welding arc and hence is used throughout while welding aluminium and magnesium. Generally, the HF unit causes radio interference which has to be eliminated to avoid public disturbance.

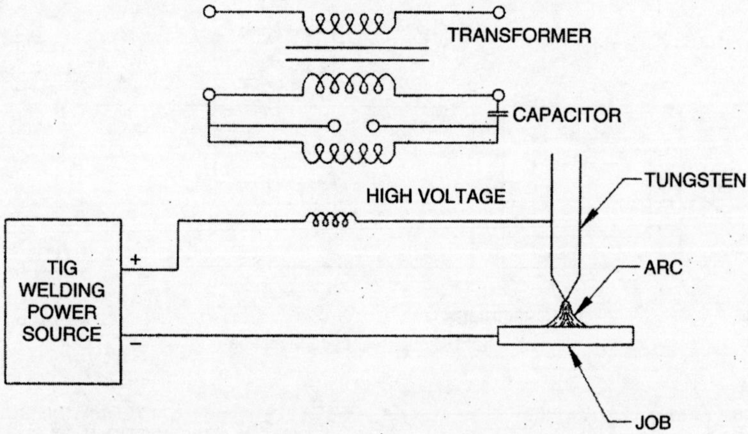

Fig. 18.19 TIG welding power source with high frequency unit.

❑❑❑

19

CHAPTER

WELDING CONSUMABLES

❑ **INSIDE THIS CHAPTER**

19.1 Welding gases, 19.2 Electrodes and filler materials.

19.1 WELDING GASES

19.1.1 Oxygen and Nitrogen Production

All the oxygen used commercially today is obtained from atmosphere by air seperation or what is frequently termed as the "liquid air process". Air is compressed and cooled so that it can be liquified. The components present in the liquid air are mainly liquid oxygen, liquid nitrogen and liquid argon and all these have different boiling points. Liquid air plants are designed to produce both high purity oxygen and high purity nitrogen. The liquid air is first partially rectified in one column. A nitrogen rich liquid mixture from the top of this column and an oxygen rich liquid mixture from the bottom are then passed onto a second column operating at a lower pressure where the final rectification takes place. Oxygen and nitrogen are withdrawn from the rectification column as cold gas and then compressed separately into steel cylinders as separate gases.

19.1.2 Oxygen Cylinders

The normal oxygen cylinder is 5 feet (1.5 m) tall including cap and 9 inches (229 mm) in diameter. It contains 244 cft (6.9 m^3) of oxygen at a pressure of 2000 psi as measured at 20°C. The oxygen cylinders are painted black and stamped as DOT3A2015—where DOT is department of transportation, 3A stands for the specification of the cylinder and 2015 for service filling pressure.

19.1.3 Nitrogen Cylinders

The nitrogen cylinders are of the same size and contain full pressure of 2000 psi at 20°C. They are painted in grey.

19.1.4 Acetylene production

Industrial acetylene is produced by the reaction of water on calcium carbide, which is a compound of calcium and carbon manufactured by the fusion of lime stone and coke in an electric furnace. Acetylene for gas welding, oxygen cutting, brazing and allied processes can be produced on a large scale, purified and charged into cylinders filled with porous mass and acetone. Acetylene filled in a cylinder is usually known as "dissolved acetylene." For site generation of acetylene, a small

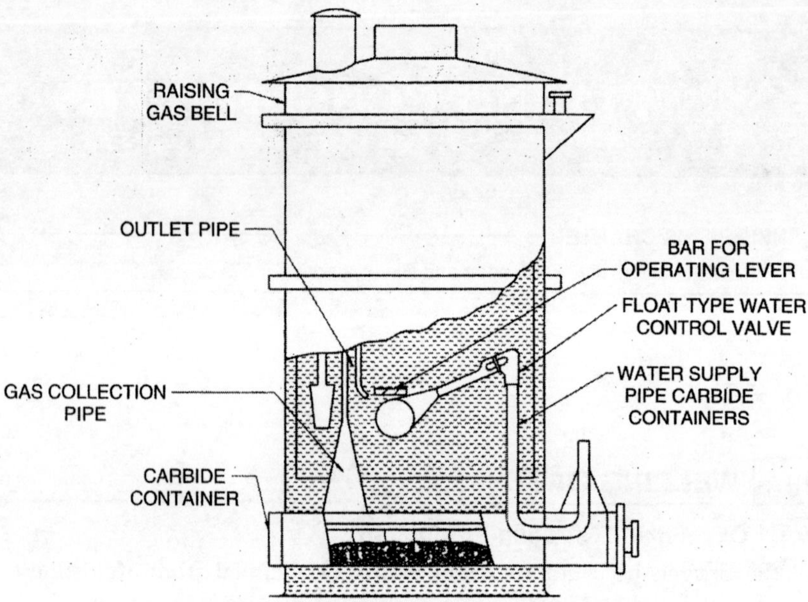

Fig. 19.1 Typical water-to-carbide generator.

type of acetylene generator is used which may be either carbide to water or water-to-carbide type. Figure 19.1 shows the commonly used water-to-carbide type acetylene generator.

19.1.5 Acetylene Cylinders

The acetylene cylinder is made from deep drawn steel plate and seam welded. A foot ring and a head ring are attached and a forged steel valve 'spud' in welded in place at the top of the cylinder. The cylinder is packed with acetone filler in a moist condition and placed in an oven for several days until a weight check shows that all the moisture has been baked out of the filler (Fig. 19.2). Finally the cylinder valve and safety devices are installed and a metered amount of acetone is pumped into the cylinder. The exact amount of acetone which may be added depends on the internal volume of the cylinder shell and the percent porosity of the filler. Free acetylene in unstable, acetylene dissolved in

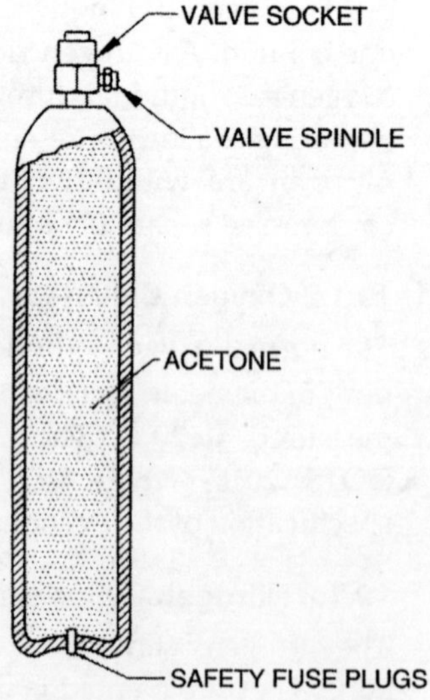

Fig. 19.2 Acetylene cylinder.

acetone can be retained in normal atmospheric pressure. Dissolved acetylene (DA) cylinders are filled to a maximum pressure of 16-17 kg/cm^2 at 20°C. They are equipped with fusible metal safety plugs or with valves having channels filled with fusible metal. The fusible metal will melt at approximately 100°C. If the cylinder is exposed to fire, the safety devices will melt and allow the acetylene and acetone to escape and burn gradually. Acetylene cylinders are painted maroon.

Important things to remember about disolved acetylene cylinders :

(i) Do not withdraw acetylene from a cylinder when it is not in an upright position. If it is so done, some of the acetone may be forced out as liquid droplets through the cylinder valves. This is not safe as it may cause erratic operation of the torch. Therefore, always store the acetylene cylinders in the upright position. It the cylinder was stored horizontally let is stand upright for a while before with drawing any acetylene.

(ii) Do not withdraw acetylene from the cylinder at a rate in excess of one-seventh the normal full capacity of that cylinder.

From a large cylinder with a capacity of 10-11 m^3 the withdrawl rate should not exceed 2 m^3/hr except for very short periods. If acetylene is withdrawn too rapidly, i.e., more come with vapour or droplet form, and the cylinder may cool down so much that it might affect the torch flame. The acetone should also to be replenished more frequently.

19.1.6 Helium Production

Helium is extracted from helium bearing natural gas. Helium is separated from natural gas by cooling the same to a temperature below the liquefaction point of its ordinary constituents but above that of helium. The gas must be treated chemically to remove small quantities of carbon-di-oxide and water vapour that will solidify at the low temperature. Natural gas is the main source of helium. The liquidified portion of the natural gas is withdrawn from the bottom at a pressure of about 14 kg/cm^2 and run back through the heat exchanger. This mixture is compressed in four stages to 190 kg/cm^2, cooled with water and dried with activated alumina, and passed to another low temperature processing unit. Here without substantial reduction in pressure, consecutive phase separation occurs producing 98.2% helium. The helium is drawn off as a gas and it passes through many stages of purifiers and heat exchangers and is stored directly into high pressure containers for storage or shipment.

Helium Cylinders :

Cylinder body colour	: Brown
Physical form	: Gas
Pressure	: 2000 psi
Specific Volume	: 6.04 m^3/kg at 20°C 1 Bar
Specification	: 99.995% pure

19.1.7 Carbon-di-oxide Production

Carbon-di-oxide is available as a by-product of the fermentation processes in the production of alcohol. Oil, coke or semi-anthracite fuels are burned with excess oxygen to give a rich CO_2 gas. The gas is stripped of its CO_2 content, which is then progressively dried, purified and converted into the liquid form by compression and cooling. CO_2 gas is also available as a by-product from several industrial processes such as production of ammonia. Carbondioxide is available in the forms of gas, liquid and solid. Liquid is the most common form available from bulk tanks and cylinders painted in black with a white neck. Syphon cylinders have now largely replaced the conventional gas cylinder for welding applications. CO_2 cylinders contain liquid and gaseous carbon dioxide in equilibrium at high pressure.

CO_2 Gas Cylinders : To ensure a constant low moisture content in the CO_2 used for welding, syphon cylinders are used. These cylinders are fitted with a metal syphon tube which extends to the bottom of the cylinder and ensures that CO_2 is always drawn from the liquid phase. As some liquid CO_2 is withdrawn, more liquid will evaporate to maintain the head of pressure. This slightly increases the concentration of moisture in the liquid phase. Since it is liquid which is actually being removed, this slightly increased moisture gets averaged over the liquid bulk and will remain substantially constant until the cylinder is exhausted. When no more liquid remains, gas is removed which is considerably drier than the CO_2 obtained from the liquid phase. Thus, whereas the moisture content of CO_2 withdrawn from the gas phase of the normal 'gas' cylinder may exceed five times its original value in a syphon cylinder, the moisture content will never rise to more than about one and half times its initial level.

Because the CO_2 is being removed as liquid no appreciable cooling of the cylinder and its contents occurs and thus freezing up in the cylinder is impossible. It is necessary, however, to pass only gaseous CO_2 into the pressure reducing valve, and an electrically heated, thermostatically controlled vapouriser unit rated at 150-250 watts must be provided between the cylinder valve and the reducing valve. It is used to prevent the partial freezing of CO_2 into ice at the pressure regulator, when converting the liquid CO_2 to gaseous CO_2 in the heater arrangement. The cylinder used for welding should never be completely exhansted. When the gas flow ceased indicating the cylinder is empty and its pressure has fallen to the reduced pressure employed, i.e., about 25 lb/in^2, in cylinder valve should be fully closed before the cylinder is disconnected. This residual gas pressure ensures that no contamination of the empty cylinder will occur during storage and transit. CO_2 can also be supplied for welding from bulk storage tanks which are filled at intervals from a road tanker. These are supplied complete with evaporator and may readily be employed in conjunction with a ring main for supplying a large number of welding units located throughout the welding shop. CO_2 is to be produced to high standard of purity with quality specification conforming to BS : 4105-1967. This states that the maximum impurity level in CO_2 from syphon cylinders for welding should be, air—0.2%, water 150 ppm.

19.1.8 Argon Production

Argon is manufactured in air separation of oxygen nitrogen plants by means of fractional distillation after the liquefaction of air. In the distillation columns liquid nitrogen is the product from the bottom of the higher pressure column, followed by liquid oxygen, containing argon. The argon rich oxygen is re-distilled and then further treated to yield very pure argon 99.95%.

Storage and Distribution : Argon is sold as gas under pressure is steel cylinders. The cylinders are painted in sky blue colour. The cylinder specifications are similar to those of oxygen and nitrogen. Table 19.1 gives essential details of various gas cylinders used in welding.

Table 19.1 : Details of gas cylinders.

S. No.	Name of gas	Physical form	Cylinder colour	Pressure	Specific volume	Specification or purity
Units				Psi	m^3/kg	%
1	Helium	Gas	Brown	2000 Psi 140 kgf/cm^2	6.04	99.995
2	Carbon-dioxide	Liquid	Black body white neck	733 Psi 150 kgf/cm^2	7.0	99.95
3	Oxygen	Gas	Black	2000 Psi 140 kgf/cm^2	6.9	99.97
4	Acetylene	Liquid	Maroon	250 kgf/cm^2	0.905	99.97
5	Nitrogen	Gas	Grey	2000 Psi 140 kgf/cm^2	7.0	99.99
6	Argon	Gas	Sky-blue	2000 Psi 140 kgf/cm^2	7.0	99.95

19.2 ELECTRODES AND FILLER MATERIALS

19.2.1 Historical Evolution of Welding Electrodes

Just prior to World War I, a Swedish engineer named Oscar Kjellberg invented a process for coating electrodes by immersing the core wire into wood plug, dipping repeatedly and drying the coating. This process is called dipping for the manufacture of coated electrodes but now it is becoming almost obsolete except for the production of some aluminium and copper electrodes. The modern method of producing electrodes is called the "Extrusion method." Very high production rates and consistency can be achieved through this process.

19.2.2 Types of Electrodes

The broad field of manual welding electrodes can be divided into many types based on two aspects as mentioned below :

(i) Depending upon the base materials which can be welded by them : mild steel electrodes, low alloy steel electrodes, stainless steel electrodes, electrodes for weather-resistant steels, non-ferrous metals etc.

(ii) Based on type of coating : Cellulosic electrodes, iron oxide-iron powder electrodes, rutile electrodes, basic coated electrodes, high deposition electrodes etc.

19.2.3 Functions of Coating

The coating of an electrode is essential for their effective performance. In shielded metal arc welding process (SMAW) which is also known as manual metal arc welding process (MMAW), the stick electrodes are being used. The stick electrode consists of a metallic core wire on which the coating has been done. The coating consists of many ingredients which are known to have specific or multiple functions and these functions can be divided into broad classes as follows :

(i) **Arc Initiators and Arc Stabilizers :** These are ionising compounds which improve the arc conductivity. Compounds of sodium, potassium, titanium and calcium present in the coating have an important function of initiating and maintaining the "plasma" column or the arc through lowering the initial ionisation potential and the operating voltage.

(ii) **Slag Formers and Gas Formers :** Slag formers are the gas formers shield the weld metal from atmospheric contamination during welding and solidifcation. When any metal melts, the atmospheric air and its constituents will try to enter the liquid weld pool, and since the resulting porosities and slag inclusions are harmful to the mechanical properties and metallurgical "cleanliness" of the weld metal, the coatings help to prevent atmospheric contamination of the weld pool by forming a slag blanket and a gas shield. Thus we avoid porosity and improve the metallurgical 'cleanliness' of the weld metal. The required mechanical properties of the weld metal can thus be obtained. Oxides of Al, Fe, Mn, Si are slag formers and calcium carbonate, dolomite and cellulose are typical gas formers.

(iii) **De-oxidisers and De-nitriders :** Electrode coatings have also de-oxidisers and de-nitriding agents which remove oxygen and nitrogen from the weld metal by selectively combining with them. These are used to obtain metallurgically 'clean' weld metal, to increase deposition efficiency and to achieve the required composition and properties.

(iv) **Metallic Powders :** Coatings may contain powdered metals like iron, chromium, nickel and other alloys such as ferro-manganese and ferror-silicon which will melt alongwith the core wire. They promote better welding characteristics like are stability, de-oxide the weld metal and increase the deposition rate, i.e., weight of weld metal deposited per hour. Alloy metal powders also impart required properties like tensile strength, hardness and wear resistance to the weld metal.

(v) **Extrusion Aids :** Modern electrode production plants have a high speed of production and hence the coating has to have sufficient quantity of extrusion aids so that it has good 'green strength' and can withstand handling without any damage to it. Coating materials have more than one function for example, liquid silicates are added as binders and extrusion aids but since they contain sodium and potassium they also help to stabilize the arc. However, clay, mica or talc are added exclusively for extrusion properties.

By controlling the slag generation and the gas evolution, the electrode coating makes it possible for us to weld in all positions, i.e., vertical up, vertical down, flat, overhead etc., which is an important technological variable in manual welding.

19.2.4 Production of Electrodes

In the production of electrodes the main raw materials are :

(i) The metallic core wire.

(ii) The fluxing materials to be coated on core wire.

In general, this core wire is of 'rimmed' quality steel. However, of late 'killed' steel wire produced in electric furnace is also used. This wire is drawn to the required diameters from the raw stock of 8 mm. diameter. This stock is produced by intergrated steel mills and is available in coils. The wire thus drawn into standard diameters viz. 6.3, 5, 4, 3.15, 2.5 and 1.6 mm. is further straightened in a straightening machine and is cut into standard lengths in a high speed cutting machine (Table 19.2). These cut lengths are fed into the extruder where the core wire is coated with flux to emerge as covered or coated electrodes. As mentioned earlier, the ingradients function as arc stabilizers, gas formers, deoxiders, denitriders, alloying elements, extrusion aids etc. These materials are first weighed accurately as per the formulation and in order to homogenise the mixture, they are mixed first in a dry mixer. In this process homogenity is achieved even though the starting materials may have been of different bulk-density values and flow properties. Then this dry mix is taken to a wet mixer where liquid silicate binder is added to it and by a combined action of kneading and 'muller' type mixing, a paste with good green-strength and good plasticity is obtained. This paste is then made into a cake in a press and this cake is loaded onto a hydraulic extruder. The extruder has the facilities for wire feeding and extrusion of the paste by hydraulic pressure through a tungsten carbide die at high pressure. The coated electrode which comes out of the die is then checked for eccentricity of the coating. If any eccentricity is noticed it

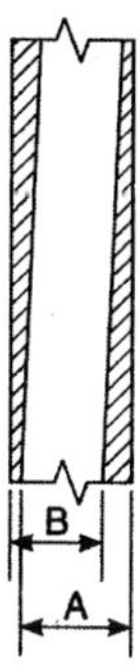

$$A\text{-}B \leq \frac{5}{100}\ \frac{A+B}{2}$$ For low hydrogen electrodes

$$A\text{-}B \leq \frac{4}{100}\ \frac{A+B}{2}$$ For low hydrogen electrodes

Fig. 19.3 Tolerance for eccentricity.

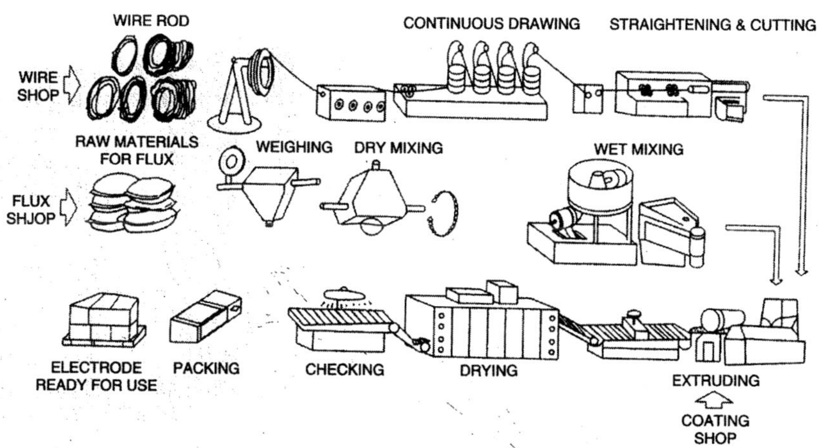

Fig. 19.4 Production line of arc welding electrode.

is adjusted to be within the predetermined limits. Then the coated electrode is delivered on to a conveyor system which has provision for holding and trimming the coating and providing a small bevel at the striking end so that arc striking with the electrodes becomes easy. Then the electrodes are dried in air and in an oven for a specified time-temperature cycle depending upon the coating types. The coating eccentricity is permitted as shown in Fig. 19.3. In Fig. 19.4, the flow chart shows the operations involved in the production of electrodes.

Table 19.2 : Standard Lengths of Electrodes

Lengths

Diameter	Mild and low alloy steel	Stainless steel
6.3	450	350
5.0	450	350
4.0	450, 350	350
3.15	450	350
2.5	350	350
1.6	350	350

All dimensions are in 'mm'.

19.2.5 Storage, Redrying and Baking of Electrodes

All electrode coatings tend to absorb moisture from the atmosphere when they are exposed. This is because of the fact that many coating ingradients may have hygroscopic or water absorbing properties and some extrusion aids may even have chemically combined water in their constitution. These materials release moisture in the arc atmosphere and as a consequence porosity, inclusions and blow holes may result in the weld metal, impairing its radiographic quality. Hence, it is always advisable to re-dry the electrodes before using them in fabrication shop. For instance, rutile coated electrodes of the AWS-E 6012 and E-6013 classifications can be re-dried upto a maximum temperature of 110°C for an hour. Care must be taken not to exceed this temperature limit as severe drying will itself cause welding defects for these classification.

Basic coated low hydrogen electrodes should always be used with short arc and these electrodes (AWS : E-7015, E-7016, E-7018 types) must be baked for atleast an hour in a baking oven. The typical baking oven in shown in Fig. 19.5. The temperature range must be 250°C to 350°C. This is to control in moisture level in the arc atmosphere and to obtain weld metal with low diffusible hydrogen values. Otherwise if the hydrogen present in the weld metal is more, there is a possibility of cold cracking (hydrogen induced cracking) occurring in the weldment.

Fig. 19.5 Electrode baking oven.

19.2.6 Filler Wires and Strips

Arc welding process like gas metal arc welding, submerged arc welding, flux-cored wire arc welding and automatic TIG and plasma arc welding processes make use of continuous filler wires of various diameters. These wires are available in coil form enabling smooth uninterrupted welding. Types of continuous filler wires are :

(i) Solid wire.
(ii) Flux-cored wire.
(iii) Continuous strip electrodes (Solid and Flux-cored).
(iv) Flux coated continuous filler wires.

All arc welding processes use solid, round shaped filler wires. Flux-cored and metal powder cored wires are used in gas shielded metal arc welding and submerged arc welding processes. Solid and flux cored strip electrodes are used in a special form of submerged arc welding, called strip cladding process.

19.2.7 Manufacture of Filler Wires

Round shaped wires are rolled and finished by drawing through dies. Flux-Cored wires are manufactured by a specialised technique of filling a sheath with cored ingradients and drawing them to smaller diameters or by folding a strip with cored ingradients properly placed inside and reducing the diameter by rolling, possibly finished, by drawing them through dies. Figure 19.6 indicates the cross-sections of various flux-cored wires. Carbon and carbon-manganese and low alloy steel filler wires are normally copper coated to prevent rusting during storage.

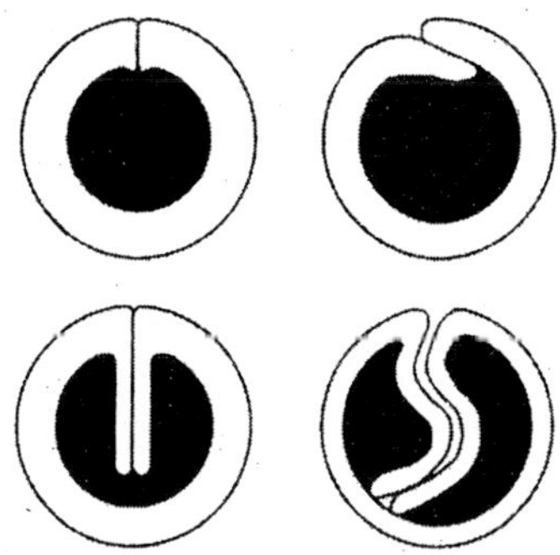

Fig. 19.6 Cross-sections of flux-cored wires.

19.2.8 Quality Control

Continuous filler wires are produced in standard sizes and the quality control of their production includes, apart from chemical analysis for various elements, the following checks :

(i) Size.
(ii) Finish and uniformity.
(iii) Cast and helix.
(iv) Coiling.

(i) Size : Filler wires are available from 0.6 mm to 6.4 mm in diameter. The normal tolerance on the finish diameter is as follows :

(a) Solid Wire :

Wire dia in mm	*Tolerance in mm*
0.6 to 1.2	0.02
1.2 to 3.2	0.05
Above 4	0.01

(b) Flux-cored Wire :

0.6 to 1.2	0.05
2 and above	0.08

(ii) **Finish and Uniformity :** The finished electrode shall have a surface which is free of defects like scratches, depressions, scales, seams or laps. The wire diameter should be uniform. The joints wherever made shall be such that they will not affect the uniform and smooth feeding of the wire during welding. For flux or metal powder cored wires, the core ingradients shall be uniformity distributed such that the required performance of the electrode and the weld metal properties will not be affected.

(iii) **Cast and Helix :** Cast (D) is the free diameter formed by one strand (free loop) of wire and helix (H) is the distance separating the free ends of one strand of wire. The cast and helix (Fig. 19.7) imparted to the wire during winding of a coil of a wire should be such that they enable the smooth feeding of the wire, ensure the positive contact of the wire with the contact tube (from which the welding current is transferred to the wire) and also enable the deposition of the weld metal at the required place in the weld joint.

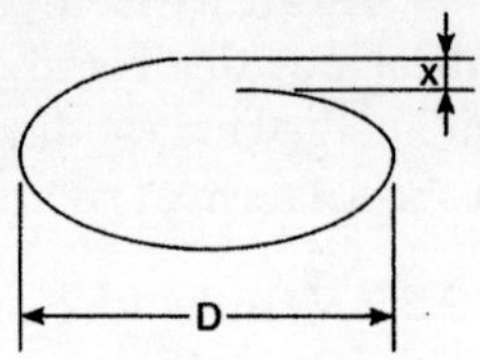

Fig. 19.7 Cast and helix.

(iv) **Coiling :** The winding of the electrodes into coils should be done such that there is no kink or bend thus ensuring free unwinding of the wire. The end of the wire to be fed into the wire feeder should be easily locatable and shall be secured to avoid unnecessary unwinding of wire spools. Wires of diameter especially above 2 mm are coiled in two ways. The coils with supports can be directly mounted on the spool holder. The coils without support have to be mounted on a coil holder and the coil holder is mounted on the machine for continuous and automatic welding applications. Wires are also packed in drums. The weight of the wire supplied in coil form varies from 1 kg to several hundred kilograms.

19.2.9 Packing

The electrode coils are suitably packed to ensure reasonable stroage life especially for carbon steel wires. The packing will be marked outside with the following information about the electrodes :

(i) Brand name and classification.
(ii) Dimentional details.
(iii) Net and gross weight.
(iv) Lot number, heat number.
(v) Supplier's name and address.

19.2.10 Storage

The wires shall be stored in a clean and dry area to avoid rusting especially carbon and low alloy steel wires. Sheath-folded type flux cored wires will not have copper coating. It is essential that these wires be stored in a proper way. High humidity and wet conditions will damage the wires quickly.

❑❑❑

APPENDIX

APPENDIX-1
QUESTION BANK

1. Explain the term 'welding' and define its classification and features.
2. Explain various arc welding processes and state their fundamental principles.
3. Explain the fundamentals of oxy-acetylene welding process.
4. Describe various types of flames used in oxy-acetyleen process and their application.
5. Describe with an illustation the working of a single-stage gas pressure rgulator.
6. Explain the various applications of gas welding.
7. Describe the setup and operations of shielded metal arc welding.
8. Write short notes on following :
   (i) Touch electrode welding.
   (ii) Bunched electrode welding.
   (iii) Multi-arc welding.
   (iv) Massive electrode welding.
9. Describe the impact of joint designs in the performancve of shielded metal arc welding.
10. What are the common defects encountered in SMAW and how are they overcome.
11. Describe briefly the functions of the coating on electrodes for SMAW.
12. What is gravity welding.
13. Describe the process fundamentals, advantages and limitations of gas tungsten arc welding.
14. What are the special power source characteristics required for GTAW of aluminium.

15. Write a short note on GTAW electrodes.
16. Explain the process fundamental and technique of plasma arc welding.
17. Describe with sketch the plasma welding torch and accessories with special mention of functions of parts.
18. Describe different methods of arc initiation and maintainence for GTAW process.
19. Write short notes on :
    (i) Pulsed current GTAW.
    (ii) Hot wire GTAW.
    (iii) Twin electrode GTAW.
    (iv) Manual programmed GTAW.
20. Explain the oeprating principle of MIG weldign process.
21. What are the different modes of metal transfer in MIG welding and their characterstics.
22. Describe the influences of shielding ass on the process behaviour.
23. Describe the features of MIG welding equipment.
24. What is electrogas welding.
25. What are the advantages of FCAW over solid wire MIG welding.
26. Differentiate between MIG and MAG welding with particular reference to shielding gas and its effects on mode of metal transfer.
27. Explain why thinner wires are more suitable for dynamic stability of GMAW process with sketches.
28. Describe the role of metal transfer on weld characteristics like bead geometry, penetration and metallurgical structure.
29. Describe circumferential weldign by SAW process. List specific applications of this type of welding.
30. Draw the schematic of Electroslag welding (ESW) setup and describe the principle and operations of process.
31. What type of power source is used in electroslag welding and why.
32. What are the special characteristics of electroslag flux.
33. What are the advantages of electroslag welding and where it should be used?
34. Describe weld structure and properties of an electrolslag weld and compare it with teh weld made by submerged arc welding?
35. Describe the principle and operations of thermit welding.
36. State the applications of thermit welding. Draw a suitable mould for one of the applications?
37. What are the different types of thermist welding?
38. What is projection welding?
39. What are teh stages involved in spot welding?
40. How the quality of spot weld is expressed?
41. Describe seam welding operation and techniques?
42. Explain the fundamentals of flash butt welding and the process techniques?

43. Describe the role of following welding variables for resitance welding methods :
    (i) Welding current.
    (ii) Weld time.
    (iii) Pressure control.
44. Write short notes on :
    (i) Roll spot and stich welding
    (ii) Mash seam welding
    (iii) Foil-butt seam welding
45. Write shorte notes on :
    (i) Series welding
    (ii) Multiple spot welding
    (iii) Pulsation welding
    (iv) Seam welding
46. Describe the process of continuous drive friction. welding with particular reference to its time-torque relationship.
47. Describe the process of explosion welding with detailed description of its principle of operation.
48. Explain the following process variables for explosion welding :
    (a) Impact velocity
    (b) Stand-off
    (c) Stand-off distance
    (d) Approach angle
49. Explain the process fundamentals and steps in achieving a friction weld.
50. Describe the process parameters of ultrasonic weld.
51. What are the advantages of diffusion bonding compared to fusion welding?
52. Describe with the help of a sketch the constructional features of an electron beam gun.
53. Compare the diode and triode type electron beam welding units and explain their field of application.
54. Explain the mechanism of key hole penetration in EBW.
55. Describe the principle and mechanism of laser beam welding operation.
56. Describe the constructional features of solid-state laser like the one that uses a ruby rod as lasing material.
57. Explain the role of beam power and mode in Laser welding.
58. Describe the hazards and safety aspects of laser welding.
59. Laser welding is supposed to be best suited for automated welding mode, why?
60. Describe the generation oflaser beam.
61. Enumerate various applications of EBW and LBW
62. Explain wet underwater welding and mention two wet welding techniques.
63. What si simple habitat welding? Draw a neat sketch of such as arrangement for pipe repair.

64. Describe what has been understood of different diving techniques that are practised in under water welding.
65. Describe thae salient features of cold pressure welding and mention its specific field of application.
66. Describe the process of welding of rails at site and compare it with flash butt welding of rails.
67. Explain 'kerf' and 'drag' in oxyacetylene cutting technique.
68. Describe the three starting techniques in oxy-acetylene cutting.
69. Describe biefly the parameters which affect the cutting quality.
70. Explain the plasma cutting process and its principle of operation.
71. What are the advantages of plasma cutting over other processes and its applications.
72. Give an account of differnt types of plasma cutting systems and the various process variables to be considered during plasma cutting.
73. What are the safety precautions to be followed during plasma cutting operation?
74. What is the purpose of gouging?
75. Explain the principle of flame arc gouging.
76. Where to use progressive gonging and spot gouging?
77. What are the voltage and current ranges of carbon arc gouging?
78. Explain the types and advantages of the different carbon gouging electrodes.
79. What are the main features of laser cutting?
80. Describe five different methods of laser cutting.
81. Explain what is understood by surfacing?
82. What are the benefits resulting from surfacing?
83. What are the major factors influencing the service environment?
84. What are the various types of wear?
85. Describe various types of corrosion.
86. List the welding processes that can be used for surfacing.
87. How does high temperature accelerates the wear?
88. What are various forms of surfacing filler materials?
89. What are factors to be considered for process selection in surfacing?
90. List the type of materials required for the following :
    (a) High impact with low abrasion resistance.
    (b) High abrasion with low impact.
    (c) High wear and corrosion resistance.
91. Describe the processes of metal surfacing and spraying with their specific fields of application.
92. Give classification and characteristics of surfacing materials with their specific fields of application.
93. Describe the steps involved in metal spraying.

94. List different methods of metallising and describe anyone of them in detail with its field of application.
95. With the help of a sketch, describe the principle of arc spraying.
96. What are the surface preparations and precautions to be dopted for spraying the first pass?
97. Mention the applications of arc spraying.
98. Explain the different kinds of atomic bonding in materials.
99. Give a brief account of the three different types of crystal structures in metals.
100. What is the significance of phase transformations in the metals? Explain the terms complete solubility and limited solubility.
101. Explain the Iron/carbon phase diagraon.
102. Explain what happens when the metal is coded from the austenitic phase. Also explain martensite and bainite.
103. Exlain time temperature transformation.
104. Describe the CCT diagram with respect to steel.
105. Describe the various types of steels and role of the respective alloying elements in the steels.
106. Name the different types of carbon steels.
107. What are the major factors that affect cold cracking in steels?
108. Why is preheating required and what is its function?
109. What is lamellar tearing? Explain with sketch.
110. Define weldability and how carbon affects weldability of steels?
111. What is stainless steel? What are the different classes of stainless steel?
112. What precautions are required in welding austenitic stainless steel? What are the procedures and parameters used?
113. How is measurement of delta-ferrite done? Mention the names of few ferrite measuring instruments.
114. How is Schaeffler diagram useful in the welding of austenitic stainless steel.
115. What precautions are required to be followed during the welding of martensitic stainless steel?
116. Name the different welding problems encountered in ferritic stainless steels and their remedils.
117. State the reasons for the following :
    (i) A short arc is to be employed for welding stainless steel.
    (ii) Stringer bead, low heat input and low interpass temperatures are employed for welding austnitic stainless steels.
    (iii) Weld decay due to inter-granular corrosion during service occurs only in unstable steels.
    (iv) Knife-line attack in stabilised stainless steel occurs if no post-weld stabilising heat-treatment is carried out.
    (v) The upper limit on delta ferrite in austenitic stainless steel weld metal depends on the service application.

118. Explain the solidification and production of the three basic types of cast irons.
119. What is the exact difficulty, metallurgically in welding cast iron?
120. Enumerate the extra care one should take while welding cast iron.
121. Make a collection of special repair welding techniques reported by various maintainence personel.
122. Highlight the important propetries of aluminium and aluminium alloys and give some typical examples of applications.
123. Give classification of aluminium alloys.
124. Describe the TIG welding of aluminium.
125. Describe the MIG welding of aluminium.
126. Enumerate the wweld defects in aluminium and the remedial steps for each.
127. Name the various common copper alloys. Compare teh thermal conductivities of these alloys with respect to oxygen-free copper.
128. Why is preheating required for copper and most of the copper alloys?
129. Can brass (copper-zinc) filler wire be used for arc welding of brass? What type of fillers are normally recommended in the arc welding of brass?
130. Why is hot peening of each pass recommended in the multi-pass welding of phosphor bronze?
131. What are the three major welding processes used to weld nickel and nickel alloys?
132. State the various techniques used and precautions necessary in nickel and nickel alloy welding.
133. Describe briefly the welding of nickel based precipitation hardenable alloys and heat-resistant alloys.
134. What are the different types of titanium alloys available?
135. What are the ELI grades of titanium alloys?
136. Why does titanium require extra inert gas protection for the hot welds?
137. Which are the welding processes normally used to weld titanium?
138. What are teh two major classifications of cobalt based alloys? State the composition of atleast two such specification from both groups.
139. What are the welding processes employed in welding stellite alloys? Describe in what way oxyacetylene welding is superior to the other arc welding processes.
140. What are the imporatnt quality aspects during deposition of stellite by gas or arc welding?
141. Describe arc discharge in a gaseous medium. How is welding arc different from this?
142. Explain the various aspects of volt-ampere characteristics of the arc.
143. What is the fundamental pricniple behind metal transfer in a welding arc?
144. List the different forces that affect the mode of metal transfer in arc welding and describe their roles in brief.

145. Classify the modes of metal transfer in arc welding and describe their characteristics and use in position welding.
146. List all the five parameters affecting the mode of metal transfer. Describe in brief teh role of them.
147. Describe in brief the modes of metal transfer in the following welding processes :
   (i) SMAW (ii) GMAW (iii) GTAW (iv) SAW
148. Describe the modes of metal transfer in GMAW for different metals using CO_2 and argon as shielding gas. Also describe the role of electrode polarity on metal transfer.
149. What are the major classification of the weld joints?
150. Draw a butt weld and a fillet weld and markout their nomenclature.
151. Draw a symbol for a double 'U' butt joint on a longitudinal seam of a vessel which is to be ground finished. The thickness of the plate is 40 m.m. The length of the seam is 2.5 meters. Also how a simple attachment of 12 m.m. thick plate weld which will be done at site should be represented by asymbol. The fillet leg length of the plate attachment weld is 12 m.m. The breadth of the plate is 150 m.m.
152. Calculate the length of weld required for a fillet of 20 m.m. equal leg length to transmit an axial load of 360 KN. The allowable shear stress of the weld material is 115 N/mm^2.
153. Describe the roles of different consittuents of welding costs.
154. Describe differnt methods of improving productivity in fabrication by welding.
155. Define the following with reference to weld fabrications :
   (i) On-cost (ii) Operator factor (iii) Standard time
156. What are the practical methods to minimise distortion during fabrication?
157. Name the materia related aprameters that influence distortion.
158. Explain the methods of hot correction.
159. Explain the effect of the welding process and heat input on distortion.
160. Enumerate the various defects encountered in welding.
161. What are the causes for porosities and blow-holes and what are the steps to be taken to avoid them?
162. What are the different types of cracks observed in welding?
163. What are the causes for :
   (i) Cold cracking
   (ii) Hot cracking
164. What is Lamellar Tearing and how to overcome this defect?
165. Explain the difference between :
   (i) Lack of fusion, (ii) Improper penetration, (iii) Root concavity
   (iv) Undercut (v) Excessive reinforcement
166. What is "slag inclusion" and how to prevent/avoid it?

167. Describe the principle underlying the following :
    (a) Liquid penetrant inspection.
    (b) Magnetic particle inspection.
168. What would be the best method for detecting surface defects on Austenitic stainless steel?
169. Which is the best NDT method for detecting slag inclusion?
170. What is understood by "Radiographic testing", what technique of RT is adopted for examining a tube butt weld, there is no access to the underside of the weld.
171. Explain the principle of "Ultrasonic testing".
172. Explain the terms "linear" and "voluminar" with reference to defects.
173. Define "tensile strength" and "yield strength" of steel. How are percentage elongation and reduction in area is measured?
174. How is toughness of steel assessed and explain how temperature affects toughness.
175. Why is a bend test conducted for welds and what is the significance of this test?
176. What type of tensile test is requied for electrode qualification?
177. Describe diffeent hazards involved in arc welding and explain in details the ones related with plasma arc welding.
178. Describe the safety precautions peculiar to electron beam and laser beam welding processes.
179. What are the preventive measures to avoid shock during welding?
180. How does the arc radiation affect the welder and what are the ways to prevent "arc eye" and "arc burns"?
181. Enumerate the cause of "back fire" in oxy-fuel cutting? What is the role of flash-back arrestor?
182. What are the annealing processes used in industries to modify metal structure? Describe each briefly.
183. Describe normalising and explain how it is useful in welding/fabrication of steels?
184. Describe the theory behind stress-relieving.
185. Describe the various resistance heating elements/pads.
186. Explain how you would setup and measure temperature of the heat treatment for post-weld stress-relieving using resistance heating pads?
187. Describe the different conceptual levels of mechanisation and automation in welding.
188. Describe the moving head systems of mechanised welding.
189. Describe the moving work systems of mechanised welding.
190. Describe fully automatic welding, its merits and demerits.
191. Write shrot notes on :
    (i) Adaptive controls in welding.

(ii) Remote control welding.

(iii) Robotic welding

192. Differentiate between manual, semi-automatic and automatic welding and give their fields of application.

193. Explain why the normal 415 V, 3-phase suply cannot be used as such for welding?

194. Explain the major divisions in the welding power source characteristics.

195. Describe the various power/current controls employed in welding transformers.

196. Describe the features of welding rectifiers and welding generators for obtaining D.C. power. What are the advantages and disadvantags?

197. Explain the major principles involved in electronically controlled power sources. HOw do they compare with the conventional ones?

198. Why are DC suppressors and high frequency units used?

199. Why are the welding electrodes to be baked and dried? Also explain the equipment and practices.

200. What are the quality control parameters to be checked while producing welding filler wires?

201. Briefly explain about the storage and handling of CO_2 gas.

❑

APPENDIX–2
SHORT QUESTIONS

1. Define welding and weldability.
2. What are the different standard sizes of coated/covered electrodes for arc welding, available in the market?
3. What is the importance of coating of electrode?
4. Name diffrent classes of electrode coatings used for welding low carbon steels.
5. What is a thermal stick and its function?
6. Differentiate between welding transformer, rectifier and motor-generator set.
7. What is duty cycle of a power source?
8. Sketch different types of static *V-I* characteritics of a welding power soruce.
9. What is an invertor power source for weldign?
10. Name different types of welds and weld joints.
11. Describe the factors accounted for in process selection for welded fabrication.
12. Describe the economic considerations for selecting a welding process.
13. What is a stringer bead?
14. What is puddling?
15. Name the different welding positions.
16. What is an arc-blow?
17. What is polarity?
18. What are dynamic *V-I* characteristics in arc welding and how are they obtained?
19. Differentiate between agglomerated and fused SAW fluxes.
20. List different methods of arc initiation in TIG welding.
21. What is the rangee of hole diameter in plasma welding torch?
22. What core material is used for making coated/covered electrodes for cast iron welding?
23. What purpose is served by mobile cathode in aluminium welding?
24. Which arc welding processes are used for welding aluminium and stainless steel?
25. Mention diffeence between TIG and plasma welding torches.
26. Difference between MIG and MAG welding.
27. Why a heater is used on CO_2 cylinder for CO_2 welding?
28. Define an autogenous weld.
29. What is series welding?
30. What is pulsation welding?
31. What is DA is gas welding?
32. What is difference between plasma cutting and flame cutting?
33. Why vacuum is needed for EBW?

34. Why all interior walls are lead-lined in EBW gun?
35. Describe hybrid friction stir welding.
36. Name some important applications of friction welding.
37. What ae beam power and mode in laser beam welding?
38. What is the role of optical fibre in laser beam welding?
39. Describe the principle of operation of a diode laser.
40. What is a thermit and what for it is used?
41. Give the principle of flame cutting.
42. What are different types of flames used in gas welding?
43. Name commonly used fuel gases for gas welding and flame cutting.
44. Name differnet welding processes used in space welding.
45. Why underwater welding is required and where?
46. Write short notes on :
    (i) Welding at low ambient temperatures.
    (ii) Welding in vacuum.
    (iii) Welding in wind.
47. Write short notes on :
    (i) Gravity welding
    (ii) Fire cracker welding
48. Why safety is required in welding?
49. What is synergic MIG welding.
50. Name important applications of Narrow gap welding.
51. What is role of fit-up in welding.
52. What do you understand by bakc-gouging?
53. Which gases are used to cover weldment in case of aluminium and copper welding?
54. Why automation has become necessary now-a-days in fabrication industries?
55. What is the main reason for stress-relieving the weldments?

❑

APPENDIX-3
METRIC CONVERSION

1. Area

1 sq cm (cm^2)	= 0.155 sq in.	1 sq in	= 6.452 sq cm (cm^2)
1 sq m (m^2)	= 10.764 sq ft.	1 sq ft	= 0.0929 sq m (m^2)
1 hectare (ha)	= 2.471 acres	1 acre	= 0.3047 hectares
10000 cm^2	= 1 sq m (m^2)	100 ares	= 1 hectare (ha)
100 m^2	= 1 are (a)	100 hectares	= 1 sq km (km^2)

2. Consumption

1 litre per hour	= 0.22 gal per lb
1 gal per hour	= 4.546 litres per hour
1 kg per second	= 3.6 tonne per hour
1 litre per min	= 0.06 m^3 per hour
1 tonne per hour	= 0.28 kg per second
1 m^3 per hour	= 16.67 litres per min

3. Density and Specific Volume

1 kg per m^3	= 0.062 lb per cu ft.	1 lb per cu ft	= 16.019 kg per m^3
1 m^3 per kg	= 16.019 cu ft per lb	1 cu ft per lb	= 0.062 m^3 per kg.
1 litre per kg	= 0.0998 gal. per lb.	1 lb per cu ft	= 10.022 litres per kg.

4. Force

1 kgf	= 2.205 pound force (lbf)
1 lbf	= 0.4536 kgf.
1 kilogram force (kgf)	= 9.81 newtons
1 newton	= 0.102 kgf.

5. Heat and Energy

1 kcal	= 3088 ft	lbf	= 427 kg fm = 3.968 B. Th. U.
1 B. Th. U.	= 778 ft	lbf	= 107.6 kgfm = 0.2520 kcal.
1 C.H.U.	= 1400 ft	lbf	= 193.6 kgfm = 0.4536 kcal.
1000 caloreis	= 1 kilocalorie (kcal)	1 kcal	= 427 kgfm
1000 kcal	= 1 therm = 4.2 kilojoules		

1 horse-power hour (metric)	= 270000 kgfm = 632 kcal.
1 kilowatt hour	= 367100 kgfm = 860 kcal.

6. Length

1 millimetre	= 0.039 inch	1 inch	= 25.5 millimetres
1 metre	= 3.28 feet	1 foot	= 0.3048 metre
1 kilometre	= 0.621 mile	1 mile	= 1.609 kilometres
1000 micron	= 1 millimetre (mm)	100 centimetres	= 1 metre (m)
10 millimetres	= 1 centimetre (cm)	1000 metres	= 1 kilometre (km)

7. Mass

1 gram	= 0.0353 ounce	1 ounce	= 28.35 grams (g)
1 kilogram	= 2.205 pounds	1 pound (lb)	= 0.4536 kg.
1 tonne	= 0.984 ton	1 ton	= 1.016 tonne (t)
1000 milligrams (mg)	= 1 gram	1000 kilograms (kg)	= 1 tonne (t)
1000 grams (g)	= 1 kilogram	100 kilograms	= 1 quintal (q)

8. Power

1 h.p. (metric)	= 542.5 ft	lb per sec.	= 0.986 h. p. (British)
1 kilowatt	= 737.6 ft	lbf per sec.	= 1.341 h.p. (British)
1 h.p. (British)	= 76.04 kg	fm per sec.	= 1.014 h. p. (metric)
1 h.p. (metric)	= 0.7355 kW		= 735.5 Joules per sec.
			= 75 kg fm. per sec.
1 kilowatt = 1.36 h. p. (metric)			= 1000 Joules per sec.
			= 102 kg fm per sec.

9. Pressure

1 kgf per cm^2	= 14.223 lbf per sq. in 1 lbf per sq in	= 0.0703 kgf per cm^2
1 atmosphere	= 14.696 lbf per sq in	= 29.92 in Hg = 33.9 ft H_2O
1 mm H_2O	= 1 kgf per m^2 1 atmosphere	= 1.0332 kgf per cm^2
10^4 kgf per m^2	= 1 kgf per cm^2	= 1.01325 bar.
1 kgf per cm^2	= 0.981 bar	= 760 mm Hg.
1 bar	= 10^5 newtons per m^2	= 10322 m H_2O

10. Volume and Capacity

1 cu cm (cm^3)	= 0.061 cu in.	1 cu in	= 16.387 cu cm (cm^3)
1 cu m (m^3)	= 35.315 cu ft.	1 cu ft	= 0.028 cu m (m^3)
1 litre	= 0.22 gallon	1 gallon	= 4.546 litres (l)
1000 cm^3	= 1 cubic decimetre (dm^3)	1000 millilitres (ml)	= 1 litre
1000 dm^3	= 1 cubic metre (m^3)	1000 litres (l)	= 1 kilolitre
1000 cm^3	= 1 litre	1 kilolitre (kl)	= 1 cu metre

11. Work

1 kgfm	= 7.233 ft lbf.	1 ft lbf = 0.1383 kgfm
1 kilogram force metre (kgfm)	= 9.81 joules	1 joule = 0.102 kgfm

❑

APPENDIX–4
TEMPERATURE CONVERSION TABLE

°C	°F	°C	°F
–273	–459	1340	2444
–200	–328	1360	2480
–100	–148	1380	2516
–30	–22	1400	2552
–10	+ 14	1420	2588
0	32	1440	2624
10	50	1460	2660
25	77	1480	2696
50	122	1500	2732
75	167	1520	2768
100	212	1540	2804
150	302	1560	2840
200	382	1580	2876
250	482	1600	2912
300	572	1620	2948
350	662	1640	2984
400	752	1660	3020
450	842	1680	3056
500	932	1700	3092
550	1022	1720	3128
600	1112	1740	3164
650	1202	1760	3200
700	1292	1780	3236
750	1362	1800	3272
800	1472	1820	3308
900	1652	1840	3344
1000	1832	1860	3380
1020	1868	1880	3416
1040	1904	1900	3452
1060	1940	1920	3488
1080	1976	1940	3524
1100	2012	1960	3560
1120	2048	1980	3596
1140	2084	2000	3624
1160	2120	2100	3812
1180	2156	2200	3992
1200	2192	2300	4172
1220	2228	2400	4352
1240	2264	2500	4532
1260	2300	2600	4712
1280	2336	2700	4892
1300	2372	2800	5072
1320	2408	2900	5252
		3000	5432

$$\frac{C}{100} = \frac{F-32}{180}$$

APPENDIX-5
SI UNITS AND CONVERSION TABLE

(A) SI-Units

S. No.	Quantity	SI Unit	Symbol
1.	Mass, m	Kilogram	kg
2.	Length, l	Metre	m
3.	Temperature, t, θ,	Kelvin, Celsius	K, °C
4.	Force, F	Newton	$N = \text{kg.m/s}^2$
5.	Work or energy, E	Joule	$J = \text{Nm}$
6.	Power, P	Watt	$W = \text{J/s}$
7.	Pressure of stress, p	Pascal or Newton per square metre	$Pa = \text{N/m}^2$
8.	Density, ρ	Kilogram per cubic metre	$\rho = \text{kg/m}^3$
9.	Resistivity, ρ	ohm metre	$\rho = \Omega\text{.m}$
10.	Thermal conductivity, λ, (k)	Watt per metre kelvin	$\lambda = \text{W/mk}$
11.	Specific heat capacity, c	Joule per kilogram kelvin	$c = \text{J/kg k}$

(B) SI-Conversion

1. 1 inch = 25.4 mm = 2.54 cm = 0.0254 m = 1000 mil.
2. 1 micron $(\mu) - 1\,\mu\text{m} = 10^{-6}$ metre $= 10^{-3}$ mm.
3. 1 lb = 0.454 kg = 454 gm = 4.45 N = 4.45×10^5 dynes. (N is Newton)
4. 1 dyne $= 10^{-5}$ N $= 0.102 \times 10^{-5}$ kg $= 0.225 \times 10^{-5}$ lb.
5. 1 kg $= 10^3$ gm = 2.21lb = 9.81 Newtons $= 9.81 \times 10^5$ dyne $= 10^{-3}$ tonne (t)
6. $1\ \text{N/m}^2 = 1$ pascal (pa) $= 1.02 \times 10^{-7}\ \text{kg/mm}^2 = 1.45 \times 10^{-4}$ psi $= 10\,\text{dynes/cm}^2$
7. $1\ \text{kg/mm}^2 = 9.81 \times 10^6\ \text{N/m}^2 = 1.426 \times 10^3$ psi $= 9.81 \times 10^7\ \text{dynes/cm}^2$
8. 1 psi $= 6.89 \times 10^3\ \text{N/m}^2 = 7.02 \times 10^{-4}\ \text{kg/mm}^2 = 6.89 \times 10^4\ \text{dynes/cm}^2$
9. 1 atmosphere = 760 mm of Hg $= 1.01 \times 10^5$ N/m= $1.03 \times 10^{-2}\ \text{kg/mm}^2$
   $= 1.03\ \text{kg/cm}^2$
10. 1 Joule (J) = 1 Nm = 1 Ws $= 10^7$ erg = 0.239 cal $= 0.625 \times 10^{12}$ eV.
11. 1 kWh $= 3671 \times 10^2$ kgfm = 860 kcal = 3.6 MJ
12. 1 kgm = 9.81 J

APPENDIX–6
HARDNESS CONVERSION TABLE

BRINELL			ROCKWELL			
Dia. in mm, 30000 kg load 10 mm ball	Hardness No.	Vickers or Firth Hardness No.	C 150 kg. load 120° Diamond Cone	B 100 kg. load $\frac{1}{16}$ " (1.6 mm) dia ball	Scleros-cope No.	Appro-ximate Tensile Strength N/mm^2
2.05	898					3050
2.10	857					2900
2.15	817					2770
2.20	780	1150	70		106	2650
2.25	745	1050	68		100	2540
2.30	712	960	66		95	2430
2.35	682	885	64		91	2330
2.40	653	820	62		87	2250
2.45	627	765	60		84	2150
2.50	601	717	58		81	2050
2.55	578	675	57		78	1980
2.60	555	633	55	120	75	1900
2.65	534	598	53	119	72	1830
2.70	514	567	52	119	70	1770
2.75	495	540	50	117	67	1700
2.80	477	515	49	117	65	1640
2.85	461	494	47	116	63	1580
2.90	444	472	46	115	61	1520
2.95	429	454	45	115	59	1460
3.00	415	437	44	114	57	1410
3.05	401	420	42	113	55	1350
3.10	388	404	41	112	54	1300
3.15	375	389	40	112	52	1250
3.20	363	375	38	110	51	1210
3.25	352	363	37	110	49	1170
3.30	341	350	36	109	48	1140
3.35	331	339	35	109	46	1100
3.40	321	327	34	108	45	1070
3.45	311	316	33	108	44	1030
3.50	302	305	32	107	43	1000
3.55	293	296	31	106	42	980
3.60	285	287	30	105	40	950
3.65	277	279	29	104	39	920
3.70	269	270	28	104	38	900
3.75	262	263	26	103	37	880

3.80	255	256	25	102	37	860
3.85	248	248	24	102	36	840
3.90	241	241	23	100	35	820
3.95	235	235	22	99	34	800
4.00	229	229	21	98	33	780
4.05	223	223	20	97	32	760
4.10	217	217	18	96	31	740
4.15	212	212	17	96	31	720
4.20	207	207	16	95	30	700
4.25	202	202	15	94	30	680
4.30	197	197	13	93	29	670
4.35	192	192	12	92	28	650
4.40	187	187	10	91	28	640
4.45	183	183	9	90	27	630
4.50	179	179	8	89	27	610
4.55	174	174	7	88	26	600
4.60	170	170	6	87	26	590
4.65	166	166	4	86	25	570
4.70	163	163	3	85	25	565
4.75	159	159	2	84	24	550
4.80	156	156	1	83	24	540
4.85	153	153		82	23	520
4.90	149	149		81	23	515
4.95	146	146		80	22	510
5.00	143	143		79	22	495
5.05	140	140		78	21	490
5.10	137	137		77	21	480
5.15	134	134		76	21	470
5.20	131	131		74	20	455
5.25	128	128		73	20	450
5.30	126	126		72		440
5.35	124	124		71		430
5.40	121	121		70		425
5.45	118	118		69		420
5.50	116	116		68		410
5.55	114	114		67		405
5.60	112	112		66		400
5.65	109	109		65		385
5.75	105	105		62		370
5.80	103	103		61		365
5.85	101	101		60		360
5.90	99	99		59		350
5.95	97	97		57		345
6.00	95	95		56		340

APPENDIX-7
INDIAN, BRITISH AND AMERICAN WELDING AND CUTTING STANDARDS – FOR REFERENCE

(A) INDIAN STANDARDS (IS)

Section 1 : Welding Consumables, Equipments, Testing, Training, Safety, etc.

S. No.	Standard No.	Title
1.	IS 812 : 1957	Glossary of terms relating to welding and cutting of metals.
2.	IS 813 : 1986	Scheme of symbols for welding.
3.	IS 814 : 1991	Covered electrodes for manual metal arc welding of carbon and carbon-managanese steel
4.	IS 817 : 1966	Code of practice for training and testing of metal arc welder
5.	IS 817 (PT1) : 1992	Manual metal arc welding
6.	IS 818 : 1968	Code of practice for safety and health requirements in electric and gas welding and cutting operations
7.	IS 822 : 1970	Code of procedure for inspection of welds
8.	IS 1179 : 1967	Equipment for eye and face protection during welding
9.	IS 1278 : 1972	Filler rods and wires for gas welding
10.	IS 1393 : 1961	Code of practice for training and testing of oxy-acetylene welders
11.	IS 1395 : 1982	Low and medium alloy steel covered electrodes for manual metal arc welding
12.	IS 2927 : 1975	Specification for brazing alloys
13.	IS 3016 : 1982	Code of practice for fire precautions in welding and cutting operation
14.	IS 3600 :	Methods of testing fusion welded joints and weld metal in steel
15.	IS 3600 (PT 1) : 1985	Cruciform fillet weld tensile test
16.	IS 3600 (PT 2) : 1985	Beam impact (charpy V notch) test
17.	IS 3600 (PT 3) : 1984	Transverse tensile test on butt welds
18.	IS 3600 (PT 4) : 1984	Longitudinal tensile test on cylindrical weld metal test pieces on butt welds
19.	IS 3600 (PT 5) : 1983	Transverse root and face bend test on utt welds
20.	IS 3600 (PT 6) : 1983	Transverse side bend test on butt welds
21.	IS 3600 (PT 7) : 1985	Longitudinal root and face bend test on butt welds
22.	IS 3600 (PT 8) : 1985	Nick break test and fillet weld fracture test
23.	IS 3600 (PT 9) : 1985	Macro and micro examination
24.	IS 3613 : 1974	Acceptance test for wire flux
25.	IS 4943 : 1968	Assessment of butt and fillet fusion welds in steel sheet, plate and pipe.

26.	IS 4972 : 1968	Resistance spot welding electrodes
27.	IS 5206 : 1983	Covered electrodes for manual metal arc welding of stainless steel and other similar high alloy steels
28.	IS 5511 : 1991	Covered electrodes for manual metal arc welding of cast iron
29.	IS 5687 : 1970	Glossary of terms relating to welding of plastics
30.	IS 5856 : 1991	Corrosion and heat resisting chromium nickel steel solid welding rods, bare electrodes
31.	IS 5857 : 1970	Nickel and nickel alloy bare solid welding rods and electrodes
32.	IS 5897 : 1985	Aluminium and aluminium alloy welding rods and wires and magnesium alloy welding rods
33.	IS 5898 : 1970	Copper and copper alloy bare solid welding rods and electrodes
34.	IS 5922 : 1970	Qualifying tests for welders engaged in aircraft welding
35.	IS 6016 : 1971	House connection for welding and cutting equipment
36.	IS 6419 : 1971	Welding rods and bare electrodes for gas shielded welding of structural steel
37.	IS 6431 : 1971	Tolerances on dimensions of plates cut by flame
38.	IS 6560 : 1972	Molybdenum and chromium-molybdenum alloy steel welding rods and bare electrode for gas shielded arc welding
39.	IS 6901 : 1988	Specification for pressure regulators for gas cylinders used in welding, cutting and related processes
40.	IS 7273 : 1974	Method of testing fusion welded joints in aluminium and aluminium alloys
41.	IS 7280 : 1974	Bare wire electrodes for submerged arc welding of structural steels
42.	IS 7303 : 1974	Covered electrodes for surfacing of metal by manual metal arc welding
43.	IS 7307 (PT 1) : 1974	Approval tests for welding procedures Part 1 : Fusion welding of steel
44.	IS 7310 (PT 1) : 1974	Approval tests for welders working to approved welding procedures Part 1 : Fusion welding of steel
45.	IS 7318	Approval tests for welders when welding procedure approval is not required
46.	IS 7318 (PT 1) : 1974	Fusion welding of steels
47.	IS 7318 (PT 2) : 1974	TIG or MIG welding of aluminium and its alloys
48.	IS 7653 : 1975	Manual blow pipes for welding and cutting
49.	IS 8363 : 1976	Bare wire electrodes for electroslag welding of steel
50.	IS 8484 : 1977	Metal powder welding electrodes
51.	IS 8666 : 1977	Copper and copper alloy covered electrodes for metal arc welding
52.	IS 8736 : 1977	Nickel and nickel alloy covered electrodes for metal arc welding
53.	IS 9495 : 1980	Test for brazeability of brazing alloys

54.	IS 190793 : 1983	Classification of defects in metallic fusion welds with explanations
55.	IS 10803 : 1984	Method of sampling and preparation of weld pad for chemical analysis of weld metal from covered electrodes for manual metal arc welding
56.	IS 10811 : 1984	Oxygen and acetylene manifold regulators for welding, cutting and related processes
57.	Is 10842 : 1984	Testing and evaluation procedure for *Y* goove crackability test
58.	IS 11368 (PT 1) : 1985	Comparison of Indian and Overseas classification and coding of welding filler materials. Part 1 : Flux coated mild steel and medium tensile steel electrodes for manual metal arc welding
59.	IS 11802 : 1986	Method of determination of diffusible hydrogen content of deposited weld metal from covered electrodes in welding mild and low alloy steels
60.	IS 13007 : 1991	Tungsten electrodes for inert gas shielded arc welding and for plasma cutting and welding
61.	IS 13043 : 1991	Covered manual metal arc welding electrodes. Determination of efficiency, metal recovery and deposition coefficient.

Section 2 : Welding Practices for Fabrication, Repair, Surfacing, and Thermal Cutting of Metals, Electroslag Welding, etc.

S. No.	Standard No.	Title
1.	SP :	Handbook of manual metal arc welding for welders
2.	IS 816 : 1969	Code of practice for use of metal arc welding for general construction in mild steel
3.	IS 819 : 1957	Code of practice for resistance spot welding for light assemblies in mild steel
4.	IS 1024 : 1979	Code of practice for use of welding bridges and structures subject to dynamic loading
5.	IS 1261 : 1959	Code of practice for seam welding in mild steel
6.	IS 2751 : 1979	Recommended practice for welding of milkd steel plain and deformed bars for reinforced construction
7.	IS 2811 : 1987	Recommendations for manual tungsten inert gas arc welding stainless steel
8.	IS 2812 : 1964	Recommendations for manual tungsten inert gas arc welding of aluminium alloys
9.	IS 3023 : 1965	Recommended practice for building up by metal spraying
10.	IS 3525 : 1983	Code of practice for use of metal arc welding for hull construction of merchant ships in mild steel
11.	IS 4353 : 1967	Recommendations for submerged arc welding of mild steel and low alloy steels
12.	IS 5139 : 1969	Recommended procedure for repair of grey iron castings by oxy-acetylene and manual metal arc welding

13.	IS 5530 : 1987	Recommendations for production, rectification and repair of steel castings by metal arc welding process
14.	IS 6916 : 1973	Code of practice for fabrication welding of steel castings
15.	IS 8987 : 1978	Recommended practice for welding of clad steels
16.	IS 9006 : 1978	Recommended practice for welding of clad steels
17.	IS 9417 : 1989	Recommendations for welding cold worked bars for reinforced concrete construction
18.	IS 9595 : 1980	Recommendations for metal arc welding of carbon and carbon-manganese steels
19.	IS 19178 : 1982	Recommended procedure for carbon dioxide gas shielded metal arc welding of structural steels
20.	IS 10186 : 1982	Recommendations for manual tungsten inert gas arc welding of copper and copper alloys
21.	IS 10234 : 1982	Recommendations for general pipeline welding
22.	IS 10801 : 1984	Recommended procedure for heat treatment of welded fabrications
23.	IS 11991 : 1986	Recommended practice for flash butt welding of tubes, rods and other sections in carbon and alloy steels
24.	SP : 6 (7) : 1972	Handbook for structural engineers—Simple welded girders
25.	SP : 12 : 1975	ISI Handbook for gas welders

Section 3 : Welding Processes Other than Arc Welding

S. No.	Standard No.	Title
1.	IS 1323 : 1982	Code of practice for oxy-acetylene welding for structural work in mild steels
2.	IS 6409 : 1971	Code of practice for oxy-acetylene flame cleaning
3.	IS 8002 : 1976	Recommended procedure for welding of flexible PVC
4.	IS 8004 : 1976	Recommended procedure for welding of rigid PVC
5.	IS 8455 : 1977	Recommended procedure for welding of polyethylene
6.	IS 9524 : 1980	Recommended procedure for repair welding of steel components by aluminothermic process

Section 4 : Testing of Weldments

S. No.	Standard No.	Title
1.	IS 1182 : 1983	Recommended practice for radiographic examination of fusion welded butt joints in steel plates
2.	IS 2417 : 1977	Glossary of terms relating to ultrasonic testing
3.	IS 2478 : 1991	Glossary of terms relating to industrial radiography
4.	IS 2595 : 1978	Code of practice of radiographic testing
5.	IS 2598 : 1966	Safety code for industrial radiographic practice
6.	IS 2953 : 1985	Glossary of terms used for interpretation of welds and castings radiographs
7.	IS 3415 : 1980	Glossary of terms used in magnetic particle flaw detection

8.	IS 3657 : 1978	Radiographic image quality indicators
9.	IS 3658 : 1981	Code of practice for liquid penetrant flaw detection
10.	IS 3664 : 1981	Code of practice for ultrasonic pulse echo testings by contact and immersion methods
11.	IS 3703 : 1980	Code of practice for magnetic particle flaw detection
12.	IS 4260 : 1986	Recommended practice for ultrasonic testing of butt welds in ferritic steel
13.	IS 4853 : 1982	Recommended practice for radiographic inspection of fusion welded butt joints in steel pipes
14.	IS 4904 : 1990	Calibration blocks for use in ultrasonic non-destructive testing
15.	IS 5334 : 1981	Code of practice for magnetic particle flaw detection of welds
16.	IS 6410 ; 1971	Specification for magnetic flaw detection inks and powders
17.	IS 7343 : 1986	Code of practice for ultrasonic testing of ferrous welded pipes and tubular products
18.	IS 7810 : 1975	Code of practice for radiographic examination of resistance spot welds on aluminium and its alloys
19.	IS 9902 : 1982	Recommended practice for leak testing
20.	IS 10543 : 1983	Method for dry powder magnetic particle testing
21.	IS 12710 : 1989	Glossary of terms used in acoustic emission testing
22.	IS 12965 : 1990	Glossary of terms used in electromagnetic (eddy current) testing

Section 5 : General Standards Related to Welding

S. No.	Standard No.	Title
1.	2879 : 1975 (1992)	Mild steel for metal arc welding electrode core wire (second revision)
2.	IS 7739	Code of practice for preparation of metallurgical specimens
3.	IS 7739 (PT 3) : 1975	Aluminium and its alloys and their examination
4.	IS 7739 (PT 4) : 1975	Copper and its alloys and their examiantion
5.	IS 7739 (PT 5) : 1975	Iron and steel and their examiantion
6.	IS 7739 (PT 8) : 1975	Nickel and its alloys and their examination

(B) BRITISH STANDARDS (BS)

General

S. No.	Standard No.	Title
1.	BS 499	Welding Terms and Symbols
	Part 1 : 1991	Glossary for welding, brazing and thermal cutting
	Part 2 : 1980	Specification for symbols for welding
	Part 2C : 1980	Wallchart of welding symbols

2.	BS 679 : 1989	Specification for filters, cover lenses and backing lenses for use during welding and similar operations
3.	BS 1542 : 1982	Specification for equipment for yee, face and neck protection against radiation arising during welding and similar operations.
4.	BS 2653 : 1955	Specification for protective clothing for welders
5.	BS 1140 : 1980	Specification for resistance spot welding of uncoated and coated low carbon steel
6.	BS 1723	Brazing
	Part 1 : 1986	Specification for brazing
	Part 2 ; 1986	Guide to brazing
	Part 3 : 1988	Methods for non-destructive and destructive testing
	Part 4 : 1988	Methods for specifying brazing procedure and operator approval testing
7.	BS 1724 : 1990	Specification for bronze welding
8.	BS 1821 : 1982	Specification for class I oxy-acetyleen welding of ferritic steel pipe work for carrying fluids
9.	BS 2630 : 1982 (1991)	Specification for ressitance projection welding of uncoated low carbon steel sheet and strip using embossed projections
10.	BS 2633 : 1987	Specification for class I are welding of ferritic steel pipework for carrying fluids
11.	BS 2640 : 1982	Specification for class II oxy-acetylene welding of carbon steel pipe work for carrying fluids
12.	BS 2971 : 1991	Specification for class II are welding of carbon steel pipe work for carrying fluids
13.	BS 2996 : 1958	Specification for the projection welding of low carbon steel studs, bosses, bolts, nuts, annular rings
14.	BS 3019	TIG Welding
	Part 1 : 1984	Specification for TIG welding of aluminium, magnesium and their alloys
15.	BS 3571	MIG Welding
	Part 1 : 1985	Specification for MIG welding of aluminium and aluminium alloys
16.	BS 4204 : 1989	Specification for flash welding of steel tubes for pressure applications
17.	BS 4515 : 1984	Specification for welding of steel pipelines on land and offshore
18.	BS 4570 : 1985	Specification for fusion welding of steel castings
19.	BS 4677 : 1984	Specification for arc welding of austentic stainless steel pipe work for carrying fluids
20.	BS 5135 : 1984	Specification for arc welding of carbon and carbon manganese steel
21.	BS 6223 : 1990	Specification for friction welding of joints in metals
22.	BS 6265 : 1982 (1991)	Specification for resistance seam welding of uncoated and coated low carbon steel

23.	BS 6944 : 1988	Specification for flash welding of butt joints in ferrous metals (excluding pressure piping applications)
24.	BS 6990 : 1989	Code of practice for welding on steel pipes containing processes fluids or their residuals
25.	BS 7123 : 1989	Specification for metal-arc welding of steel for concrete reinforcement
26.	BS 7475 : 1991	Specification for fusion welding of austenitic stainless steel
Equipment		
27.	BS 638	Arc Welding Power Sources, EQuipment and Accessories
	Part 1 : 1979	Specification for oil cooled power sources for manual, smei-automatic and automatic metal-arc welding and for TIG welding
	Part 2 : 1979	Specification for air cooled power sources for manual metal-arc welding with covered electrodes and for TIG welding
	Part 3 : 1979	Specification for air cooled power sources for semi-automatic and automatic metal arc welding
	Part 4 : 1979	Specification for welding cables
	Part 5 : 1988	Specification for accessories
	Part 6 : 1984	Specification for safety requirements for cosntruction
	Part 7 : 1984	Specification for safety requirements for installation and use.
	Part 8 : 1984	Specifciation for electrode holders and hand held torches and guns for MIG, MAG and TIG welding
	Part 9 : 1990	Specification for power sources for manual arc welding with limited duty
	Part 10 : 1990	Specification for safety requiremnts for arc welding equipment : welding power sources
28.	BS 807 : 1955	Specification for spot weldign electrodes.
29.	BS 1389 : 1986	Specification for hose connections and hose assemblies for equipment for gas welding, cutting and related processes
30.	BS 3065 : 1982	Specification for rating of resistance welding equipment
31.	BS 3067 : 1984	Specification for dimensions of seam welding wheel blanks
32.	BS 3856 : 1986	Specificationf or slots in platens for projection welding machines
33.	BS 4215	Specification for Resistance Spot Welding Electrodes, Eelctrode Holders and Ancillary Equipment
	Part 2 : 1987	Specification for straight resistance spot welding eletrodes
	Part 3 : 1987	Specification for the dimensions of electrode taper fits
	Part 4 : 1990	Specification for taper plug gauges and taper rign gauges
	Part 5 : 1987	Specification for electrode back-ups and clamps
	Part 6 : 1987	Specification for electrode adaptors, female taper 1 : 10

	Part 7 : 1987	Specification for male electrode caps
	Part 8 : 1987	Specification for insulation caps and bushes
	Part 9 : 1990	Specification for electrode adaptors, male taper 1 : 10, conical fixing 1 : 10
	Part 10 : 1990	Specification for electrode adaptors, male taper 1 : 10, parallel shank fixing for end thrust electrodes
	Part 11 : 1990	Specification for electrode holders, taper fixing 1 : 10
	Part 12 : 1990	Specification for electrode holders, Morse taper fixing
	Part 13 : 1990	Specification for electrode holders, parallel shank fixing for end thrust
	Part 14 : 1990	Specification for cooling tubes
	Part 15 : 1991	Specification for insulated pins for use in electrode back-ups
34.	BS 4577 : 1970	Specification for materials for resistance welding electrodes and ancillary equipment
35.	BS 4819	Specification for Resistance Welding Water-Cooled Transformers for Press-Package and Portable Types
	Part 1 : 1989	Specification for press-package transformers
	Part 2 : 1989	Specification for portable transformers
36.	BS 5120 : 1987	Specification for rubber hoses for gas welding and allied processes
37.	BS 5741 : 1979	Specification for pressure regulators used in welding, cutting and related processes
38.	BS 5924 : 1989	Specification for safety requirements for the construction and installation of equipment for resistance welding and allied proceses
39.	BS 6158 : 1982	Specification for safety devices for fuel gases and oxygen or compressed air for welding, cutting and related processes
40.	BS 6503 : 1984	Specification for handheld blowpipes, mixers and nozzles, using fuel gas and oxygen, for gas welding, cutting and related processes
41.	BS 6572 : 1986	Specification for Bourdon tube pressure gauges used in welding, cutting and related processes
42.	BS 6942	Design and Construction of Small Kits for Oxy-Fuel Gas Welding and Allied Processes
	Part 1 : 1988	Specification for kits using one or more non-refillable gas containers for oxygen and fuel gas
	Part 2 : 1989	Specification for kitss using refillable gas containers for oxygen and fuel gas
43.	BS 7125 : 1989	Specification for transformers for resistance welding machines
44.	BS 7278 : 1990	Specification for gas tightness of equipment for gas welding and allied processes
45.	BS 7279 : 1990	Specification for graphical symbols for resistance welding equipment

46.	BS 7329 : 1990	Specification for materials for equipment used in gas welding, cuttng and allied processes
47.	BS 7418 : 1991	Code of practice for isolation of the welding circuit in arc welding plant
Welding Consumables		
48.	BS 639 : 1986	Specificaton for covered carbon and carbon manganese steel electodes for manual metal arc welding
49.	BS 1453 : 1972 (1987)	Specification for filler materials for gas welding
50.	BS 1845 : 1984	Specification for filler metals for brazing
51.	BS 2493 : 1985	Specification for low alloy steel electrodes for manual metal-arc welding
52.	BS 2901	Specification for Filler Rods and Wires for Gas-Shielded Arc Welding
	Part 1 : 1983	Ferritic steels
	Part 2 : 1990	Specification for stainless steels
	Part 3 : 1990	Specification for copper and copper alloys
	Part 4 : 1990	Specification for aluminium and aluminium alloys and magnesium alloys
	Part 5 : 1990	Specification for nickel and nickel alloys
53.	BS 2926 : 1984	Specification for chromium and chromium-nickel steel electrodes for manual metal-arc welding
54.	BS 4165 : 1984	Specification for electrodes, wires and fluxes for the submerged arc welding of carbon steel and medium-tensile steel
55.	BS 5465 : 1987	Specification for electrodes and fluxes for the submerged arc welding of austentic stainless steels
56.	BS 6693	Diffusible Hydrogen
	Part 1 : 1986	Method for determination of hydrogen in manual metal-arc weld metal using 3 day collection
	Part 2 : 1986	Method for determining of hydrogen in manual arc weld metal
	Part 3 : 1988	Primary method for the determination of diffusible hydrogen in manual metal-arc ferritic steel weld metal
	Part 4 : 1988	Primary method for the determination of difusible hydrogen in submerged-arc ferritic steel weld metal
	Part 5 : 1988	Primary method for the determination of diffusible hydrogen in MIG, MAG, TIG or cored electrode steel weld metal
57.	BS 6787 : 1987	Method for determination of ferritic number in austenitic weld metal deposited by covered Cr-Ni steel electrodes.
58.	BS 7084 : 1991	Specification for carbon and carbon-manganese steel tubular cored welding electrodes
59.	BS 7384 : 1991	Guide to laboratory methods for sampling and analysis by particulate matter generated by arc welding consumables

60.	BS EN 26848 : 1991	Specification for tungsten electrodes for inert gas shielded arc welding and for plasma cutting and welding
Testing and Inspection		
61.	BS 709 : 1983	Methods of destructive testing fusion wleded joints and weld metal in steel
62.	BS 1295 : 1987	Code of practice for training in arc welding skills
63.	BS 2600	Radiographic Examination of Fusion Welded Butt Joints in Steel
	Part 1 : 1983	Methods for steel 2 mm up to and including 50 mm thick
	Part 2 ; 1973 (1983)	Methods for steel over 50 mm up to and including 200 mm thick
64.	BS 2704 : 1978 (1983)	Specification for calibration blocks for use in ultrasonic falw detection
65.	BS 2910 : 1986	Methods for radiographic examination of fusion welded circumferential butt joints in steel pipes
66.	BS 3451 : 1973 (1981)	Methods of testing fusion wleds in aluminium and aluminium alloys
67.	BS 3923	Methods for Ultrasonic Examination of Welds
	Part 1 : 1986	Methods for manual examination of fusion welds in ferritic steels
	Part 2 : 1972	Automatic examination of fusion welded butt joints in ferritic steels
68.	BS 3971 : 1980 (1985)	Specification for image quality indicators for industrial radiography (including guidance on their use)
69.	BS 4069 : 1982	Specification for magnetic flaw detection inks and powders
70.	BS 4129 : 1990	Specification for welding primers and weld-throguh sealant, adhesives and waxes for resitance welding of sheet steel
71.	BS 4206 : 1967	Methods of testing fusion welds in copper and copper alloys
72.	BS 4331	Methods for Assessing the performance Charactersitics of Ultrasonic Flaw Detection Equipment
	Part 1 : 1978 (1989)	Overall performance : on-site methods
	Part 2 : 1972 (1987)	Electrical performance
	Part 3 : 1974 (1987)	Guidance on the in-service monitoring of probes (excluding immersion probles)
73.	BS 4870	Specification for Approval Testing of Welding Procedures
	Part 1 : 1981	Fusion welding of steel
	Part 2 : 1982	TIG or MIG welding of aluminium and its alloys
	Part 3 : 1985	Arc welding of tube to tube-plate joints in metallic materials
	Part 4 : 1988	Specification for automatic fusion welding of metallic materials, including welding operator approval

74.	BS 4871	Specification for Approval Testing of Welders Working to Approved Welding Procedures
	Part 1 : 1982	Fusion welding of steel
	Part 2 : 1982	TIG or MIG welding of aluminium and its alloys
	Part 3 : 1985	Arc welding of tube to tube-plate joints in emtallic materials
75.	BS 4872	Specification for Approval Testing of Welders when Welding Procedure Approval is not Required
	Part 1 : 1982	Fusion welding of steel
	Part 2 : 1976	TIG or MIG welding of aluminium and its alloys
76.	BS 5044 : 1973 (1987)	Specification for contrast aid paints used in magnetic particle flaw detection
77.	BS 5289 : 1976 (1983)	Code of practice for visual inspection of fusion wedded joints
78.	BS 6072 : 1981 (1986)	Method for magnetic particle flaw detection
79.	BS 6084 : 1981	Method of test for comparison of prefabrication primers by porosity rating in arc welding
80.	BS 6443 : 1984	Method for penetrant flaw detection
81.	BS 6691	Fume from Welding and Allied Processes
	Part 1 : 1986	Guide to methods for the sampling and analysis of particulate matter
	Part 2 : 1986	Guide to methods for the sampling and analysis of gases
82.	BS 7009 : 1988	Guide to the application of real-time radiography to weld inspection
83.	BS 7257 : 1989	Methods for radiographic examination of fusion welded branch and nozzle joints in steel.
84.	BS 7363 : 1990	Methods for controlled thermal severity (CTS) test and bead-on-plate (BOP) test for welds
85.	M 42 : 1972 (1983)	Methods for non-destructive testing of fusion and resistance welds in thin gauge materials
86.	PD 6493 : 1991	Guidance on methods for assessing the defects.

(C) AMERICAN WELDING SOCIETY AWS–STANDARDS

S. No.	Standard No.	Title
1.	A1. 1-89	Metric Practice Guide for the Welding Industry
2.	A2. 1-86	Welding Symbols Chart
3.	A2. 4-86	Standard Symbols for Welding, Brazing and Nondestructive Examination
4.	A3. 0-89	Standard Welding Terms and Definitions
5.	A4. 2-91	Calibrating Magnetic Instruments to Measure the Delta Ferrite Content of Austenitic Stainless Stel Weld Metal, Standard Procedure

6.	A4. 3-86	Determination of Diffusible ydrogen Content of Martensitic, Banitic and Ferritic Steel Weld Metal Produced by Arc Welding, Standard Procedure
7.	A5. 01-87	Filler Metal Procurement Guidelines
8.	A5. 1-91	Specification for Carbon Steel Electrodes for Shielded Metal Arc Welding
9.	A5. 2-88	Carbon and Low Alloy Steel Rods for Oxyfuel Gas Welding, Specification
10.	A5. 3-88	Aluminium and Aluminium Alloy Electrodes for Shielded Metal Arc Welding, Specification
11.	A5. 4-81	Covered Corrosion-Resisting Chromium and Chromium-D Nickel Steel Welding Electrodes, Specification
12.	A5. 5-81	Low alloy Steel Covered Arc Welding Electrodes, Specification
13.	A5. 6-84	Covered Copper and Copper Alloy Arc Welding Electrodes, Specification
14.	A5. 7-84	Copper and Copper Alloy Bare Welding Rods and Electrodes, Specification
15.	A5. 8-89	Filler Metals for Brazing, Specification
16.	A5. 9-81	Corrosion Resisting Chromium and Chromium-Nickel Steel Bare and Composite Metal Cored and Standard Welding Electrodes and Welding Rods, Specification
17.	A5. 10-88	Bare Aluminium and aluminium Alloy Welding Electrodes and Rods, Specification
18.	A5. 11-90	Nickel and Nickel Alloy Wleding Electrodes for Shielded Metal Arc Welding, Specification
19.	A5. 12-80	Tungsten Arc Welding Elecetrodes, Specification for,
20.	A5. 13-80	Solid Surfacing Wleding Rods and Electrodes, Specification
21.	A5. 14-89	Nickel and NIckel Alloy Bare Welding Electrodes and Rods, Specification
22.	A5. 15-90	Welding Electrodes and Rods for Cast Iron, Specification
23.	A5. 16-90	Titanium and Titanium Alloy Welding Electrodes and Rods, Specification
24.	A5. 17-89	Carbon Steel Electrodes and Fluxes for Submerged ARc Welding, Specification
25.	A5. 18-79	Carbon Steel Filler Metals for Gas Shielded Arc Welding, Specification
26.	A5. 19-80	Magnesium Alloy Welding Rods and Bare Electrodes, Specification
27.	A5. 20-79	Carbon Steel Electrodes for Flux Cored Arc Welding, Specification
28.	A5. 21-80	Composite Surfacing Welding Rods and Electrodes, Specification
29.	A5. 22-80	Flux Cored Corrosion-Ressiting Chromium and Chromium-Nickel Steel Electrodes, Specification

30.	A5. 23-90	Low Alloy Steel Electrodes and Fluxes for Submerged Arc Welding, Specification
31.	A5. 24-90	Zirconium and Zirconium Alloy Welding Electrodes and Rods, Specification
32.	A5. 25-91	Specification for Carbon and Low Alloy Steel Electrodes and Fluxes for Electroslag Welding
33.	A5. 26-78	Specification for Carbon and Low Alloy Steel Electrodes for Electrogas Welding
34.	A5. 27-85	Copper and Copper Alloy Rods for Oxyfuel Gas Welding, Specification
35.	A5. 28-79	Low Alloy Steel Filler Metals for Gas Shielded Arc Welding, Specification
36.	A5. 29-80	Low Alloy Steel Filler Metals for Gas Shielded Arc Welding, Specification
37.	A5. 30-79	Consumable Inserts, Specification for, (Mild Steel, Cr-Mo Steel, Cr-Ni Stainless Steel, Nickel Alloys and Cu-Ni Alloys)
38.	B1. 10-86	Guide for the Nondestructive Inspection of Welds, (Visual, Penetrant, Magnetic Particle, Radiography, Ultrasonic, and Eddy Current Inspection)
39.	B1. 11-88	Guide for the Visual Inspection of Welds
40.	B2. 1-84	Welding Procedure and Performance Qualification Standard
41.	B2.1.001-90	Shielded Metal Arc Welding of Carbon Steel (M-1/P-1, Group 1 or 2), 3/16 through 3/4 inch, in the As-Welded Condition, with Backing, Standard Welding Procedure Specification
42.	B2.1.002-90	Gas Tungsten Arc Welding of Carbon Steel (M-1/P-1, Group 1 or 2), 3/16 through 7/8 inch, in the As-Welded Condition, With or Without Backing, Standard Welding Procedure
43.	B2.1.003-90	Gas Metal Arc Welding of Galvanized Steel, 10 through 18 Gage, in the As-Welded Condition, With or Without Backing, Standard WeldingProcedure Specification
44.	B2.1.004-90	Gas Metal Arc Welding of Carbon Steel, (M-1, Group 1), 10 through 18 Gage, in the As-Welded Condition, With or Without Backing, Standard Welding Procedure Specification
45.	B2.1.005-90	Gas Metal Arc Welding of Austenitic Stainless Steel, (M-8 or P-8), 10 throguh 18 Gage, in the As-Welded Condition, With or Without backing, Standard Welding Procedure Specification
46.	B2.1.006-90	Gas Metal Arc Welding of Carbon STeel to Austenitic Stainless STeel, (M-1 to M-8 or P-8), 10 through 18 Gage in the As-Welded Condition, With or Without Backing, Standard Welding Procedure

47.	B2.1.007-90	Gas Tungsten Arc Welding of Galvanized STeel, 10 through 18 Gage, in the As-Welded Condition, With or Without Backing, Standard Welding Procedure Specification
48.	B2.1.008-90	Gas Tungsten Arc Welding of Carbon Steel, (M-1, Group 1), 10 through 18 Gage, in the As-Welded Condition, With or Without Backing, Standard Welding Procedure
49.	B2.1.009-90	Gas Tungsten Arc Welding of Austenitic Stainless Steel, (M-8/P-8), 10 through 18 Gage, in the As-Welded Condition, With or WithoutBacking, Standard Welding Procedure
50.	B2.1.010-90	Gas Tungsten Arc Welding of Carbon Steel to Austenitic Stainless Steel, (M-1 to M-8 or P-8), 10 throguh 18 Gage, in the As-Welded Condition, With or Without Backing, Standard Welding Procedure
51.	B2.1. 2-91	Brazing Procedure and Performance Qualification, Standard
52.	B4. 0-85	Mechanical Testing of Welds, Standard Methods for, (Bend, Tension and Fracture Toughness Testing as well as Fillet Weld Break and Shear Testing)
53.	C1. 1-66	Resistance Welding, Recommended Practices
54.	C1. 3-70	Resistance Welding Coated Carbon Steels, Recommended Practices
55.	C2. 2-67	Metalizing with Aluminium and Zinc for Protection of Iron and Steel, Recommended Practices
56.	C2. 14-74	Corrosion Tests for Flame Sprayed Coated Steel, 19-Year Report,
57.	C2. 16-78	Thermal Spray Operator and Equipment Qualification, Guide
58.	C3. 2-82	Strength of Brazed Joints in Shear, Standard Method for Evaluating
59.	C3. 3-80	Design, Manufacture and Inspection of Critical Brazed Components, Recommended Practices
60.	C3. 4-90	Specification for Torch Brazing
61.	C3. 5-90	Specification for Induction Brazing
62.	C3. 6-90	Specification for Furnance Brazing
63.	C3. 8-90	Recommended Practices for Ultrasonic Inspection of Brazed Joints
64.	C4. 1-G	Oxygen Cutting Surface Roughness Gauge
65.	C4. 1-WC	Criteria for Describing Oxygen-Cut Surfaces
66.	C4. 2-90	Operator's Manual for Oxyfuel Gas Cutting
67.	C4. 3-83	Gas Heating Torch Operation, Operator's Manual
68.	C5. 1-73	Plasma Arc Welding, Recommended Practices
69.	C5. 2-83	Plasma Arc Welding, Recommended Practices for,
70.	C5. 3-91	Recommended Practices for Air Carbon Arc Gouging
71.	C5. 4-84	Stud Welding, Recommended Practices

72.	C5. 5-80	Gas Tungsten Arc Welding, Recommended Practices
73.	C5. 6-89	Gas Metal Arc Welding, Recommended Practices
74.	C5. 7-89	Electrogas Welding, Recommended Practices
75.	C6. 1-89	Friction Welding, Recommended Practices
76.	D1. 1-90	Structural Welding Code-Steel,
77.	D1. 2-90	Structural Welding Code-Aluminium,
78.	D1. 3-89	Structural Welding Code-Sheet Steel,
79.	D1. 4-79	Structural Welding Code-Reinforced Steel,
80.	D1. 5-88	Bridge Welding Code
81.	D3. 5-85	Steel Hull Welding, Guide
82.	D3. 6-89	Underwater Welding, Specification
83.	D3. 8-90	Aluminium Hull Welding, Guide
84.	D5. 2-84	Welded Steel Elevated Tanks, Standpipes, and Reservoirs for Water Storage, Standard
85.	D8. 5-66	Automotive Portable Gun Resistance-Spot Welding, Recommended Practices
86.	D8. 7-88	Automotive Frame Weld Quality-Resistance Spot Welding, Recommended Practices
87.	D8. 8-89	Automotive Frame Weld Quality-Arc Welding, Specification
88.	D9. 1-90	Sheet Metal Welding Code
89.	D10. 4-86	Austenitic Chromium-Nickel Stainless STeel Piping and Tubing, Recommended Practices
90.	D10. 6-91	Titanium Piping and Tubing, Gas Tungsten Arc Welding of, Recommended Practices
91.	D10. 7-86	Chromium-Molybdenum Steel Piping and Tubing, Recommended Practices
92.	D10. 8-86	Chromium-Molybdenum Steel Piping and Tubing, Recommended Practices
93.	D10. 9-80	Qualification of Welding Procedures and Welders for Piping and Tubing, Specification
94.	D10. 10-90	Recommended Practices for Local Heating of Welds in Piping and Tubing
95.	D10. 11-87	Recommended Practices for Root Pass Welding of Pipe Without backing
96.	D10. 12-89	Welding Carbon Steel Pipe, Recommended Practices for,
97.	D11. 2-89	Guide for Welding Iron Castings
98.	D14. 1-85	Industrial and Mill Crane and Other Material Handling Equipment, Specification for Welding,
99.	D14. 2-86	Metal Cutting Machine Tool Weldments, Specification
100.	D14. 3-82	Earthmoving and Construction Equipment, Specification for Welding,
101.	D14. 4-77	Machinery and Equipment, Classification and Application of Welded Joints
102.	D14. 5-80	Presses and Press Components, Specification for Welding

103.	D14. 6-81	Rotating Elements of EQuipment, Specification
104.	D15. 1-86	Railroad Welding Specification
105.	F1. 1-85	Sampling Airborne Particulates Generated by Welding and Allied Processes, Methods
106.	F1. 2-87	Measuring Fume Generation Rates and Total Fume Emission for Welding and Allied Processes, Laboratory Method
107.	F1. 3-91	Evaluating Contaminants in Welding Environment/A Sampling Strategy Guide
108.	F1. 4-87	Methods for Analysis of Airborne Particulates Generated by Welding and Allied Processes
109.	F1. 5-87	Methods for Sampling and Analyzing Gases from Welding and Allied Processes
110.	F2. 1-78	Electron Beam Welding and Cutting, Recommended Safe Practices
111.	F2. 2-89	Lense Shade Selector
112.	F3. 1-89	Welding Fume Control, Guide
113.	F4. 1-88	Preparation for Welding and Cutting of Containers that Have Held Hazardous Substances, Recommended Safe Practices
114.	F6. 1-78	Sound Level measurement of Manual Arc Welding and Cutting Processes, Method
115.	Zs49. 1-88	Safety in Welding and Cutting
116.	AWN	Arc Welding and Cutting Noise
117.	AWS	Arc Welding Safety
118.	FGW	Fumes and Gases in Welding Environemnt

❑

APPENDIX–8
WORKSHOP EXERCISES

A. Fillet Weld Lap Joint (Horizontal Position)

Objective : To develop skill for producing quality fillet welds on the lap joint in the horizontal position.

Material : 2.5 mm. mild steel plates –2 pcs.
Filler rod : AWS RG–45 (2 mm)
Gas pressure setting :
Oxygen : 0.4 kg/cm^2
Acetylene : 0.4 kg/cm^2
Tip of size : No. 4

Setting of plates : Set the plates one over teh other so as to cover half the width of the plate on each side and tack both ends of the plates without filler wire as shown in fig. 1 (a).

(i) Place the work piece on five bricks. Position the torch under a 90° work angle.
(ii) Position the torch with a 90° work angle and a 40°–45° travel angle, push at the root of the joint.
(iii) Hold the filler rod at 45° in opposite direction to the torch, Fig. 1 (b).
(iv) Filler should be given on teh ege of the upper plate to assure complete fill on top side and to prevent overfill on bottom plate.
(v) Be sure to melt out 3 m.m. of top plate to assure complete penetration.
(vi) Weld bead should have equal legs evenly rippled and penetrate into both plates.

Note : For heavy steel plates the torch travel angle will be around 80°–85°.

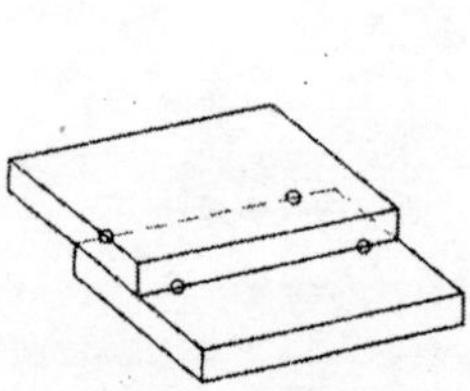

(a) Tack welding of plates for lap joint

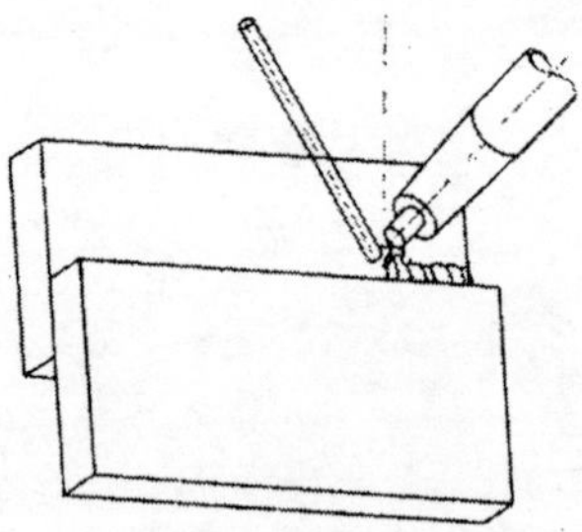

(b) Relative positions of torch and filler rod

Fig. 1

B. Square Butt Joint-Flat Position

Obejctive : To develop skill to produce quality welds with comlete penetration in the flat position. Key hole technique to be used for achieving penetartion.

Material : 2.5 m.m. mild steel plates–2 pieces.
Filler metal : RG–45, Q. 2 m.m.

Gas pressure setting :

Oxygen : 0.4 kg/cm ^{2}

Acetyleen : 0.4 kg/cm ^{2}

Tip size : No. 4

(a) Position two plates to form butt joint (Fig. 2)

(i) Set square butt joint with 3 m.m. root gap using 3 m.m. spacer wire.

(ii) Tack at one end of the work piece.

(iii) After tack welding the root opening will be larger due to expansion. Realign the joint with 4.5 m.m. root gap and tack the other end (Fig. 2a).

(iv) Place the work piece on the fire bricks.

(b) Deposit weld bead

(i) Position the torch with 90° work angle and 40°–45° travel angle.

(ii) Hold filler wire at 40° from vertical with similar work angle (Fig. 2–b, c,).

(iii) Preheat plates and start 5 m.m. behind the tack weld at the right end of the joint.

(iv) Move the torch evenly along adding filler metal to the leading edge of the weld puddle.

(v) Be sure that the torch is centered over the sqaure groove.

(vi) Prior to fusing into the tack weld, move the torch into circular motion to improve tie in. (Fig. 2-d).

(vii) After the edges of the tack weld and the plates have been sufficiently melted to form a small hole, add filler metal until the proper bead size is attained.

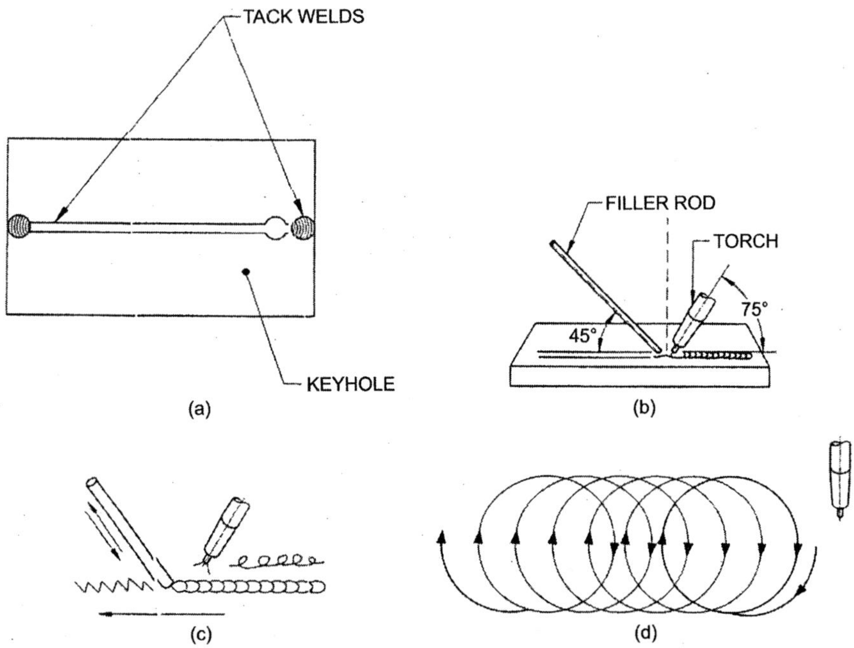

Fig. 2 Setting up plates for square butt weld.

(viii) The weld face should be 2 m.m. wider than the groove and should be flat.

(ix) Weld should complely fuse into sides of joint and completely penetrate to form a root reinforcement of 1.5 m.m.

(x) If the weld bead is excessively concave, add a secnd pass to completely fill the joint.

C. Single 'V' groove butt joint (Flat, horizontal, vertical and overhead positions)

Objective : To develop skill to weld in 'V' groove butt joints in flat, horizontal, vertical and overhead positions.

Material : 3.5 m.m. mild steel plates–2 pcs.
Filler rod : RG45 (ϕ 3.15 mm)
Gas pressure setting :
Oxygen : 0.5 kg/cm^2
Acetylene : 0.5 kg/cm^2
Tip size : No. 4

Preparation of Material

(i) Bevel the plate edges to 30° by machining or grinding.
(ii) Position two plates to form butt joint with 3.15 m.m. root opening.
(iii) Tack both teh ends of the joints.
(iv) Place work piece on fire bricks in flat position.

Flat downhand (Fig. 3a)

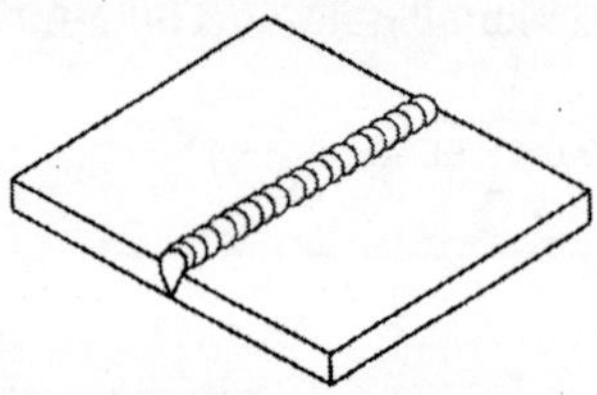

Fig. 3(a) Single 'V' groove butt joint in flat downhand position.

First weld bead deposition :
(i) Position the torch with 90° work angle and 40°–45° travel angle.
(ii) Hold filler rod at 35°–40° angle in opposite direction to the torch.
(iii) Start the weld and move torch evenly on the eges of the plates in forehand progression.
(iv) Add filler rod at the outer edge of molten metal.

Make a key hole, melted out of weld puddle centre to assume good penetration, weld should melt each of joint. Bead should be slightly concave on the back side ensuring complete penetartion.

Second weld bead deposition :
(i) Use the same torch angle.
(ii) Travel in opposite direction (back hand progression).
(iii) Use alternating semi circular motion with torch and filler rod. The rod should be in continual contact with weld puddle.

(iv) When torch is at one side of weld puddle, filler rod should be at opposite edge.

(v) Be sure weld penetrates well into root bead and side of the joint.

Horizontal (Fig. 3 (b))

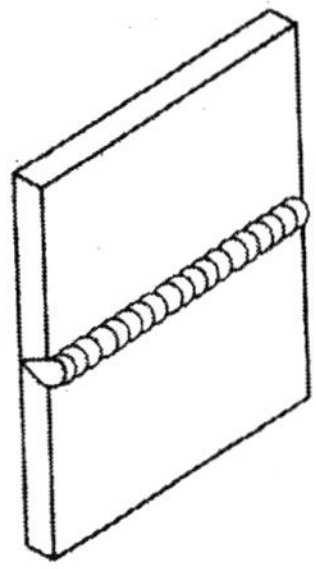

Fig. 3(b) Single 'V' groove butt joint in horizontal position.

First weld bead deposition :

(i) IIold the torch with an 80° 85° work angle and a 5°–10° travel angle, push.

(ii) Hold filler rod 45° opposite from torch and 5° up from horizontal.

(iii) Move torch evenly along joint adding filer metal to the top edge of the joint.

(iv) Travel just fast enough to form key hole and control weld puddle.

Second bead deposition :

(i) Hold the torch and filler rod same as first weld bead and travel back hand.

(ii) Weld puddle must be long and steeply angled from vertical so lower side will form a small shelf to hold up molten metal.

(iii) Rod should be fed slowly into the weld puddle with continuous semicircular movement. Torch should move within lower two thirds of weld puddle in oval motion so molten metal will stay chilled and in plastic state.

(iv) The finished bead face should be slightly convex and approximately 12 m.m. wide.

Vertical position up (Fig. 3-c)

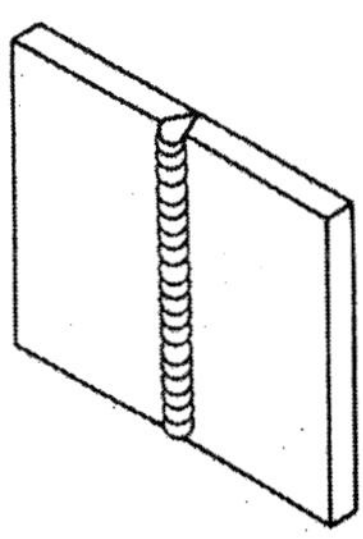

Fig. 3(c) Single 'V' groove butt joint in vertical position up.

(i) Tack weld the plates as earlier done and clamp it on the work table in vertical position.

(ii) Position torch with a 90° work angle and a 20°–30° travel angle, push.
(iii) Hold filler rod 20°–25° in opposite direction with no side angle.
(iv) Start the weld from bottom of the joint and feed the filler rod in and out of the leading edge of puddle.
(v) The finished weld bead should be slightly convex completely penetrate joint and properly fuse into the joint edges.

Overhead position (Fig. 3-d)

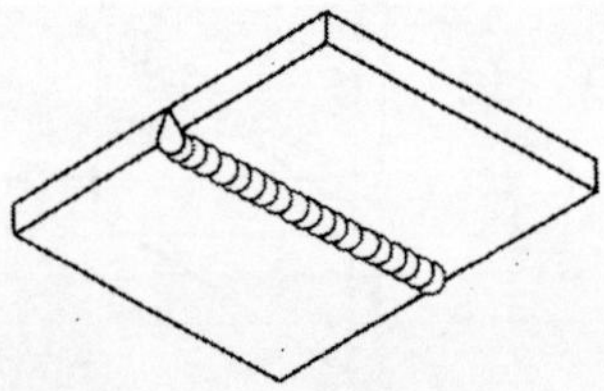

Fig. 3(d) Single 'V' groove butt joint in overhead position.

(i) Tack the plates as earlier done and clamp it on the work table in overhead position.
(ii) Hold the torch with a 90° work angle and a 20° travel angle, push.
(iii) Hold filler rod 45° horizontal from the plate and no side angle.
(iv) Move the torch evenly along the joint adding filler metal to the leading edge of the puddle.
(v) Completely fuse the filler metal into the joint edges and fully penetrate the joint to form a convex bead face.

D. Single 'V' groove butt joint pipe welding horizontal fixed positions

Objective : To develop skill to produce quality welds in the single 'V' pipe butt joint in the horizontally fixed positions.

Material : Carbon steel pipe ϕ 63.5 m.m. × 3.5 m.m.
Filler metal : RG–45, ϕ 2.5 m.m.
Gas settin pressure :
Oxygen and acetylene : 0.5 kg/cm^2 each.
Tip size : No. 6

Prepare work piece

Prepare 35° bevel on the edges of the tube with no root face.

Tack weld work piece

(i) Place two pieces in jig with 2.5 m.m. root opening.
(ii) Use 2.5 m.m. filler rod to make root gap in the pipe joint.
(iii) Deposit 5 m.m. length tack welds at four places in the pipe by giving equal spaces between the tack welds.
(iv) Clamp the assembled pipe joint in the horizontally fixed position with tack welds at 12, 3, 6, 9 O'clock positions (Fig. 4a).

Root bead depostion

(i) Start the weld 10 m.m. back from the edn of the 6 O'clock tack weld.
(ii) Position torch with 35° work angle vertically and without side angle.

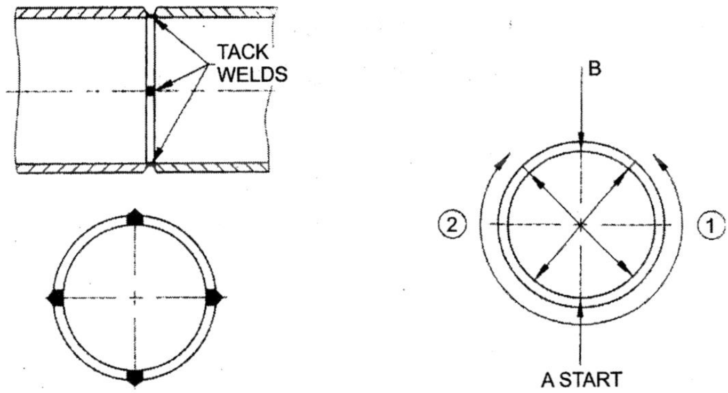

(a) Tack welds in pipe welding (b) Pipe joint horizontal fixed torch movement

Fig. 4 Pipe joint horizontal fixed torch movement.

(iii) Hold filler rod 10° from verticla with no side angle.

(iv) Keep the filler rod behind the inner core of teh flame. The torch should be moved forward and rod should be followed along with the torch flame.

(v) Add filler metal to teh leading edge of the weld puddle as you move the torch evenly along.

(vi) Be sure to maintain a key hole wider than the root opening.

(vii) Use a circular torch motion to properly fuse the edges of the pipe. Teh filler rod should be immersed in the molten metal. The molten metal should be projected slightly inside the tube by pushing the filler rod for getting uniform penetration.

(viii) After reaching the weld to 12 O'clock position, draw back the torch to 6 O'clock again to start weld on the other sideo f the pipe.

(ix) Weld 6 to 9 and 9 to 12 O'clock position as done on the other side, *i.e.*, 6 to 3 and 3 to 12 O'clock. (Fig. 4-b)

(x) The root bead should penetrate the joint and fill almost 2/3rd of the V-groove.

Deposition of the second weld bead

(i) Use same torch and filler rod angles.

(ii) Travel from 6 to 12 O'clock.

(iii) Oscillate the travel using semi circular weaving motion.

(iv) The filler should be held in the centre of the molten metal. The torch should be forward and the rod should follow the torch, *i.e.*, back hand motion should be used.

(v) Weld completely around the pipe.

(vi) The finished face should be slightly convex and about 12 m.m. wide.

APPENDIX–9
WORKSHOP CALCULATION

1.0 Length

Length is defined as the linear physicla charactersitc of a body in a given direction. It is the distance betweeen any two given points along a straight line or along a defined path. The standard unit of a length is mtre.

Example :

(A) In Fig. 1, the length of the weld is 400 m.m.

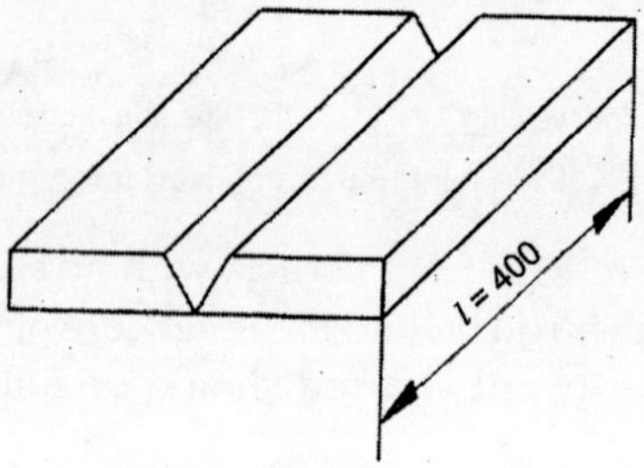

Fig. 1

(B) In Fig. 2, the length of the weld = circumference of the circle with radius R

$$= \pi \times D = \pi \times 2 \times R = \pi \times 2 \times 300$$
$$= 1884.96 \text{ m.m.}$$

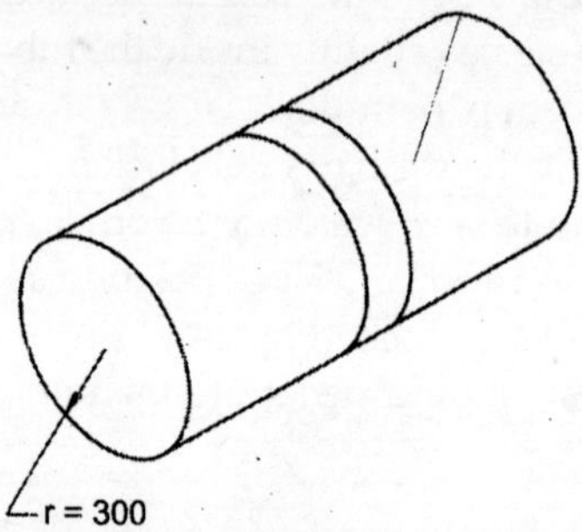

Fig. 2

2.0 Mass, Gravity and Weight

Mass : Mass is a measure of the amount of matter in a body. It may also be considered as a quantitative measure of the inertia of a body, *i.e.*, the property causing resistance to change in the motion of the body. Teh standard unit in MKS system is kg.

Gravity : Earth attracts all the substances towards its centre. This phenomenon is called gravitational pull. This causes all substances to exert a pressure on which it is resting. Due to the gravity, the freely falling bodies gain an acceleration of 9.81 m/sec^2.

Weight : Weight is the force due to gravitational attraction on the mass of a body. The unit of weight is kilograme force.

Here, we must understand that the weight of a body is not the same as its mass. Weight is related to mass as per the following equation :

$$\text{Weight} = \text{Mass} \times \text{acceleration due to gravity}$$

Comparison of mass and weight is given in table 1.

Table 1 : Comparison between mas and weight

Mass	Weight
1. Quantum of material contained in a body.	Force with which the body is pulled towards the earth.
2. Constant at all places.	Varies from place to place.
3. Measured by comparing known masses.	Measured using spring balances or in a common balance.
4. Scalar quantity.	Vector quantity.

3.0 Density and Specific Gravity

Density : Density of a body is its mass per unit volume. Its unit is kilograme per cubic metre.

Specific gravity : Specific gravity of any substance is the ratio between the density of the substance to the density of water. Density of water is 1 gm/cm^3.

Example : Calculate the mass of a steel plate of the dimensions 500 m.m. × 500 m.m × 20 m.m. Assume the density of steel is 7850 kg/m^3.

$$\text{Volume of the plate} = 500 \times 500 \times 20 \text{ m.m.}^3$$

$$= \frac{500}{1000} \times \frac{500}{1000} \times \frac{20}{1000} \text{ m}^3$$

$$= \frac{1}{2} \times \frac{1}{2} \times \frac{1}{50} = \frac{1}{200} \text{ m}^3$$

$$\text{Density of the steel} = 7850 \text{ kg/m}^3$$

Therefore, the mass of the steel plate = volume × density

$$= \frac{1}{200} \text{ m}^3 \times 7850 \text{ kg/m}^3$$

$$= \frac{7850}{200} = 39.25 \text{ kg.}$$

4.0 Time

Time is the measure of the duration between the start and end of any event. The standard unit of time is second.

Example : Calculate the time required for a single pass welding of a 9 m long pipe if the speed of welding is 300 m.m. per minute.

Length of pipe $= 9 \text{ m} = 900 \text{ m.m.}$

Speed of welding $= 300 \text{ m.m./min.}$

So, time required for welding pipe = length/speed of welding

$$= \frac{9000 \text{ m.m.}}{300 \text{m.m./min}} = \frac{9000}{300} \text{ minutes}$$

5.0 Velocity

Velocity is the rate of change of displacement, *i.e.*, distance moved by an object per unit time. It is also called the speed of travel. The unit of velocity is metre per second.

Example : A plate of 6 m length is welded using gas metal arc welding. If the total time taken for welding is 30 min., find the velocity of the welding head (speed of welding).

Length of plate $= 6 \text{ m} = 6000 \text{ m.m.}$

Time taken for welding $= 30$ minutes.

Velocity of welding head, *i.e.*, speed of welding $= \dfrac{6000}{30}$

$= 200$ m.m./minute.

6.0 Acceleration

Acceleration is the rate of change of velocity with time. If the sign is negative then it is called deceleration. The unit of acceleration is meter per seconds.

7.0 Force

Force is defined as the product of mass and acceleration. Force can be caused either by gravitational or by external source. Its unit is kilogram force.

Example : A job having a mass of 5 tons is resting on the machine table. Find the force exerted on the bearing of the rotary table.

Mass of job $= 5 \text{ T} = 5000$ kgs.

Acceleration due to gravity $= 9.81 \text{ m/sec}^2$

Force exerted on the bearing = Mass × acceleration

$= 5000 \times 9.81$

$= 49050$ newtons

As 9.81 newton = 1 kgf

So, force $= \dfrac{49050}{9.81}$

$= 5000$ kgf.

8.0 Momentum

Momentum is the product of mass and velocity. Its unit is kilogram metre per second.

Example : A body of mass 100 kg is moving with a velocity of 5 m/sec, find the momentum of the body.

Mass of the body $= 100$ kgs.

Velocity $= 5$ m/sec.

So, momentum $= m \times v$

$= 100 \times 5 = 500$ kg m/sec.

9.0 Work

The work done is the energy spent to move a body over a distance. This is calculated by the product of force and distance moved. Its unit is kgf-m.

Example : Find the work done if a steel plate is moved over a distance of 500 m.m. by applying a force of 90 kgf.

Work done $= 90 \times 500$ kgf - m. m. $= 45$ kgf-m.

10.0 Power

Power is the work done per unit time. The unit of power is watts or kilowatts. Electric power is given by $V \times I$, where V is the voltage and I is the current in amperes.

Example : Calculate the power required by an electric arc during welding, operating at 20 volts adn 150 amps.

$$\begin{aligned} \text{Power} &= V \times I \\ &= 20 \times 150 \\ &= 3000 \text{ watts} \\ &= 3 \text{ kW} \end{aligned}$$

11.0 Energy

Energy is the ability to do work. There are many sources of energy like electrical, mechanical, chemical etc. The electrical energy is given by the product of current, voltage and time and it is denoted by unit watt-sec.

Example : Find the energy spent in a welding arc operating for 15 minutes at 25 volts and a current of 200 amps.

$$\begin{aligned} \text{Energy spent} &= \text{voltage} \times \text{current} \times \text{time (in seconds)} \\ &= 25 \times 200 \times 15 \times 60 \\ &= 45{,}00{,}000 \text{ watt-sec} \\ &= 4500 \text{ kw seconds} \end{aligned}$$

Mechanical Energy : It is mainly of three types.

A. Potential Energy : It is the energy stored in an object due to its height from the earth surface.

Potential energy stored in an object

= mass of the body × height above the earth's surface × acceleration due to gravity

$= m \times h \times g$

B. Kinetic Energy : Kinetic energy is the energy acquired by a body due to its movement.

$$\begin{aligned} \text{Kinetic energy} &= \frac{1}{2} \times \text{mass of body} \times (\text{velocity})^2 \\ &= \frac{1}{2} mv^2 \end{aligned}$$

C. Strain Energy : Strain energy is due to the deformation occured to a body.

12.0 Heat

Heat is a form of energy. Mechanical, electrical, chemical energies can be convereted into heat energy.

The basic unit of heat is calorie or kilocalorie. One calorie is the amount of heat required to raise the temperature of 1 gram of water through one degree centigrde.

$$1 \text{ calorie} = \text{mass of water (1 gram)} \times \text{specific heat (1 for water)} \times \text{raise in temperature (1 degree centigrade)}$$

$$1000 \text{ calories} = 1 \text{ kilocalorie}$$

13.0 Temperature

It is a measure of hotness or coldness of a body. It is expressed as degree celsius or degree Farenheit. It is measured using a thermometer or other sophisticated systems.

Absolute temperature : –273°C is known as 0 kelvin. In this system temperature is expressed by adding 273°C to the temperature on the celsius scale. For example 27°C is equal to 27 + 273 = 300 K. This method of representing temperature is called absolute temperature.

14.0 Specific Heat

It is the amount of heat required to raise. The temperaure of 1 gram of a substance through degree centigrade. Specific heat is expressed as calories/gram/degree centigrade or cal/gr/°C or kcal/kg/°C.

Example : Calculate the heat energy required to raise the temperature of one kilogram of steel to 100°C. Specific heat of steel is 420 J/kg°C. Assume room temperature to be 30°C.

Weight of steel	= 1 kg
Room temperature	= 30° C
Temperature to which 1 kg of steel is to be raised	= 100° C
Rise in temperature	= 100 – 30 = 70° C

Heat energy required = mass of the body × specific heat × temperature rise

$$= 1 \text{ kg} \;\; 420 \times 70$$
$$= 29400 \text{ joules}$$
$$= 29.4 \text{ kJ}$$

15.0 Heat Input

In arc welding operation, heat is applied to the job through an electric power. The quantum of heat generated by the direct electric current is equal to the arc voltage multiplied by the arc current.

Heat generated $Q = V_{arc} \times I_{arc}$

Heat input per unit length of the weld is an important factor, in determining the weld quality. This is calculated as follows :

$$\text{Heat input per unit length of weld} = \frac{\text{Quantum of heat input}}{\text{Welding speed}} = \frac{V \times I}{S}$$

16.0 Welding Speed

The welding speed is the speed (velocity) at which the arc moves relative to the job. This is expressed in metres/minute (m/min).

Example : Calculate the heat input per unit length in submerged arc welding using the following welding parameters :

Arc voltage $= 30$ V

Arc current $= 500$ amps.

Welding speed $= \frac{500}{60}$ mm/sec.

Heat input per unit length of weld $= \frac{V \times I}{S}$

$$= 30 \; 500 \times \frac{60}{500} \text{ J/m.m.}$$

$$= 1800 \text{ J/m.m.}$$

17.0 Pressure

Pressure is defined as the force per unit area, acting on a given surface. The unit of pressure is kgf/mm^2. One atmosphere is the pressure exerted by the atmosphere and it is equal to 1 kgf/cm^2.

18.0 Fluid Flow

Fluid flow can be steady or unsteady. When the fluid velocity V at any given point of time is constant, the fluid motion is said to be steady. This condition can be achieved at low flow speeds. In unsteady fluid flow, the velocities are a function of time. This means that the fluid velocities vary erratically from point to point as well as from time to time. The fluid flow is expressed in centimeter cube per minute or litres per minute.

$$1 \text{ litre} = 1000 \text{ centimeter cube}$$

Fluids can be compressible or incompressible. Liquids can usually be considered as incompressible whereas gases will be compressible. Similarly, fluid flow can be viscous or non-viscous. Viscosity intoduces tangentical forces betwen layers of fluid in relative motion and results is dissipation of mechianical energy.

19.0 Stress, Strain and Modulus of Elasticity

The materials on which an external force acts will deform. But in the process material offers an internal resistance to the external force in such a way that the sum of the external forces acting in any direction will be equal to the sum of the internal resistance forces. The internal resistance offered by the material will be distributed over an area and if the cross-section is uniform then the internal resistance will be uniform over the section. This internal resistance offered by the material against deformation is called stress. In the case of a rod with a uniform cross-sectional area subjected to an axial load P at the two ends, the stress in the midsection will be calculated as

$$S = \frac{P}{A}$$

Example : What is the stress developed if a steel plate of 10 sq. cm. is subjected to an external force of 100 kgf.

External force applied $= 100$ kgf

Area of the palte on which load is applied $= 10$ cm^2

Therefore, stress developed in the plate $= \dfrac{\text{Froce}}{\text{area}}$

$$= \frac{100}{10} = 10 \text{ kgf/cm}^2$$

Due to the application of the load, the length of the member changes from L to $L + dL$. The ratio of the change in length of the original length of the member is called strain.

$$e = \frac{dL}{L}$$

For external loads of smaller magnitudes, the material deforms as long as teh external load is pressent and once the load is removed it will return to the original length. This material behaviour is called "elastic behaviour". The material exhibits the elastic property upto a dcertain level of stress level excees the elastic limit, then on unloading, the material will not come back to the original length. The material is then said to be plastically deformed.

It is observed that when a material is loaded within elastic limit, the ratio between the stress to the corresponding strain is constant and this ratio is called "Young's Modulus" or "Modulus of Elasticity" which is a property of a material.

Example : A load of 1000 kgs acts on a steel rod having an area of cross-section of 50 sq. m.m. and length of 1500 m.m. Calculate the stree, strain and the increase in lengfth of the bar under the action of hte load. Young's Modulus of the steel may be taken as 20000 kgf/mm^2.

Load acting on the member $= 1000$ kgf

Area of cross-section $= 50$ sq. m. m.

Stress experienced $= \dfrac{1000}{50} = 20$ kgf/sq. m. m.

$$\text{Young's Modulus} = \frac{\text{Stress}}{\text{Strain}}$$

Hence, $\text{Strain} = \dfrac{\text{Stres}}{\text{Young's Modulus}}$

$$= \frac{20}{20{,}000} = 0.001$$

But we know that strain $= \dfrac{\text{Increase in length}}{\text{Original length}}$

So, increase in length = strain × origianl length

$$= 0.001 \times 1500 = 1.5 \text{ m. m.}$$

□□□

REFERENCES

1. Gold Man : Electric Arcs in Argon, Physics of the Welding Arc Welding Institute, London, UK 1966.
2. Jackson C.E. : The Science of Arc Welding.
3. A.C. Davies : The Science and Practice of Welding, 8th edition, Cambridge University Press, UK 1984.
4. P.T. Houldcraft : Welding Process Technology, Cambridge University Press, UK 1977.
5. Jean Cornu : Advanced Welding Systems.
6. E.C. Rollason : Metallurgy for Engineers.
7. R.A. Higgins : Engineering Metallurgy.
8. Ivan H Griffin and Edward M Roden : Welding Processes.
9. The Procedure Handbook of Arc Welding : Lincoln Electric Company, Ohio, USA.
10. James A. Pender : Welding Processes.
11. Submerged Arc Welding Handbook, Union Carbide Corporation, New York.
12. Electroslag Process : B.E. Paton, Kiev.
13. P.L. Chuloshnikov : Resistance Welding.
14. Dr. A.S. Bahrani : Modern Welding Process : The Queen's University of Belfast, UK.
15. Welding Methods and Metallurgy : M.D. Jackson, Griffin, London.
16. Electron Beam Welding : Principles and Practice Book, A.H. McGraw Hill, England.
17. Underwater Welding : IIW Conference.
18. Welding Skills and Technology - Dave Smith.
19. Welding fundamentals and Procedures : Jerry Galyen, Garry Sear, Charles A. Tuttle.
20. Basic Welding and Practise by W. Kenyon - Itman Publications.
21. Welding Principles and Practices by Raymond J. Sacks.
22. Welding and Welding Technology by Richard L. Little.
23. Welding and Cutting of Metals by M. Khanapetov - MIR Publications.
24. Selecting the Optimum Surfacing Technique by T.S. Eyre.
25. AWS Handbook.
26. Various Technical Papers.

INDEX

□□□

Win Prizes!

ATTENTION READERS

We request you for your frank assessment regarding some of the aspects of the book given as under:

WELDING TECHNOLOGY

By

Er. G.D. Garg

Please fill up the given space in neat capital letters.

(i) What topics of your syllabus that are important from you examination point of view are not covered in the books ?

....
....
....
....

(ii) What are the chapters wherein the treatment of the subject matter is not systematic or organised or updated ?

....
....
....
....

(iii) Have you come across misprints/mistakes/factual inaccuracies in the book? Please specify the chapters and the page numbers.

....
....
....
....

(iv) What other books on the same subject have you found/heard better that the present book? Please specify the terms of price and quality.

....
....
....
....

(v) Further suggestions and comments for the improvement of the book:

....
....
....
....

(PS: If need be please attach a separate sheet to write your views.)

OTHER DETAILS

(i) Who recommended you this book ? (Please tick (✓) in the box near the option relevant to you.

☐ Teacher ☐ Friends ☐ Bookseller

....

....

....

....

(ii) Name of the recommending teacher, his designation and address :

....

....

....

....

(iii) Name and address of the bookseller you purchased the book from :

....

....

....

....

(iv) Name and address of your institution (Please mention the University or Board as the case may be)

....

....

....

....

(v) Your name and complete postal address :

....

....

....

....

(vi) Course you are enrolled in :

....

....

....

....

The best assessment will be awarded each month: The award will be in the form of our publications as decided by the Editorial Board.

Please mail the filled up coupon at your earliest to :

S.K. KATARIA & SONS®

4885/109, Prakash Mahal, Dr. Subhash Bhargav Lane,

Opposite Delhi Medical Association, Daryaganj, New Delhi–110002 (INDIA)

Phone: +91-11-43551243, Mobile: +91-9871775858, 7678425209

e-mail: katariabook@gmail.com, katariabooks@yahoo.com

Website: www.skkatariaandsons.com